SALTWATER

SALTWATER

A Natural History of the Sea

ASHLEY WARD

ITHAKA

First published in the UK in 2026 by Ithaka Press
An imprint of Bonnier Books UK
5th Floor, HYLO, 105 Bunhill Row,
London, EC1Y 8LZ

A CIP catalogue record for this book is available from the British Library.

Hardback ISBN: 978-1-78658-297-3
Trade Paperback ISBN: 978-1-78658-298-0

Also available as an ebook and an audiobook

1 3 5 7 9 10 8 6 4 2

Design and Typeset by Envy Design Ltd
Printed and bound in Great Britain by CPI (UK) Ltd, Croydon CR0 4YY

FSC
www.fsc.org
MIX
Paper | Supporting responsible forestry
FSC® C013604

The authorised representative in the EEA is
Bonnier Books UK (Ireland) Limited.
Registered office address:
Block B, The Crescent Building
Northwood, Santry
Dublin 9, D09 C6X8
Ireland
compliance@bonnierbooks.ie

www.bonnierbooks.co.uk

To the researchers, volunteers and advocates
who dedicate themselves to understanding
and protecting our seas.

Contents

Prologue

'The Sea, once it casts its spell, holds one
in its net of wonder forever.'
Jacques-Yves Cousteau

WHERE DOES A LIFELONG PASSION COME FROM? IT'S as though something snags on the fabric of our early existence, then catches. Thereafter, we're always in thrall to this thing, whatever it is. I remember as a small child, walking at my grandad's side on the springy turf of a headland on the Yorkshire coast, feeling the mingling thrill and dread of vertiginous cliffs just beyond our path. The summer sun was on our backs and the world felt perfect. Kittiwakes soared on the salt updrafts, looking quizzically into our faces as they emerged above the lip of the limestone escarpments, before dipping a wing and wheeling expertly away. We watched ships as they inched across the horizon and followed gannets as they quartered the waves from high above in search of prey. Perhaps at the sight of an incautious fish, a bird transformed in a moment from graceful aeronaut to a missile, wings folded tight against its body as it arrowed into the sea, leaving a legacy of bubbles on the surface. Following its lead, a squadron of other gannets dived into the water, and each time one disappeared beneath the waves, my grandad's hand would tighten fractionally about mine.

Later, we wound our way down the cliff path to the shore itself and, as the rest of my family ate sandwiches, I scampered off to look for animals. The flood tide trickled through gaps in the rocks, filling pools and reclaiming the land. I watched as the rising waters inundated a glistening jewel, rich ruby red stark against its chalk boulder setting, and to my astonishment the jewel responded by unfurling a crown of tentacles, seeming to bloom like a flower. I saw the blur of fish torpedoing to safety as my shadow trespassed on their domain and sensed the urgency of crabs shuttling between hiding places as the tide returned to embrace them. In the confines of a thimble-sized pool, a barnacle encased in its own tiny rock castle opened the door to its keep and reached out jerkily with pinkish arms to beckon me. A moment later, it withdrew sharply, as if shy, but then seemed to reconsider and returned to gesticulate again. Gradually, the incoming tide drew its watery veil over these wonders, leaving me marooned and bereft on the beach. Still, the eight-year-old version of me was spellbound. When we left the coast to head back home the next day, I took with me a collection of treasures – a whelk's empty egg case, a discarded crab claw, mussel shells with their peerless, pearlescent interiors, and my greatest prize, a mermaid's purse from the strandline.

Over time, I learned that the purse hadn't belonged to a mermaid but instead had once held an embryonic shark, that crabs could regrow lost claws, and that the barnacle wasn't waving to me. Rather, it was essentially lying on its back, kicking out its feathery legs, which mesh into a kind of net that combs the water for morsels of food. These new insights didn't dispel the magic, they intensified it. The more I found out, the more I craved to know. The sea had snagged and caught me on that visit, and I've been captivated by it ever since. When I sat on the transom of a boat in Tonga so many years later, marvelling at the memory of cavorting humpbacks, it was with the same wonder I'd felt as a child, and with a sustained and sustaining love of the sea and all the life it contains.

My connection to the sea, though personal, is far from unique. It casts a powerful spell on so many of us, one that echoes across cultures and represents the most profound of all our relationships with the natural world. The sea feeds us, both literally and spiritually, and it draws us for our holidays and recreation, to paddle in the shallows, to swim or perhaps simply to savour the unique magnificence of the coast. While there, the sensory embrace of the sea percolates through us, registering at a level far beneath our immediate consciousness, tuning and recalibrating our deepest physiology to leave us refreshed and renewed.

A sea view unhitches us from the mundane concerns of everyday life, somehow placing our existence in context. You can look at the same expanse of sea a thousand times and never see it the same twice. It is at once both familiar and enigmatic, a dimension that we can visit but never truly inhabit, sometimes hostile, other times soothing. Standing on the sandstone cliffs at Sydney's Coogee Bay, looking out over the Pacific, I'm confronted by an incredible panorama of beryl, cobalt, aquamarine and topaz, with diamonds of sunlight glinting across the undulating surface. Waves approach the shore, each at the climax of an epic but unheralded journey that might have encompassed a thousand kilometres or more. For a fleeting moment, each rears and curls in its final act, the wind whipping spindrift from its peak, before plunging against the land in a chaos of sound and spume. As the spent waters of its predecessor retreat, the next in line follows suit in an ageless perpetuity, the metronomic sequence depicting the puissant heartbeat of the deep and mirroring my own.

From my perspective on the cliffs, the horizon lies perhaps 12 kilometres away; and while this expanse of water provokes a sense of awe in me, it can't begin to convey the immensity of the ocean. If I were to travel due east from here, I'd have to cross more than 11,000 kilometres of open sea before I once again hit land on the coast of Chile. Indeed, the scale of the ocean is extraordinary: all the world's landmasses could fit comfortably within the Pacific alone. That we don't necessarily perceive its vastness is due in part to our limited view from dry land, but it may also be the result of our familiarity with the Mercator projection, the instantly recognisable map pinned to classroom walls the world over. This focuses on the continents, pushing the oceans into the background and making them easy to ignore, even splitting the great Pacific in two. The antithesis, an alternative map known as the Spilhaus projection, offers a dramatic challenge to this, and one that revolutionises our perspective on our home planet. First published in 1979, the Spilhaus projection compellingly presents the world's oceans as a single, unified and unbroken expanse of water, centred on Antarctica. This interconnected ocean breaks on the shores of every continent and drives home the point that this is one global commons, shared by us all.

My bird's-eye view of the ocean is also limited to the surface, so neglecting the vastness of what lies *beneath* the waves. At its deepest point, around 300 kilometres southwest of Guam, an area known as Challenger Deep, the ocean plunges 11 kilometres below the surface: sufficient to swallow Mount Everest with over 2 kilometres to spare. A pebble dropped from the surface into this void would be in freefall for well over an hour. Indeed, the *average* depth of the sea approaches 4 kilometres, with everything below 1 kilometre wreathed in complete darkness. The oceans are a vast aquatic firmament that comprise 97% of the living space on the planet.

Despite its capricious and unforgiving nature, the ocean was the cradle of life on our planet. We still do not know for certain how the very first organisms flickered into being, but an expedition led by researchers from Woods Hole Oceanographic Institute, Massachusetts, in 1977 provided our strongest clue yet. Amid the near-freezing black of the deep-sea Galapagos Rift, 2.5 kilometres below the surface, they discovered something entirely new. In an area of the seabed where volcanic activity drives the Earth's giant tectonic plates apart, they found the first examples of what became known as hydrothermal vents. Similar discoveries followed, identifying vent fields populating the deep ocean floor, spewing superheated water at a temperature great enough to melt zinc. The unique confluence of chemical and thermal energy at these sites fosters the conditions needed to catalyse the complex reactions that are essential to allow organic molecules to aggregate and then replicate.

The first steps, in other words, towards life itself.

Over three billion years on from this seminal transition from chemistry to biology, the ocean remains the wellspring of life on our planet. It is home to a huge diversity of creatures, from the hordes of tiny, unheralded algal cells that are the lungs of our planet to the majestic blue whale, the largest animal on Earth. And it is home to multitudes. There are thought to be something in the order of a quadrillion – a million billion, if you prefer – of the finger-length fish known as bristlemouths in the deep sea. If you were, by some strange means, to line up all these bristlemouths nose to tail, they would reach from Earth to the sun and back eighty times over. It's extraordinary to reflect that despite their prodigious superabundance, few people have ever heard of a bristlemouth, let alone seen one. They were not even known to science until 1958.

If it's the small size of bristlemouths that allows us to overlook them, the same cannot be said for the coelacanth, a fish that reaches approximately the same size as us, and which can live for over a century. Coelacanths were thought to have gone extinct along with the dinosaurs, but a chance discovery of one in the nets of a local fisherman just before Christmas 1938 off the east coast of South Africa revealed that these so-called living fossils had, in fact, existed alongside us throughout our entire history. They're extraordinary-looking creatures, sturdy, thickset and entirely lacking the hydrodynamic finesse of more modern fish. However, their true significance lies in their place in the story of life on this planet. The coelacanth is an ancient representative of a taxonomic group that had their heyday in the Devonian period, and who, millions of years ago, made the epoch-defining transition from the water to the land. In doing so, these pioneers established a line that eventually gave rise to all terrestrial vertebrates, and ultimately, to us.

Between them, the bristlemouth and the coelacanth exemplify not only the incredible biological potential of the sea but also its mystery. It's a facet of the human psyche that we're fearful of the unknown; accordingly, our ancestors conjured fantastical and terrifying creatures in their imaginations, such as the kraken, sea serpents and the hydra, that lurked unseen in the depths. Our enlightenment is a relatively recent thing. For the longest time, the ocean resisted our attempts to explore and understand it. In the Mediterranean and in the East, small numbers of divers have been crossing the frontier into the marine environment for millennia in search of sponges or pearls, but water robs our eyes of the ability to focus, and a single breath can take us only so far. Thus, when strange creatures washed up on beaches, or were hauled to the surface, crumpled in a seine that became their shroud, it served only to tantalise us. The challenges of remaining submerged for any amount of time, let alone being able to leave the uppermost regions of the sea, proved immense. We developed the capacity to build vessels capable of sustaining us underwater only relatively recently and, as so often, it was the twin pursuits of wealth and of warfare that drove us to innovate. Diving bells had their heyday in the seventeenth century when they were used to reclaim gold from sunken ships. Meanwhile, it was not until the mid-nineteenth century that submarines, in a recognisably modern form, were developed and deployed by the French Navy, providing the inspiration for the *Nautilus* in Jules Verne's 1870 novel, *Twenty Thousand Leagues Under the Sea*. Now we could penetrate the ocean, but not yet in a way that would afford us an understanding of the world beneath the waves. It was the development of the aqualung in the middle of the twentieth century that revolutionised our relationship with the sea. Finally, we could cut our links to the surface, at least for a time; we could breathe underwater, we could move freely, even balletically, and we could encounter living marine organisms on equal terms. We could participate in the ocean for the first time.

As Jules Verne was captivating the reading public with his adventures of Captain Nemo, groundbreaking plans were being made in Britain for a voyage that would transform our understanding of the sea. With the support of the Royal Society of London, Charles Wyville Thomson, professor of natural science at the University of Edinburgh, secured a vessel from the Royal Navy with the aim of conducting a voyage of exploration across the world's oceans. The ship was stripped down, its guns replaced by scientific instruments, its quarters reconfigured into laboratories and its crew drawn from some of the leading academics of the age. Sailing from Portsmouth in December 1872, HMS *Challenger* travelled more than 100,000 kilometres throughout the world's oceans for the next three and a half years, collecting samples, cataloguing new species, and meticulously recording observations as she went. The richness and abundance of the data that they amassed was such that it would be a further 19 years before all the findings were published, with the final report amounting to almost 30,000 pages, spread over 50 hefty volumes. The quality of the work was of such a high standard that it remains in use even today. The legacy of the voyage was immense, not least in the discovery of over 4,000 new species, but it was the unprecedented wealth of new physical, chemical and biological discoveries that have led to the *Challenger* expedition being regarded by many as the founding moment of both oceanography and marine biology as modern scientific disciplines.

In the aftermath of *Challenger*, research institutes sprang up around the world, dedicated to the systematic analysis of marine ecosystems. These have been joined by government agencies, set up to tackle key questions and to inform policy decisions, and universities, which have devoted ever more space in their laboratories and their syllabuses to the study of marine biology. While it remains the case that most marine biology is conducted at the margins of the sea and in shallow waters, technological advances have allowed researchers ever greater scope to explore far from land and to dive deeper. Much work remains to be done, of course, but in the 150 years that have passed since the *Challenger* expedition, our understanding of marine biology has blossomed. The question is: what have we learned?

For a start, everything's different in the ocean. What we comfortably presumed to be the natural order of things from our terrestrial perspective are morphed and transmogrified in the marine environment. Our assumptions of life, sex and death are all, so to speak, at sea. There are marine mammals that live for centuries, corals that live for millennia and jellyfish that are effectively immortal. As for sex, how about the male anglerfish who essentially merges with his mate, or the flatworm rivals that fight duels with rapier-like penises.

The undersea world is a bewildering, dazzling, fascinating tableau of life; it's as though Mother Nature had a fever dream, as if evolution broke its shackles and ran riot. Take, for instance, goblin sharks that can catapult their jaws beyond the extent of their mouths to seize unwary prey, or the spookfish, which gains a panoramic view of its surroundings by dint of having a transparent head. And, of course, the giant squid, a charismatic but little understood behemoth of the deep sea that may be longer by a third than a double-decker bus. The sea encompasses some of the most hostile environments on the planet, such as the crushing, frigid conditions of the deep ocean. Yet even here, there are animals whose solutions to the challenges of life are so extraordinary that they test the very limits of biology. There are horrors beneath the waves, such as the 3-metre-long bobbit worms that lie buried, ready to ambush the unwary in their viciously keen, trap-like jaws. And there's mesmerising beauty: an immense shoal of silvered fish, darting this way and that in unison as though choreographed by some unseen hand among submarine gardens of fantastical, labyrinthine corals.

We've learned that the life in the ocean has extraordinary resilience, that it has an enormous capability to support and sustain us, and we have also learned that it is more vulnerable than we could ever have feared.

Even now, when discussing life beneath the waves, we sometimes do so in a manner that suggests it is separate and distinct from terrestrial organisms, and indeed from us. Nothing could be further from the truth. The ocean is the world's life support system. As the pioneering marine biologist and oceanographer Sylvia Earle succinctly put it: 'No water, no life. No blue, no green.' Notwithstanding the incalculable value of marine biota in protecting and providing for us, such as the oxygen produced by phytoplankton, the reefs and mangroves that act as a bulwark against storms, and the millions of tonnes of seafood that feed us every year, as a few examples among many, the ocean moderates our planet to an extent that we scarcely realise. Our profligacy with fossil fuels continues to change the character of our atmosphere, charging the air with greenhouse gases and knocking the Earth's climate out of kilter. Over the last half-century, the ocean has soaked up around a quarter of the excess carbon dioxide that we've belched into the atmosphere. At the same time, it has absorbed well over 90% of the extra heat that the remainder has driven. It's estimated that without this, the planet would be around 36°C warmer than it is now. To put it another way, without the ocean's intervention, we'd already be history.

But the heat that the ocean soaks up doesn't simply disappear; the spectre of global warming looms as large beneath the waves as above them. Marine organisms are straining under the burden of change that comes at a pace that evolution is not equipped to match. As the mercury rises, entire reefs bleach to bone while kelp forests wither to leave rock barren and legions of species retreat poleward in search of relief from the heat. The extra carbon dioxide, too, spells trouble. As it dissolves in seawater, it reacts to form carbonic acid, reducing the pH and inexorably changing the chemical character of the oceans. This has many implications, but one of the most concerning is that it becomes ever more difficult for organisms to sequester carbonate ions, which are the basic constituent of the bodies of shell builders, such as molluscs, corals and even plankton. We're already seeing zooplankton sheathed in pitifully thin shells that are only a quarter as substantial as their pre-industrial forebears, and it's estimated that if current trends continue, the pH levels anticipated by 2100 would cause animals such as the entrancing and delicate sea butterfly to simply dissolve.

Despite our professed love for the sea, and our reliance upon it, we've consistently used the marine environment as our dumping ground and sewer. The great agglomerations of floating plastic debris that litter the oceans today are testimony to this, the most famous of which, the Great Pacific Garbage Patch, is a stain the size of France on the surface of the sea. Whales, turtles and fish are poisoned or choked with the trash that we so carelessly discard. Submersibles exploring and mapping virgin stretches of the seabed all too often report that previously uncharted deep-sea plains are littered with refuse. We're being beaten to these marine wildernesses by our own junk. Some of this might be put down to carelessness, but not all. The ocean has proven too tempting a solution to the problem of how to dispose of toxic waste. In 2021, a seafloor dump twice the size of Manhattan was identified off the coast of Los Angeles, containing thousands of discarded barrels of the banned pesticide DDT. We have little idea of how many of these noxious middens exist around the world, hidden beyond our view, but we can see their effect. As chemicals leach out of corroded drums and into the surrounding water, they poison the environment, infiltrating the bodies of marine organisms and accumulating throughout the food web. Mammals as well as fish have been found with ulcers, tumours and congenital deformities, and those are just the survivors.

Our rapacious quest to strip anything of value from the sea also leaves terrible scars. The insatiable demand for concrete means that sand is now the world's largest extraction industry, with something like 50 billion tonnes mined every year, much of which comes from the ocean.[1] The net result of this is the transfiguration of coastlines and the loss of priceless habitats. Drilling for oil and gas, meanwhile, not only results in sporadic and catastrophic spills, but the seismic surveys used to locate new deposits beneath the seafloor generate their own problems. The sound impulses used can exceed

1 Desert sand, it turns out, is near useless for building because of the shape of the sand grains.

200 decibels – which equates to an intensity over a thousand times greater than that which would burst a human eardrum – and the result is a swathe of strandings among marine mammals, who are disorientated and deafened by the sonic barrage. Even the deep sea, for the longest time beyond our reach, is now in the crosshairs of prospectors who can see dollar signs amid the darkness. The technological advances that make abyssal minerals accessible to wealthy mining consortia threaten to usher in a new gold rush, and one that all of us will surely pay for.

Alongside the removal of materials, fishers empty the sea's bounty in a frantic race to the bottom, the tragedy of the commons embodied in each cast of their nets. The collapse of the Grand Banks cod fishery in the late twentieth century stands as grim testament to the fragility of marine resources. Where once it was claimed that the fish off Newfoundland were so uniquely plentiful it might be possible to walk across their backs to shore, decades of industrialised fishing concluded in the annihilation of the world's most productive fishery. The denouement was witnessed in 1990 by a Canadian scientist, George Rose, aboard a fisheries research vessel. The ship's echo sounders reported the arrival at the Banks of the last great cod shoal in the western Atlantic, and George could only watch as the fish swam straight into waiting nets of a host of greedy trawlers. If cod are the tragic icon for our rapine attitudes towards the marine, they are by no means alone; more than 80% of the world's commercial fish stocks are fully exploited or approaching collapse.

For anyone who cares about the sea, these are troubling times. For too long, we've regarded the sea as being infinitely capable of feeding us, of swallowing our rubbish and of withstanding our abuse. In essence, we've consistently thought of the sea and its bounties as being limitless. As recently as 1955, Francis Minot, director of the Woods Hole Oceanographic Institution, co-authored the book

The Inexhaustible Sea, in which it was confidently predicted that we would one day come to appreciate that the sea's capacity to provide was illimitable, that its resources were endless. He could not have been more wrong. Nevertheless, this is an illustration of what was, until very recently, the prevailing view, the result of which is a tragic litany of despoilation. Even now, our greater understanding hasn't yet translated into a change in our behaviour. Our relationship with the sea is paradoxical and troubling: we're utterly reliant upon a healthy marine environment for our continued existence, and yet we seem intent on its destruction. For all that, however, now is not the time to give up. There is hope, as I'll show in these pages, both because of the incredible resilience of the ocean and its ability to bounce back, if given a chance, and because of a groundswell of committed people who care about the sea and are willing to fight for it.

My own affection for the sea was kindled in early life and fostered during my during my 25-year career as a research biologist. For others, especially those who seldom interact with the sea, a connection to the marine environment may seem less obvious and less immediate. And so, while attitudes are changing in the wider populace and among our leaders, they are doing so at a pace that feels painfully slow. This pace needs to increase, and my aim in writing this book is to lay before people the incredible wonders of the sea, as well as to uncover the shameful and very often covert practices that damage it, in the hope of inspiring them to join in the task of safeguarding the future of this, the planet's most important realm.

This is a book about the sea, the life that flourishes within it and our essential relationship with both. It is both a love letter to the sea and a call to arms, and its writing has been the most demanding as well as the most rewarding thing that I have ever done. Demanding because the scope of marine life is as awe-inspiring as the scale of the oceans, and because what we call marine biology is a composite of many different disciplines and specialities, all of which could present a challenge to incorporate into a single volume. My approach, therefore, is to draw together different strands and to emphasise the commonalities. And although it is true that the various regions of the global marine environment have their own qualities and peculiarities, each part of the sea connects intimately with every other and the living creatures it supports do not recognise the arbitrary boundaries that we impose.[2] Each chapter centres on a particular habitat, but it does so in a way that accentuates the interrelation and interdependency that exists throughout the marine world. This is not a catalogue of marine creatures, or of biomes, but a series of interconnected stories that weave a path across and through the oceans. I have avoided jargon in favour of a more inclusive approach, and rather than clog the text with scientific nomenclature, I tend towards common names. This is addressed in the species glossary. I hope that you will enjoy reading this book as much as I enjoyed writing it, and that in celebrating the sea, we can help to bring about a change in perspective that will allow future generations to enjoy its mesmerising spectacle.

---

2 Cartographers divide the global marine environment into five oceans and dozens of seas; however, in truth, there are no sharp divisions, just one interconnected whole. I hope that the reader will forgive my rather cavalier approach to the words 'sea' and 'ocean', which I use interchangeably.

1.
Constellations of the Deep

The deep sea is a mysterious and challenging biological frontier

IT'S A WARM SPRING MORNING IN THE PACIFIC, 15 kilometres north of the Tongan island of Foi'ata. For the past half hour or so, ever since the whale research vessel slowed to begin tracing a wide figure-of-eight search pattern, I've been casting anxious glances up at the crew on the flybridge as they scan for movement among the waves with their binoculars. There's an air of anticipation on board, conversation is hushed, all eyes fixed on the rolling surface of the ocean. Suddenly, there's a shout. Heads turn in the direction indicated by the skipper's outstretched arm. A spout, and then another, moving this way. The boat glides into place.

Now the motor is cut, the mechanical thrum giving way to the peaceful sounds of water plashing against the hull. My heart is pounding; the stilling of the engines signals that the moment has come to enter the whales' world. Clamping my mask to my face, and with camera in hand, I slip off the stern of the boat and into the ocean. As the curtain of bubbles clears from my field of view, I'm confronted by a breathtaking sight. Two gargantuan male humpbacks in pursuit of a smaller female, all three arrowing through the water towards me. The males joust for the female's attention; there's a white flash as one of the males drives his calloused pectoral fin, 5 metres long and more than a tonne in weight, against his rival. Dodging the boisterous males, the female banks to her left and accelerates with a flex of her mighty tail. Immediately, her suitors follow suit, arcing balletically into a sharp turn. Neither has forgotten his opponent, however. Abruptly, one dives beneath the other then rises, his enormous mouth gaping wide in what feels like a show of cetacean bravado. The pair corkscrew around each other, both straining to keep pace with the female. I am a spectator to the whales' cavorting for perhaps only 30 seconds before the trio rocket away, their huge forms merging into the blue at the limits of my vision.

I'm here in the Tongan Archipelago, one of the world's great humpback breeding grounds, to study the social behaviour of these whales. Each year, the returning ice compels humpbacks to leave Antarctic waters, where they've fed throughout the Austral summer, and to travel 6,000 kilometres north, to court, to mate and to give birth in the benign seas of the tropics. The exhilarating cetacean cavalcade that I witnessed was part of this: though the female seems to be fleeing the harassment of the males, in reality she solicits their attention. In doing so, she encourages them to vie for the chance to mate with her in what's known as a heat run. The females do this to winnow out all but the strongest in an exhausting pursuit that may last for hours, ensuring that the victorious male possesses the vigour to make a worthy father.

I haul myself back aboard the boat and sit at the stern, feeling electrified by the encounter. The size of these oceanic leviathans is incredible; 30 or perhaps even 40 tonnes of bulky might, slicing through the ocean in absolute and paradoxical silence. More than this, however, it's a vivid impression of potency, of beauty and of grace, accompanied by the sense of being in the presence of a parallel consciousness, that of a creature whose experience is so alien to our own.

This first encounter with the whales left me greedy for more, and I had the good fortune over the days and weeks that followed to observe a panoply of humpback behaviour. Sometimes the whales were urgent; I felt the water around me shudder as one female thundered past, metres below the tips of my fins. At other times, they were restful, lolling at the surface and perhaps enjoying the feel of the Tongan sun on their slate dark backs. I was careful to keep my distance out of respect for the whales, unthreatening though they are. Nevertheless, the curiosity of these giants sometimes led to them approaching me. One, a male, materialised by the side of the boat, his huge head lifting vertically from the waves in a behaviour known as spy-hopping, regarding me for a thrilling moment with his rheumy, grapefruit-sized eye before descending back into his watery domain. Equally memorable was a female who cruised slowly past me with her offspring tucked tight against her flank in the manner of so many whale mothers. It might be hard to conceive an animal as a newborn when it is already 4 metres long, but that fact was given away by its softly furled dorsal fin and the folds in its flanks that spoke of months curled tightly in its mother's womb. It was a wonderful sight, a calf born into a generation of whales that will live beyond the shadow of the harpoon. In time, the calf's fin will stiffen, and its birth folds will disappear. It will grow fat on its mother's rich milk and, in a few weeks, embark on the first of many oceanic journeys that are the humpbacks' inheritance.

As one life begins, another is drawing to a close. Out to sea, far beyond these Tongan shores, an elder humpback labours at the surface, her exhaled breath swirling amid the spray of squally waves. She has travelled almost a million kilometres across the oceans since she was born some 80 years back, and instinct, migratory memory, pushes her on towards equatorial breeding grounds. She last calved long ago, and her womb, like her belly, is empty. She swims alone, perhaps finding solace in the distant song of her kind that carries to her across the void. Once she was fleet; she had to be to evade the explosive danger attending the hunting ships that stalked her youth. Now her heart, a 300-kilogram miracle of organic engineering, is weakening, and she with it. She is aware that, unseen in the dark waters beyond, her frailty has drawn a menacing escort of predators. She is too tired to dive, too weary to charge as once she might; she knows, perhaps, that her end is near. As dawn begins to break over the Pacific, her giant heart beats its last.

The extinguishing of a life, especially that of a creature as majestic as a humpback, is always sombre. Yet her death offers opportunity to others. Her escort of sharks close in, wary at first, but then with a quickening bloodlust. Opportunistic seabirds settle on her giant flanks, their questing beaks gouging plugs of skin and blubber. The whale's body represents a rich bounty of food in a watery desert, but it is ephemeral. As the air gradually leaves her lungs, she begins to sink, embarking on a final journey, the ocean reclaiming its own.

With every extra 10 metres that the whale descends, another atmosphere of pressure is brought to bear, squeezing and compressing her giant body. As she reaches 200 metres, she enters the mesopelagic, known more familiarly as the twilight zone, the gateway to the deep sea. Her backdrop is a scene of shadowy gloom. As her stately descent continues, she enters the bathypelagic, the midnight zone, and leaves the sunlight behind for the last time. The only luminance beyond this point is that created by living organisms, who offer a silent serenade of cold, sapphire flares as she sinks. She sinks further still, the water now hovering just above freezing and stirred only by the gentlest of currents. She slides through the darkness until she meets the seabed, her final resting place.

Life in the deep sea is one of paucity, of hunger and of scarce resources. No plants grow in the darkness, so the abyssal food web lacks the foundation of primary production that supports virtually every other habitat on Earth. Instead, most of the energy flows into the deep from above. One such comes in the form of particles of organic matter that fall in a perpetual slow drift down from the surface waters. The naturalist and deep-sea pioneer William Beebe likened this gentle fall to marine snow and the name stuck, but if the name conjures ideas of freshness and purity, think again. Marine snow is an amalgam of dead plankton, mucus, bacteria and turds, and though it represents a crucial flow of calories to the deep, it's thin fare. The arrival at the seabed of the carcass of our humpback, what marine biologists refer to as a whale fall, represents a bonanza that is the equivalent of 2,000 years' worth of marine snow.

The arrival of such a banquet rapidly draws diners, attracted by the sound of a heavy body touching down and the enticing odours of decomposition. Sleeper sharks, their 7-metre bodies lacking the streamlined finesse of their surface-water cousins, are among the first to the fleshy table. They tear into the carcass, twisting and pirouetting in a macabre dance to gouge great hunks. They're joined at their repast by herds of hagfish, creatures whose fascinating biology far outweighs their unprepossessing appearance.

Hagfish have been around for well over 300 million years, a feat of evolutionary longevity exceeding that of any other vertebrate and one that speaks to a winning strategy. Their body plan is a simple one, at least superficially, consisting of a simple tube, much like an eel, though their loose, wrinkled skin calls to mind a badly fitting pink sock. They are jawless, but this presents little obstacle to them when it comes to feeding. Rows of wickedly sharp teeth are mounted on two dental plates, arranged not in the horizontal manner of our own mouths, or those of every other vertebrate, but vertically. These plates, looking like medieval instruments of torture, can be projected beyond the mouth to grasp onto the flesh of the whale before being drawn back inside, causing them to clamp together with a force that exceeds that wielded by most of their jawed cousins. Still, the half-metre lampreys lack the bulk of their feeding companions, the sleeper sharks, and powerful though their bite is, it's a challenge to contend with the robustness of a fresh whale. To overcome this, the hagfish does something so stupendously odd that if it were invented as part of a fictional bestiary, it would seem preposterous. Specifically, the hagfish ties itself in a knot and slips that knot up along its body until it braces against the whale and provides leverage to liberate a gobbet of flesh from the huge cadaver.

It might reasonably be assumed that the presence of dozens, if not hundreds, of hagfish at the whale fall might encourage the huge sleeper sharks to diversify their diet, but the sharks know better than to trifle with these animated frankfurters. Hagfish are exquisitely defended: when provoked, a battery of glands within the creature's body respond in a fraction of a second, squirting out slime from pores arrayed along each of their flanks. Upon contact with water, the slime expands dramatically and almost instantaneously to produce a welter of oozing gel.

If this seems underwhelming, that's because you're not a fish. Until fairly recently, the value of the hagfish's defences was a matter of speculation, and then, in 2009 and 2010, video footage taken from baited deep-sea camera stations off New Zealand showed just how potent the slime can be. Two large predators – a seal shark and a wreckfish – made the mistake of attacking a hagfish only to very visibly regret the decision moments later. In both cases, the predator released the hagfish, unharmed, immediately, and could be seen performing the fish equivalent of coughing or retching. The slime is an effective weapon against fish because it doesn't just fill their mouth cavity with goop, it goes on to clog their gills and poses the risk of choking them. It does this by dint of being composed of both mucus and long strands of hair-like proteins, which hold the whole mess together, potentially tangling the delicate structures of the gills. Hagfish have even been known to repurpose their slimy defences for attack, using their gooey superpower to fill the seabed burrows of slender creatures known as bandfish, thereby suffocating them. With such an effective weapon at their disposal, there's always a risk that hagfish can sabotage themselves, since they too have cloggable gills, but they avert the danger by using their amazing self-knotting technique to slide and squeegee the gunk from themselves.

The sleeper sharks and hagfish don't get to enjoy the windfall whale alone; they're soon joined by swarms of other hungry deep-sea denizens. Giant isopods, carnivorous snails, squat lobsters and octopus, all drawn by the smell of decay as chemicals with evocative names such as putrescine and cadaverine leak from the carcass into the surrounding waters. With so many diners at the feast, it's little surprise that the more obviously edible parts of the body are depleted rapidly. After a few months, or possibly a year, depending on just how deep the whale's resting place is, the flesh and the soft tissues have been stripped away, leaving the skeleton. Yet while the flesh eaters now move on to pastures new, there remains much to titillate the appetites of other, more specialised consumers.

Whale bones differ from those of land mammals by being not just larger but also less dense; after all, it makes sense not to have a skeleton so heavy that it pins you to the seabed. Accordingly, whales evolved over time to have bones that have a spongy core, full of marrow and, crucially, fats, which make up more than half of their weight. The result of this is that their bones are more buoyant and better able to resist pressure than those of their terrestrial relatives. This serves the whale well in life, and after death the succulent bones conceal a wealth of energy-rich sustenance to any creature that can gain entry to their interior. Plenty are adapted to precisely this, including *Rubyspira* snails, bone specialists of the deep sea that rasp away at the skeleton with hardened, abrasive mouthparts. And perhaps the most celebrated of all, the extraordinary zombie worms of the genus *Osedax* that, in time, colonise the skeleton and give it a pink, fluffy appearance. This carpeting of fuzz is generated by the feathery, blood-rich gills of thousands of worms, the only part of their bodies that protrudes from the bone surface. Beneath this respiratory plumage, the less showy parts of each worm are safely ensconced in its own burrow, a hole that they make in the bone by secreting acid. Drilling into the skeleton in this way gives them access to the heart of the bone and its bounty of oil and marrow. Oddly, for us at least, in most *Osedax* species, the largest worms are all females, who maintain a harem of perhaps a hundred or so microscopic males within their bodies. The size difference between the sexes is dramatic – females can be tens of thousands of times larger than their consorts. The equivalent would be if a human married a blueberry.

The incursions made in the bone allow not only creatures like *Osedax* to mine the riches within, they also allow bacteria to penetrate the interior. Once inside, these bacteria thrive on the fatty residues and subsequently release hydrogen sulphide, a gas with a signature smell of rotten eggs. This is chemosynthesis and, to the deep-sea food web, it's the equivalent of the photosynthesis that occurs in the shallows; rather than being powered by sunlight, the deep is driven by chemical energy. Chemosynthesis forms the basis for a rich and complex community of animals, comprising hundreds of different species. Dense beds of mussels and clams use symbiotic bacteria to convert the plentiful hydrogen sulphide to food, while worms and limpets graze on the mats of bacteria that coat the skeletal remains. These, in turn, draw predatory whelks and crabs, with the end result being one of the most diverse assemblages of animals to be found anywhere in the deep. And if this carnival of biology seems alien, then consider this: in the substrates around a whale fall, microbes such as *Firmicutes* and *Bacteroidetes* thrive. These are the two dominant types of bacteria of our own gut flora; hence they play a crucial part in the digestive processes of both the deep sea and your body.

This sulphur-based food web can endure for decades at a large whale fall. And even after the last vestiges of energy are stripped from the bones, leaving nothing but the stony minerals behind, it remains invaluable to those creatures, such as anemones, barnacles and even corals, that rely on having a hard substrate to attach to amid the soft mud of the ocean floor. Long after the living whale perished, she transforms into a reef, becoming in a sense her own deep-sea memorial garden.

Whale falls are islands of plenty amid the barrens of the deep ocean, and although they are ephemeral, they aren't necessarily rare. Even as whales recover from the hunting that took some to the brink of extinction, it's estimated that there may be the best part of a million active whale fall communities scattered around the oceans. Evenly distributed, this would mean that the separation between these oases is only around 12 kilometres. Near enough, in other words, for not only mobile scavengers to manage the commute but also for the roving larvae of the more sedentary whale fall animals. This ability to colonise new sites when an existing one begins to fail allows us to picture whale falls as a series of stepping stones for deep-sea organisms to navigate their way through the hardships of their environment.

The discovery of whale falls, and in particular the complex, thriving and long-lived communities that they support, transformed our understanding of the deep sea. It came about in a golden era of exploration, spanning the closing decades of last century, when a new breed of submersibles yielded up the secrets of the abyss. Along with the revelation of whale falls came the discovery of two further phenomena: seeps and vents, where mineral-rich water gurgles or rushes from the seabed, providing a home for creatures quite unlike anything we had previously known. Tantalisingly, there are suggestions of a close evolutionary link between all three, giving rise to the idea of a network of deep-sea habitats, founded not on the light of the sun, but on raw chemical energy.

* * *

Belching out black smoke, a forest of chimneys towers over their surrounds, calling to mind the mill towns of my childhood. Unlike the factory chimneys, however, these are not regimented or regular. Each is unique and idiosyncratic, Gaudiesque in its whorls and flourishes. These hydrothermal vent stacks rise from the seabed, thousands of metres below the surface, at the imperfect margins of tectonic plates where molten magma strives for release. At such fracture zones, two elemental forces meet as freezing abyssal water percolates down through fissures in the seabed to converge with the geological furnace of the mantel. The water, supercharged by Hadean extremes of temperature, cannot boil due to the crushing pressure, but expands with the heat and inexorably forces its way upwards until it jets out of the crust and into the near-freezing water. Here its burden of minerals rapidly precipitates out of solution, creating the impression of smoke and, in time, forms stacks that might stand as much as 20 storeys above the seabed.

Magnificent as these geological phenomena are, for most of our history we were ignorant of them. Their discovery was tied in with a scientific theory that emerged early in the twentieth century to explain the perplexing similarities between rock formations and fossils between far distant landmasses. This theory, plate tectonics, described the way that huge slabs of the Earth's crust are engaged in a slow dance, sliding across the planet's mantle, separating some provinces and uniting others. While our focus is on the movements of continents, most of the action takes place out of sight, beneath the sea, at the boundaries between plates. By the 1960s, the theory was beginning to gain widespread acceptance, and the race was on to make direct observations of spreading zones and subduction zones, where the plates respectively move apart or fold together.

The 1960s saw a dramatic increase in deep-sea exploration; expeditions set out to uncover evidence of arguably the planet's most important geological phenomenon. By the 1970s, the pieces of the puzzle were coming together, with one peculiar exception: the geothermal heat that was expected to accompany the process of spreading was missing. With magma chambers containing molten rock at temperatures of up to 1400°C lying just beneath the spreading zones, the ridges at the epicentre of the process should have been much warmer than they were. Had there been some mistake? If the instruments were right, maybe the heat was escaping, but if so, how? One proposal conjectured the existence of localised hot water vents, the marine equivalents of geysers. These could arguably act like oceanic exhaust pipes, spewing the heat out in concentrated patches and explaining that missing from elsewhere. This wasn't an idea that gained universal acceptance though; sceptics argued that while the eruptions of terrestrial geysers are driven by water boiling deep underground, the same thing couldn't happen on the ocean floor because of the massive pressures.

Getting to the bottom of this mystery brought together teams of geologists and oceanographers from across the world. Between 1971 and 1974, a French-American collaboration dubbed Project FAMOUS sent crewed submersibles down almost 3 kilometres into the rift between two gigantic plates at the Mid-Atlantic Ridge. There, in a deep-sea valley larger than the Grand Canyon, they were the first to witness an epicentre of tectonic activity. Despite the successes of Project FAMOUS, however, it failed to locate any hot water vents. Undaunted, the scientists turned their focus to the Galapagos Rift, a seafloor spreading zone 400 kilometres northeast of the Galapagonian island of Santa Cruz, and one so active and volatile that its spasms regularly result in mass kills of deep-sea fish. If vents were to be found, an area where the parboiled corpses of fish regularly decorated the ocean surface seemed to be the perfect place to look. A preliminary expedition offered encouragement, identifying a region of chemical and thermal anomalies near the floor of the rift itself. What was needed now was to move beyond inference to hard evidence, to locate the last piece of the puzzle and to find a hydrothermal vent.

So it was that in February 1977, a team of geologists and oceanographers set out to investigate the Galapagos Rift aboard the research vessels *Knorr* and *Lulu*, the latter the mothership to the most famous submersible in the annals of deep-sea research, DSV *Alvin*.[3] The speedier Knorr arrived first at the site and deployed a 2-tonne sledge, bristling with camera equipment and sensitive instruments, that could be towed along at walking pace just above the seabed. For six hours, the sledge reported back a

3 Deep Submergence Vessel *Alvin* might seem like a strange moniker for such a high-tech, cutting-edge vessel, especially since Alvin and the Chipmunks were a popular novelty band at the time it was launched. Actually, *Alvin* is named for Allyn Vine, who led the development of the submersible. Students at Woods Hole have their own name for it: The Pregnant Guppy.

monotonous sequence of unchanging data – two degrees, two degrees, two degrees – and then, suddenly, the temperature spiked. This spike lasted for less than three minutes before returning to the near-freezing baseline. On board the excitement was palpable; it seemed as though they'd found the very thing they'd come looking for: a hydrothermal vent. Proof positive could only be gained by recovering the camera and, this being before the age of digital photography, developing the film.

Having retrieved the camera from the frigid waters of the seabed, 2.5 kilometres below the ship, the team were impatient to see what they'd found. There was a risk of damaging the film, however, if they didn't allow the camera to warm up gradually. Two agonising hours of waiting followed before the camera roll could finally be recovered, and then there was a further delay while the onboard lab developed the film. At long last, the photographs were ready, and scientists pored over the sequence of thousands of images. Picture after picture showed fresh, solidified lava that they described as resembling mounds of petrified toothpaste, but nothing could have prepared them for the 13 photos taken as the sledge passed through the hotspot. Gone was the stark, lifeless moonscape of the previous photos, replaced by huge beds of clams and mussels, clustering amid swirls of misty water. They'd found the hydrothermal vent that they'd been looking for and something else, something entirely unexpected: a riot of life thriving on the hot springs. It was the zoological find of the century.

By now *Lulu* had arrived, bringing *Alvin* to the site, and with it the exciting prospect of a first crewed voyage to a hydrothermal vent. At first light the next morning, *Alvin* dipped beneath the waves and navigated down to the precise coordinates recorded by the sledge. There, in the submarine's spotlights, the crew were greeted by the sight of hot water shimmering out of cracks in the rock, and of hordes of the shellfish whose portraits had earlier been captured by the sledge. The entire patch was just a few tens of metres across, a deep-sea metropolis in a vast and barren plane. *Alvin*'s robotic arm was pressed into action to scoop samples of the creatures from their volcanic bath. Back on the ship, the specimens were preserved using the only thing that the scientists had to hand: vodka. No one had expected to find any life worth cataloguing, let alone preserving, since the prevailing view was that deep-sea life was scarce, sparse and small. As a result, this groundbreaking expedition, which discovered a whole new realm of biology, ironically included not a single biologist.

In the days that followed, the crew aboard *Alvin* discovered several new sites, and many strange new animals. There were ghostly white crabs, metre-long tubeworms, peculiar, stalked creatures that looked like yellow flowers, and odd-looking pink fish. Not only was life diverse at the vents, but it was also superabundant. The obvious question was: how? What could be sustaining so many animals? Clues to this came with each biological sample that was delivered to the ship's lab. Every jar that was unsealed released the telltale eggy stench of hydrogen sulphide, causing the scientists to rush to open the portholes. Noisome as it is, the presence of this gas is fundamental to vent ecosystems. Although animals can't use the sulphides directly, bacteria and their distant, single-celled cousins, the archaea, thrive on the chemical energy bound up in the rich, mineralised water of the vents, and as a by-product of their activity, convert that energy into nutrients that sustain animals.[4] Until then, it had been thought that all life on Earth depended directly or otherwise on photosynthesis, but here was something completely

4 Specifically, the symbiotic bacteria work to oxidise chemicals in the vent waters, such as sulphide (H_2S), methane (CH_4) or hydrogen (H_2). This drives the fixation of inorganic carbon into organic matter, which in turn feeds the animal host.

new: animals nourished not by the sun, but by the chemical energy of the Earth.

Almost 50 years have passed since hydrothermal vents were first discovered and, in that time, hundreds of others have been found in a network tracing along plate boundaries throughout every ocean. In each case, fascinating communities of animals have made their homes at the vents. Some graze on the microbial mats that form around the hot springs like cattle munching grass in a paddock, whereas others take things a step further, embracing the microbes and hosting them in their own bodies. This symbiotic arrangement is used to dramatic effect by the giant *Riftia pachyptila* tubeworms. The worms are gutless, lacking anything that has even a semblance of digestive equipment. They don't even have a mouth. What they *do* have is a sac-like structure called a trophosome, stretching almost the entire length of their oversized body. This houses a bacteria farm, and they're entirely reliant on this for their food.

Wayward clusters of *Riftia* tubeworms cover vent surfaces like the pipes of a dilapidated church organ, their bright red gill plumes poking shyly out of the top of enormous, 3-metre-long cylindrical homes. Part of the reason that their tubes are so long is to give their crown of gills access to colder, oxygenated water while the lower extremities bathe in the hot, sulphide-rich vent fluids. Even though hydrogen sulphide is one of the main sources of nutrition for the animals, it is highly toxic. Absorbed by the body, it works in a similar way to cyanide, binding tightly with haemoglobin, interfering with oxygen transport and leading to asphyxiation. Tubeworms have evolved to cope with this by developing large and highly complex haemoglobin molecules that wrap up the sulphides in a chemical embrace while simultaneously retaining the crucial ability to carry oxygen. It's an extraordinary arrangement, and one that enables *Riftia* to be one of the most dominant animals at the vents.

The thickets of giant *Riftia* tubes around the vents make an undersea forest whose glades are inhabited by a host of other creatures. Mussels, limpets, snails, shrimps and crabs abound here, and these plucky invertebrates are joined by a fish, *Thermarces cerberus*, a vent specialist named for the mythical hound of hell. References to the infernal netherworld are not uncommon among our descriptors of these places, and of the organisms that live here, conjured by the scalding temperatures and the volcanic activity. The vent fish, its smooth, pink form giving it an air of naked vulnerability, cruises around among the *Riftia*, apparently indifferent to the heat, the dark, the toxicity and the pressure. It eats the shellfish that cluster here, and in doing so performs a service to the tubeworms, clearing the limpets whose presence can hamper the ability of juvenile *Riftia* to find the space to set up home here.

All of these creatures are fuelled, directly or otherwise, by a diversity of bacteria that feed on the chemical effluent of the vents. But while *Riftia* and its lodgers enjoy a little warmth, preferring water temperatures comparable to that you might find at a coral reef, there are other organisms that like it hot – very hot. Microbes in particular have managed to find a way to survive some of the most extreme conditions imaginable. One, known rather prosaically as Strain 121, can prosper in temperatures as high as 121°C – hence the name – and can actually survive short stints in even hotter water. This is an organism that'd want you to turn the heat up if you put it in a pan of boiling water. Any other cell encountering such thermal extremes, or anything even close to this, would die instantly. Exposed to this heat, most proteins denature, collapsing into a tangled heap (which is why eggs and meat change colour and texture during cooking), and the fats that make up cell walls melt. But for so-called hyperthermophiles like Strain 121, this doesn't happen. They have evolved sophisticated biochemical scaffolding in their proteins that keeps them together at high temperatures, as well as using novel materials to protect their cell walls. Aside from being an astonishing example of the adaptability of life, these hyperthermophiles have much to offer us. The PCR tests that were at the forefront of our defences against Covid and that forensics teams use to analyse suspects' DNA rely upon the enzymes from these heat-loving organisms.[5] More than this, the enzymes' robustness to high temperatures provide us with analytical techniques that have huge potential in fields as diverse as the sustainable production of biofuel, mineral extraction and medicine.

The superheated extremes of the vents are strictly the preserve of microbial life, but there are animals who push the limits, too. One such is the Pompeii worm, whose scientific name, *Alvinella pompejana*, pays homage to the legendary submersible. Pompeii worms occupy sweltering tubular penthouses at the top of hydrothermal vent chimneys. These aren't just hotter versions of *Riftia*, though. They're much smaller, at only about a finger's length, and they don't farm bacteria within their bodies. Instead, they slurp up microbes from their surrounds using crowns of rusty-coloured tentacles mounted on their heads. What really sets Pompeii worms apart from their distant relations, however, is their toughness. The temperatures inside their tubes might reach a searing 84°C, with occasional spikes even exceeding this, potentially making it the most heat tolerant of all animals. They cope in the most extraordinary manner imaginable, seeming to protect themselves using a homespun

5 Polymerase Chain Reaction (PCR) tests are used to amplify DNA, thereby making it easier to analyse.

version of the fireproof underwear worn by Formula 1 drivers, a shaggy affair consisting of millions of bacteria.

The bacteria don't just support the worms against the heat, they also detoxify the water in the interior of their shared tubular home, removing heavy metals and processing the excess hydrogen sulphide. In return, worms reward them by secreting tasty mucus from glands along their bodies for the bacteria to eat.[6] Worm mucus might not sound all that appealing to us, but to bacteria of many species it's a major draw. Not all of these are quite so well suited to the worm's requirements, however, and to discourage all but their favoured varieties, Pompeii worms produce a powerful, targeted antibiotic. Although it's unclear whether this might one day have an application in human medicine, it's a further reminder of the rich potential of the deep.

The conditions at hydrothermal vents are so challenging that you could be forgiven for thinking that only a handful of animals could rise to the challenge of living here. Yet in the five decades since their discovery changed our understanding of biology, over 600 vent species have been discovered. For context, that's about the same as the number of different birds that have ever been recorded in Britain. And with new vents being discovered regularly, these numbers continue to climb. Where it was once assumed that the zoological marvels of the Galapagos Rift were representative of vent animal communities throughout the world's oceans, the more we look, the more diversity we find, and the more incredible creatures turn up. Part of the reason for this is that the conditions that support vent communities are few and far between. Each vent field represents a tiny life-support bubble amid a huge expanse of hostile,

6 Despite the many services provided by its microbial lodgers, the Pompeii worm gives landlords a bad name by periodically, and unsentimentally, devouring some of its tenants.

near-freezing waters. The remainder comes down to the fact that no two vent systems are exactly alike.

At the fountainhead of hydrothermal discovery, the Galapagos Rift system revealed by Bob Ballard and his team in 1977, the water shimmers gently from the seabed at little more than room temperature. A little further away at the East Pacific Rise, it's a very different story. This was the site that the towering chimneys that came to be known as black smokers were first found in 1979, jetting out plumes of inky, mineral-saturated water superheated to an incredible 350°C. As one of the observers aboard *Alvin* during that first encounter, the geophysicist Jean Francheteau, remarked in awe: 'They seem connected to hell itself.' And hellish it might have been, had the submersible got too close; the first temperature probe that they used to investigate simply melted, an event made all the more concerning by the fact that the probe was made of the same material as *Alvin*'s viewing ports. Since this discovery, other black smokers have been located that exceed 400°C, and with pH levels as low as 3, similar to that of sulphuric acid.

The diverse chemical and thermal environments of hydrothermal vents support equally varied wildlife. To survive these extremes, evolution has conjured some of the most extraordinary animals on Earth. There are the pallid, mango-sized yeti crabs, discovered only in 2005, that bear a forest of luxuriant hair on their appendages, supporting the growth of bacteria that the crabs then snack upon. And there are the spangly, iridescent scale worms that cluster at some of the world's deepest hydrothermal vents, 4 kilometres down in the Gulf of California. They look more like bristly woodlice than worms, and they're lent a glamorous air by their flamboyant rainbow colouration.

Few creatures are as exceptional as the scaly-foot snail, a golf ball-sized gastropod native to just three hydrothermal vents in the

Indian Ocean. What makes it so special is its cloak of overlapping iron pyrite scales, giving it the mien of a medieval knight. Indeed, it's the only iron-clad animal. Surprisingly, though, the purpose of its attire isn't to stymie predators but to funnel toxic sulphurous waste out of its body. What's good for the snail may also be good for us; iron pyrites are used in the production of all manner of green technologies, from solar cells to lithium batteries to water-splitting catalysts, but our traditional approaches to fabricating them rely on lots of heat energy. Evolution has equipped the snail with a much more efficient means of doing this, which promises to revolutionise the way that we produce pyrites.

If you were to keep travelling eastwards from the home of the metal snail, to arrive at the vents along the Mid-Atlantic Ridge, you wouldn't find any of the giant *Riftia* tubeworms that are so successful in the East Pacific vents. Mussels and clams abound here, as well as huge swarms of *Rimicaris* vent shrimps. These chipolata-sized crustaceans gather on vent chimneys in densities of up to 3,000 per square metre. Like many vent animals, they grow their own food, in their case tucking millions of bacteria in the space beneath their shells and absorbing the organic matter produced by those bacteria straight through their skin, which saves them the bother of eating in the normal way. But the weirdest thing about the vent shrimps is that, as adults, they lack eyes in the place you'd expect to find them: the head. Instead, they have a giant eye in the middle of their back. Scaled to the size of you or me, it'd be about the size of a can of soup. This eye doesn't form images, though, it's more of an infrared detector that it uses to find the heat of a hydrothermal vent in the darkness, and to avoid the hottest parts of its environment. Not that it's flawless; surveys of the shrimps on the Mid-Atlantic Ridge vent fields report a high proportion of individuals with burn marks, which likely represents an occupational hazard for an animal that inhabits such underwater infernos.

* * *

Aside from the threat of being accidentally poached, or succumbing to sulphurous toxins, vent animals face another fundamental challenge: instability. Hydrothermal vents might last for decades, possibly even centuries, but they can be destroyed in the blink of an eye by lava belching from the mantle. The Rose Garden, an area of the original Galapagos Rift system named for its multitude of swaying giant *Riftia* tubeworms whose bright-red gill plumes called to mind a field of flowers, was obliterated by a lava flow early this century. But long before this cataclysmic event, the tubeworms had been ousted by mussels and clams. The spreading seafloor shuffles vents along, changing their flow and creating new conditions in which some of their inhabitants prosper and others succumb.

As one vent dies, another gushes into life elsewhere. Success for a vent specialist is therefore about more than just adapting to the conditions in these ephemeral pockets of habitat, it demands that they find a way to cross a cold and unwelcoming void to colonise new vent territories. This is a task that is beyond the ability of the adults of most species to negotiate; with limited mobility, their lives are inextricably tied to their chosen vent and when it expires, so do they. The challenge of breaking new ground therefore falls to their offspring. It's for this reason that the young of hydrothermal vent animals are radically different to their parents. This difference isn't just about appearance, it encompasses their characteristics and preferences, their physiology and diet. Expelled as eggs from the vents to begin their development in the icy waters of the deep, these are functionally different animals.

With hundreds, or even thousands of kilometres separating vent fields, the youngsters' journey can take an extremely long time. The eggs of heat-loving Pompeii worms enter a state of suspended animation as they travel through the blackness, calling to mind the 'cryosleep' of interstellar astronauts in sci-fi movies. Rather than being deadly, as it might for their adult equivalents, the cold preserves vent larvae, slowing down their maturation and, in effect, expanding the range over which they're able to travel in the slow deep-sea currents. Some vent mussels can spend more than a year simply drifting, while the young of giant tubeworms are similarly unhurried – they don't even develop a mouth until they're several weeks old. The likelihood of any of them actually finding a hydrothermal vent are infinitesimal, so for their parents, this is a numbers game. Producing millions of eggs gives them more tickets in the lottery.

Even so, a strategy that relies on relentlessly churning out vast numbers of energetically expensive eggs and sperm into the unforgiving deep like broken slot machines isn't ideal, so some species have developed ways to tilt the odds in their favour by channelling their inner gophers. Beneath the vent fields, another ecosystem has recently been revealed, one that calls to mind the famous catacombs below the boulevards of Paris. Here, among the cracks and fissures of the basalt, reside larvae that have found an alternative to open-water wayfaring. The youngsters can disperse through pores in the crust in relative safety, avoiding exposure to an army of hungry seabed predators, and make their way towards new vent sites. Cunning as this tactic is, however, it's one that is likely best suited to migrating between nearby vents.

At every established hydrothermal vent field that we've so far discovered, even ones that are huge distances from their nearest neighbours, our cameras have recorded a profusion of animals. And in many cases, the virgin lava of new vents is colonised surprisingly rapidly. Does this mean that the world's oceans are saturated with long-distance vent larvae just floating around and waiting for a lucky break? It seems unlikely. As far as we can tell, there are hundreds of vent fields around the world, but these collectively amount to a total area of about 50 square kilometres, the equivalent of a small city. Finding a needle in a haystack would be child's play in comparison to locating a vent field in the immensity of the world's seas. So perhaps the network of vents is connected by a series of stepping stones, truck stops on the oceanic highway where vent animals might find refuge. Remarkable as hydrothermal vents undoubtedly are, they're not the only places in the sea that provide the kinds of fuel needed to support the bacterial partners of the animals that live on them.

Cold seeps are areas of the seabed where methane, hydrogen sulphide and other foul-smelling chemicals bubble up from cracks in the bedrock. These chemicals are manna to a host of chemosynthetic bacteria and, in turn, the animals who depend on the microbes' chemical-converting capabilities for their own food. There are perhaps ten times as many cold seeps on the ocean floor than hydrothermal vents, and this abundance is potentially what makes them so valuable as stepping stones. Add in the hundreds of thousands of whale falls worldwide where methane-munching and sulphur-scrunching bacteria also live, and suddenly the distances separating the vent fields seem far more manageable.

Although hydrothermal vents were discovered only recently, their finding still predates those of cold seep and whale fall communities. Consequently, our understanding of these peculiar deep-sea assemblages is in its infancy, and the question of whether individual animals hop from vent to seep to whale fall remains open, not least because tracing the movements of microscopic larvae is challenging. What we can say is that the species that live at these kinds of habitats have close evolutionary links. Some animals, like the sea cucumber, *Chiridota*, happily live at all three types of habitat. Others have made a more decisive jump to a hot-water lifestyle from an ancestral seep existence. Both habitats are rich in *Bathymodiolus* mussels, who entertain chemosynthetic bacteria in their gills that translate the chemicals on offer in these habitats into organic matter for their hosts. Similarly, the giant tubeworms of the vents have close relatives living at seeps, and while *Riftia* lives a turbocharged life, bathed in warm water, and is one of the fastest growing of all animals, its kinsworms, *Escarpia* and *Lamellibrachia*, take things a little slower. Indeed, some individual Escarpians alive at seeps today are older than the United States by half a century or more. Though these creatures are indisputably animals, some of their characteristics make them seem like deep-sea trees. Aside from their stupendous longevity, they have roots that anchor them to their substrate and draw in nutrients, and they grow in thick groves as tall as an average human, making a miniature forest on the ocean floor, inhabited by shrimps, sea cucumbers and lobsters. Few predators have much of an appetite for these tubeworms; steeped in sulphides, their blood would likely be toxic to anything that ate them, so they live their slow, serene existence, decade after decade, and century after century.

Seeps, vents and falls are unique environments, but it'd be wrong to think of them as being entirely anomalous, standing apart from the rest of the deep-sea biome. Everything in the ocean is interconnected and these habitats play a disproportionate role in the ecology of the deep sea. With so few nutrients percolating down from above, vent and seep organisms, dislocated from a reliance on solar energy, provide huge amounts of food for other animals. Many different kinds of predators visit seeps and vents, gaining a substantial share of their diet from such places. Meanwhile, crowded beds of seep clams broadcast enormous numbers of eggs into the wider environment; every square metre of these clams produces the best part of a billion eggs each year, the overwhelming majority of which end up in the bellies of deep-sea creatures.

The importance of the vents and seeps doesn't end here, however, and nor is it limited to the deep sea. The processing of methane by bacteria at seeps plays a critical role in preventing this potent greenhouse gas from reaching the atmosphere. In addition, the plumes of hot water that gush from hydrothermal vents rise high in the water column and disperse over thousands of kilometres. The cargoes of dissolved metals carried by these plumes act as giant, ocean-scale fertilisers, providing crucial fuel for surface phytoplankton, and supporting both the foundations of the shallow-water food chain and the sequestration of a vast amount of carbon dioxide from the atmosphere.

It's less than half a century since we discovered vents and seeps, and we're still only beginning to appreciate their importance, and that of the deep sea in general, to the healthy functioning of our planet. Only in recent decades have we had the technology to penetrate these mysteries, and there is a long way to go before we can claim a full understanding of this, the last wilderness on Earth. The pursuit of knowledge, however, is shadowed by another pursuit, that of money. The very deepness of the sea has, to now, largely protected it from our depredations. That is now changing, and the dividends of the depths are now firmly in the sights of mining companies.

* * *

Strewn across the seabed, largely at depths beyond 4 kilometres, are huge numbers of strange, rounded nuggets, dark grey to brown in colour and ranging from microscopic to melon-sized. These are polymetallic nodules, accretions of dozens of different minerals, rich in manganese, iron, copper and nickel, among much else. The quantity of nodules dispersed across the ocean floor almost defies belief. By some estimates, there's more than a *trillion* tonnes of them, representing a treasure trove of raw materials. Even harvesting a small proportion of this would, according to mining advocates, lead a manufacturing revolution that would allow a rapid transition to green technology, and provide our best opportunity to solve the energy crisis and arrest global warming.

Such a boost to our efforts to pivot away from fossil fuels is appealing. Adding impetus to this was a recent report by the International Energy Agency, which warned of a looming mineral supply crisis in the face of steepling demand for battery components. Extracting these materials on land is a process dogged by pollution and contamination, and in some places, human rights abuses. Moreover, collecting nodules from the deep sea is likely to be incredibly lucrative. At an estimated price of $500 a tonne, with nodules lying thick like windfall apples on the ocean floor, fortunes are right there for the taking. The incentives – environmental, moral, financial – are considerable. Collecting polymetallic nodules from the sea seems like a panacea. So, what's stopping us?

It's an unavoidable fact that deep-sea mining would obliterate ecosystems that have developed and emerged over the course of millennia in the slow-paced world of the deep. The effects spread beyond the immediate operational area, since the hulking extractive machines proposed would kick up mountainous plumes of sediments as they scour the seabed, choking and despoiling millions of square kilometres of the deep sea. This fact isn't contested – the areas in question are referred to, unsentimentally, as sacrifice zones, and within them, life would be annihilated. We've hitherto taken for granted the regenerative powers of nature, its capacity to bounce back from the body blows that we deliver. This, however, is different. Even if the deep ocean could recover from mining, it would do so on a completely different timescale to that which we're used to. It is, in effect, irreplaceable.

The nodules at the heart of the controversy are the result of a strange phenomenon in which metals precipitate around a particle of hard material on the seabed, building up gradually, layer upon layer. It's the geological equivalent of pearl development in oysters, but it's a breathtakingly slow process; the gradual accretion that generates a potato-sized nodule takes millions of years. The existence of a large majority of the creatures in the deep are intertwined inextricably with these structures, both because they provide a hard substrate for these animals to attach to and because the nodules have recently been found to act as 'geobatteries', electrolysing water molecules and generating precious oxygen to support the deep-sea environment.

Some might say that they're unconcerned by the fate of a community of weird, abyssal animals. This misses the point on several levels, the first of which relates to a little self-interest. Even if I can't convince such people that these animals have intrinsic value in their own right, then consider what we've already gained, and stand to gain, from similar marine organisms. Novel drugs for treating conditions such as cancer, hypertension and HIV, to name but a few, have been discovered in exactly the kinds of creatures that these people might loftily dismiss, including sponges, crinoids and tunicates. The ocean, and especially the deep sea, is a drugstore of enormous potential, and it's one that's largely untapped. The extraordinary challenges of life in the deep sea have given rise to equally extraordinary biochemical solutions. By most estimations, the chances of finding new pharmaceutical compounds in the sea are several hundred times greater than in the terrestrial environment. There's a strong, and obvious, argument not to destroy the deep before we've effectively explored it.

It isn't just medicine that stands to lose. Bioinspired materials have the promise of revolutionising the way we live. For instance, the humble glass sponges of the deep have evolved a unique structural framework whose strength and lightness significantly outperforms those designed by human engineers. Meanwhile, studies of the skin and tissues of animals including squid and jellyfish are driving new developments in artificial materials that can respond adaptively to their environment, with applications for everything from smartphones to smart windows, and from thermally sensitive blankets to infrared movement detection. The deep sea can certainly yield metals, but it has so much else to offer.

At the same time, the ocean floor stores unimaginable amounts of carbon, locked away in the sediments. Disturbing this in the process of collecting nuggets carries the risk of destabilising the marine carbon cycle and releasing massive reservoirs of hitherto trapped carbon dioxide and methane into the atmosphere. Interred alongside the greenhouse gas in the thick mud of the deep-sea floor are other elements, including radioactive materials. Liberating these, and sending them spiralling into the surface waters, runs a strong risk of compromising the marine food chain, and with this, the safety of the seafood that we rely on.

Throughout our history, we've played a kind of ecological Jenga, whipping out the lowest block and trusting to fate that the whole stack doesn't come crashing down. This has never been more so than in the case of deep-sea mining. Though there is no doubt that we need the materials, it would represent an incredibly high-risk strategy to sacrifice deep-sea ecosystems when we simply don't fully appreciate what's down there, or what holds the ecology of the habitat together. The deep sea provides the foundations of the entire marine ecosystem, which is, in turn, the planet's life support system – and ours. At a time when we have no idea what we stand to lose, I'd argue that we'd be insane to take the gamble.

The only thing that's holding the deep-sea gold rush back is a fragile consensus under the aegis of the International Seabed Authority (ISA), a body established under a United Nations treaty, the Convention on the Law of the Sea. All the world's major nations, with the exception of the United States, which has never ratified the treaty, participate in the ISA, whose remit is to safeguard the deep sea and to regulate mining activities. State-sponsored and private companies are pushing hard to begin mining, however, and the ISA, which is sympathetic to mining interests, has at times looked perilously close to firing the starting pistol on nodule extraction. At the time of writing, some 40 countries have called for a moratorium on deep-sea mining. This includes the UK, much of the EU, Canada and many Central and South American countries, as well as some of the Pacific nations. From others – China, the US, India, Russia and Australia, for instance – the silence is telling. How ironic it would be if a part of our planet that has often been suggested as the cradle of life itself becomes simply another page in the catalogue of destruction that blights our civilisation.

* * *

The deep sea has an allure that sets it apart from any other place on Earth. This allure exists despite its inaccessibility and its hostility, residing instead in its place as a proving ground for biology and as a triumph of evolution. It was once confidently predicted that no creature could exist beyond 300 fathoms, or about 550 metres, but we now know that life is present at every point from the surface to the very deepest point in the sea, more than 11 kilometres below the waves. The extraordinary demands imposed by the pressure, the cold and the darkness of the deep might seem to conspire to stifle the spark of existence, yet to virtually every challenge posed, life has conjured an answer. The results can be seen in the images of phantasmagorical animals captured by cameras mounted on submersibles. Photographs and video footage are one thing; direct experience is quite another. It has always been my dream to explore our planet's inner space and to see for myself, so when the opportunity arose, I took it eagerly.

Roatan is the largest of the Bay Islands, a scattering of peaks that rise from the Caribbean 50 kilometres north of Honduras. Lush tangles of tropical vegetation cling to the flanks of the ridge that runs lengthways along the island, while palms and mangroves claim the ground at sea level and picket the pale ribbon of sands at its margins. Scarlet macaws glide among the forest canopy, crimson comets among the green, and hummingbirds quarter the undergrowth, their darting, buzzing flight startling the watchful agoutis. Diverting as the island's birds and mammals are, however, it is not these that I've come halfway around the world to see.

A narrow shelf of corals and seagrass meadows girds the island, but travel just a short distance beyond the shore and the seafloor falls steeply away. On the north of the island, the Cayman Trench descends precipitously. It was this proximity of deep water to land that brought submarine expert Karl Stanley to Roatan years ago. His base at Half Moon Bay, once a pirate stronghold, is just a couple of hundred metres away from the edge of the trench.

After a couple of weather delays, I finally get the go-ahead on a blisteringly hot Caribbean afternoon. The shore swarms with giddy local children, leaving wet footprints as they scale the unforgiving volcanic rock to dive into the cool sea. Amid the excitement, suspended above the water at the end of a wooden dock, is Karl's extraordinary creation: the submarine *Idabel*. It's a small craft, not much larger than a regular family car, with a dome mounted on top as a conning tower. *Idabel*'s bright-yellow livery and rounded profile, in combination with its large portholes, reminded me irresistibly of one of the Minions from the film *Despicable Me*. The comical imagery, however, belies the fact that *Idabel* is a masterpiece of engineering and a veteran of almost 2,000 dives into the darkness and mystery of the deep. Now, on its latest voyage, I am to venture into the abyss myself.

Any trepidation I'd felt in the run-up is replaced by excitement as I thread myself through the narrow hatch and squeeze myself into the submarine. Part vehicle, part life-support system, the interior is crammed with an array of equipment, placing free space at a premium. Despite the confines of *Idabel*, however, I have no sense of claustrophobia. In front of me is a large viewing window. Right now, it's offering a split-screen panorama, the vivid blue of the sky giving way to shoals of silvered slivers that dart to collect unseen morsels from the water's surface. Karl boards the vessel behind me and readies it for departure. I can see him only from the knees down as he stands in the central section, but I listen to his disembodied voice as he commentates on the preparations.

Finally, we're off, the faint whirr of the propellers drowned by the sloshing of wavelets against the hull of the submarine. We make stately progress across a shallow, sunlit meadow of seagrass for a few minutes before the waters ahead darken forebodingly. The darkening tells us that we've reached the lip of the trench. Beyond this, the seabed drops away to a depth of 7 kilometres; were anything to go wrong, there's no coming back from that. As the tanks begin to fill with seawater, the nose of the submarine tilts forwards and we slide beneath the waves.

The descent is slow, controlled. There are no points on which I can take a bearing in this seemingly limitless aquatic firmament, so it feels almost as though we're not moving. But if I crane my head to look up through the upper surface of the window, I can see the faint glow from the surface receding far above. Around the sub, the waters dim almost imperceptibly as the sunlight yields to the depths. I watch with a thrill as the needle on the depth gauge tips below 200 metres: at last, I'm in the deep sea. At this depth, little light remains, and beyond the viewing port is an expanse of diffuse blue-black. The temperature has dropped, too, and a film of condensation forms on the cool metal of the hull and begins a slow rain on my head. Happily, that's the only water inside *Idabel*.

Though I'm unaware of it as I sit cross-legged in the steel bubble of the submarine, the pressure is ratcheting up as we descend. With each additional 10 metres, the pressure on the viewing window alone equates to another 7 tonnes of water. By the time we reach our destination, a shelf on the canyon's flank, 610 metres below the sea's surface, the window will have to withstand a stupendous amount of pressure, the equivalent of 45 fully grown African elephants standing in a precarious tower on its domed surface.

I try to keep these worrisome calculations from my mind by concentrating on looking out for life. Every few seconds, some new curiosity emerges onto the stage before me. Squid, startled by our appearance in their realm, jet away leaving clouds of brownish ink. The extraordinary colonial animals known as pyrosomes, or more informally, sea pickles, drift languidly by, consumed by the business of consuming plankton. Deep-sea shrimps charge past, trailing enormous antennae far longer than their own bodies, while fish zigzag into the darkness beyond the pool of *Idabel*'s lights. Among the cast of creatures, Karl points out a bristlemouth, momentarily frozen in the spotlight. The spindly, insubstantial body of this fish terminates in disproportionately large jaws, lined with a forest of fine, hair-thin teeth that give them their name. Unheard of by most, there may exist as many as a million bristlemouths for every person alive today. Our lack of awareness of the existence of such a superabundant animal emphasises better than anything the mystery of the deep sea. Recovering its composure, the bristlemouth twitches and glides beyond the glare of the lights, returning to the anonymity of the blackness.

Another 20 minutes of careful descent takes us to the deepest part of the dive. We're approaching the shelf over half a kilometre beneath the waves. Shapes and structure begin to emerge from the darkness, and it's a view beyond anything I could have imagined. Basalt boulders are coated with a profusion of life. Huge, intricate yellow fan corals rise like the sails of sunken galleons, catching not wind, but particles of drifting food. In the delicate rigging of the coral, snake stars wind and twist, hoping to intercept the coral's food. A ghostly, white garden of *Lophelia* coral grows thick across the surface of the rocks. This is the mainstay of deep reefs around the world, looking like a wintry forest of Lilliputian trees, studded with batteries of cup-shaped polyps. Here and there, whip corals grow, some coiled like pigs' tails, others trailing out obliquely into the darkness like discarded aerials.

It can come as a surprise to some that there are corals in the deep at all, given our association of them with shallow, tropical waters, yet more than half the world's corals live in waters like these, far beyond the sun. The cold, in combination with the lack of food, means that they have a slow pace of life. Some of the corals I'm seeing here are certainly hundreds of years old, possibly even thousands. In 2009, a deep-sea black coral collected off Hawaii was dated to an age of over 4,000 years. These might grow as little as 5 micrometres per year, at which rate it would take 15 years just to enlarge by an amount equivalent to the width of a human hair. The hefty corals before me must be ancient at that rate; it's exhilarating to be among organisms that could have been alive at the time of Tutankhamun.

Bestriding feather-like cascades of soft gorgonian corals, squat lobsters respond defiantly to *Idabel*'s intrusion, their claws primed and raised in our direction like gunslingers from the Wild West. A cluster of brittle stars gathers at the edge of an outcrop, their legs reaching outwards into open water in the hope of embracing a meal. Vase-like glass sponges, named for their spiky silica scaffolding, have found a berth here among the crowds, and they're joined by fat, furry-looking sea urchins that ease their way through the throng. There are sea lilies, their long stems bearing aloft a crown of feathery branches. They look more like palm trees than lilies; in fact, they're not plants at all, but animals – close relatives of urchins. Oddest of all is the fact that should they tire of their surroundings, some sea lilies can detach themselves and scamper along the seabed using the root-like structures that normally anchor them as makeshift legs, like miniature marine Ents.

Below the restive lilies, a gang of brachiopods cluster beneath an overhang. Better known as lamp shells, they resemble clams on sticks. Seeing these is like time travelling; they're throwbacks to a distant age, reminders of how the deep is a refuge for biological museum pieces. Hundreds of millions of years ago, these were common throughout the world's oceans, but competition from arriviste molluscs, which they resemble but are not related to, consigned them to permanent exile here among the darkness. Collectively, the vista is almost overwhelming in its vibrancy and exuberance. I'd expected life in the deep sea to be sparse, but this cornucopia of creatures has taught me that this isn't always so.

Just like a shallow reef, this deep-water coral garden supports a host of other, more mobile animals. At our approach, a pair of slimeheads slink into the shelter of the corals and overhangs. Rebranded by marketers as the orange roughy, this species is a motif for our troubled relationship with the deep sea. Diffuse as life may be in much of the deep, it's nevertheless a biome that occupies a seemingly limitless space, so when roughies congregate to spawn, their shoals can be enormous. Regrettably, once the presence of such bonanzas became known, it was only a matter of time before commercial interests took notice. Beginning in the late 1970s in New Zealand, the roughy rush brought huge catches and equally huge profits. By 1990, something like 100 million orange roughies were being hoovered out of the deep each year. Worse still, the bottom trawling used to catch them bulldozed through thousand-year-old corals in the blink of an eye, smashing the delicate deep-sea ecosystem.

The subsequent collapse of the fishery was predictable, especially given the nature of our characteristic 'shoot first, ask questions later' approach. When we began targeting orange roughies, we knew almost nothing about their life history. The assumption was that it'd be comparable to other commercial fish. It isn't. Roughies grow slowly, maturing only when they're 30, or possibly even 50 years old. Even then, their fecundity is perhaps only about 10% of that of an equivalent, shallow-water species. Oh, and they live considerably in excess of two centuries; a roughy served in a restaurant today might well have been a teenager at the time of Waterloo.

Powerful lobbyists are at work to try and convince consumers of the sustainability of the remnants of this fishery, but what might ultimately save the roughy is a healthy dose of pragmatism. The longevity of the species means that they have the time to accumulate all manner of toxins in their flesh. Mercury, in particular, is a problem, with concentrations up to it's actually 50 times higher than in sardines times those found in other food fish. Targeting orange roughies is a terrible idea for a host of different reasons, but it's a lesson that we haven't fully learned yet. As I watch one of these charming fish, safely ensconced in the midst of a rich and diverse deep-sea reef, its downturned mouth lending it an air of pathos, I can only hope that the trawlers don't come to wreak their destruction around Roatan.

Among the plethora of peculiarities down here, however, one creature really stands out: the dumbo octopus. It's named for the pair of ear-like appendages on either side of its body that it flaps, winglike, to propel its way through the murk, like the flying elephant of Disney's imaginings for which it is named. The dumbo's peculiar means of getting about, as well as its short arms and dark, mournful eyes, give it a very different appearance to other octopuses, but like them, it's a hunter. It cruises a short distance above the seabed, seeking prey such as shrimps and prawns and, having found a victim, pounces to subdue it. When the hunter becomes the hunted, however, the dumbo has a strange trick up its many sleeves – it inflates itself in a manner known as ballooning. Each arm fills rapidly with water, becoming distended in the process, and suddenly the octopus is transformed, as though it were an extra in 'Cinderella', into a turgid pumpkin. Who's to know what a potential predator makes of this (surprise would seem likely to be the least of it), but one thing's for certain – its augmented size would make it difficult for any except the largest of predators to deal with.

Megafauna substantial enough to take on a defensive dumbo are relatively rare in the deep; nevertheless, there was one such that I hoped to see on my next descent in *Idabel* the following night. To have any chance of seeing this creature, we needed to tilt the odds in our favour, and the best way to lure a predator is with food. We were using meat, a mix of offal and roadkill, and lots of it. In preparation for our dive, Karl had arranged to deploy the bait from a boat an hour or so before our dive. This done, all I could do was keep my fingers crossed while I watched the distant street lights of West End disappear from view as I dipped below the waves aboard *Idabel* for a second time.

Less than half an hour later, Karl expertly found the cache he'd dropped from the surface, 400 metres above us, but a mist of sediment and the scattered remains of the bait announced that our quarry had beaten us to it. We began to circle the area, peering into the blackness for any signs of our mystery diner. I was losing heart when Karl called out; he'd spotted something off to starboard and once he'd turned the sub, I got my first sight of a deep-sea behemoth: a female sixgill shark, coasting along the edge of the undersea cliff. At over 5 metres in length, she's longer than the submersible, and her apparent indifference to our presence allows a close view. As she passes by, she seems to peer into the sub, weighing us up. I look into her phantom green eyes, a hallmark of deep-sea sharks, and share a moment of mutual contemplation.

Her cavernous mouth hangs slightly agape, providing a clear view of her battery of saw-like teeth, and slightly further back is the array of six parallel gill slits that give the creature its name. Most modern sharks boast only five gill slits, and this extra pair on my new acquaintance is a trait of ancestral – some would say primitive – sharks; sixgills have barely changed in the last 200 million years. She is no throwback, however; she's the top predator here, a survivor whose evolutionary stasis speaks of consummate success.

Finding little to reward her curiosity, the enormous shark swims languorously away, her sinuous, metre-high tail fin arcing slowly from side to side as she merges into the sepulchral scene. Her retreat signals our time to depart. Gently, we lift off from the bottom, ascending in our halo of light while the darkness reclaims the ocean floor. Once we're in open water, Karl has a surprise in store. He cuts the engines, and the lights. Blackness descends, unbroken but for the faint glow of the instruments. All is silent and for a split second I wonder what's happening, but then the show starts. It's one of the most awe-inspiring things I've ever seen, a biological aurora generated by a cast of what seems like millions out in the waters beyond. As far as I can see, sparking pulses of brilliant blue flare and fade against a backdrop of velvet black as *Idabel* drifts slowly upwards through inner space.

* * *

The spectacular display that I experienced during *Idabel*'s ascent was provided by masses of bioluminescent animals. This ability to produce light is a feature throughout the oceans, but here in the deep it's pronounced. More than 90% of all the animals down here are bioluminescent. Most produce blue light, though across the gamut of species every colour of the spectrum, from violet to red, is represented. The light serves as a means of communication and one that can be used to attract or to deter, to inform or to deceive. It's ubiquity in the zoological lexicon of the deep, and diversity of colour dialects that exist among species, have earned light its title: the language of the deep.

The eyes of deep-sea animals are testament to the importance of bioluminescence. Rather than follow the evolutionary lead of subterranean and cave-dwelling species in abandoning vision, plenty of deep-sea creatures have disproportionately huge eyes. Size, however, is only one part of the equation. Deep-sea fish have eyes that are equipped with powerful lenses and superbly adapted retinas comprising several stacked layers of light-sensitive cells to harvest every stray photon that comes their way. Some, like the sixgill, have reflective layers at the back of their eye to recycle what little light there is. The pay-off of this hardware is impressive: it's been calculated that deep-sea fish have vision that is hundreds of times more sensitive to light than our own, so the scene that looked to me like pitch blackness at 400 metres is, to them, perhaps no more than an overcast afternoon. Impressive as it is, the ability of their eyes to capture light is only the beginning.

Anyone who's ever peered into an aquarium is familiar with the eyes of a typical fish. In the deep, however, things are a little different. For a start, many have yellow-tinted eyes. Confusingly, under the bright lights of a submarine, these can look not yellow but green. Regardless of that, the benefit to the fish is that it improves their ability to distinguish bioluminescence from any residual sunlight that makes it down this far. A handful of fish have evolved what amounts to a pair of binoculars, each eye forming into a tubular shape that provides an extremely detailed image. Hatchetfish have tubular eyes that point upwards, allowing them to distinguish the bodily outlines of their favoured planktonic prey against minute amounts of light that well down from the surface. The tubular eyes of telescope fish, in contrast, point forwards, again giving them an edge when hunting. The problem with tubular eyes, of course, is that while they offer excellent vision, they provide only a narrow field of view, leaving a substantial blind spot. So, the adoption of tubular eyes required an evolutionary plug-in to deal with this – light-sensitive cells on the outside of their eyes to warn of approaching menace.

When one of us ventures out at night into an unlit area, the scene appears monochromatic like an old film; the colours have vanished, replaced by a black-and-white analogue. This is due to our visual specialisation to daytime, and specifically because the cones that provide our perception of colour only function above a threshold of light. This leaves our rod cells in sole charge after the sun has set, and though they're sensitive to small amounts of light, on their own they're capable of providing only a simplified, colourless image of the world. The eyes of deep-sea fish, however, are much better equipped to navigate the dark and to pick up on the movements of those they share it with. Not content with the single, workaday rod photopigment gene that we make do with, some of them have expanded things dramatically. The silver spinyfin, a disc-shaped and lugubrious-looking fish, is the master of this. It boasts an impressive

38 different rod photoreceptor genes, enhancing its ability to perceive prey animals in the darkness as well as allowing it to distinguish among a swathe of different colour signals, and particularly the blues and greens that are the mainstay of bioluminescence.

The light itself comes from a chemical reaction involving a compound known as luciferin and an enzyme known as luciferase. Oddly, it's thought that the original driver behind the evolution of bioluminescence had nothing to do with light and everything to do with a deadly new substance that began to accumulate in the atmosphere about 2.5 billion years ago: oxygen. The ancient microbes that existed at the time needed to find a way of detoxifying oxygen and one way of achieving this was to lock it up in a new, less harmful molecule. This process 'excites' the molecule, as chemists put it, and as it calms down, it emits light. Nevertheless, this light was only ever an unintended by-product and of little interest to the microbes themselves as they struggled to avoid being poisoned by oxygen.

It wasn't long, however, before evolution began to harness the possibilities offered by this quirky means of generating light. Its value is suggested in the fact that bioluminescence has evolved independently at least 40 and perhaps as many as 100 times, numbers suggested by the diversity of different types of luciferin found throughout the tree of life. Building on the microbes' early work, animals have either invited them into their bodies as light-producing symbiotic partners or cut out the middleman by getting the luciferin through their diet. Though we have no means of generating bioluminescence ourselves, we've made historical use of it, from miners using the weak light emitted by bacteria thriving on fish skins to navigate underground to early submarines lit using luminous fungi. The Japanese have long used small marine crustaceans known as *umihotaru*, or sea fireflies. By rubbing dried *umihotaru* on their hands and mixing it with a healthy amount of spit, they could co-opt the animals, light and use it for signalling or even, at a push, reading on the darkest nights. The most valuable application of bioluminescence to us, however, came more recently, when green fluorescent protein (GFP) was isolated from the crystal jellyfish. When GFP is introduced to cells, it can be used like a microbiological highlighter pen to identify cellular activity, from gene expression to protein synthesis, which has led to huge improvements in the diagnosis and treatment of cancer, alongside much else.

The crystal jellyfish is just one organism among thousands that engage in luminous communication in the oceans. The magical light show that I was treated to during *Idabel*'s ascent isn't a particularly natural state of affairs – essentially the disturbance we created set off battalions of biological burglar alarms. Nevertheless, even under normal conditions, there's plenty of information being transmitted in this light language. For flashlight fish, bioluminescence potentially provides a means to attract their planktonic prey and, at the same time, allows them to keep in touch with one another. The light itself comes from a bean-shaped pocket under each of their eyes. The pocket is crammed with bacteria that do the job of providing the light, while the fish holds up its end of the bargain by supplying the bacteria with food and lodgings. The helpful microbes don't simply run free within the pocket; they're packed into miniature tubes, allowing the fish to keep control of when their light is visible to others. To make the lights shine out, the fish rotates these tubes so that they're pointing straight out of the pocket like an array of tiny torches. To switch the light off, the fish simply turns its tubes inwards. This happens quickly, often more than once per second, and in this way, the fish looks like it's blinking, but the effect is much more intentional than this. The blink patterns change according to context, producing a kind of semaphore that keeps the gang of flashlight fish in close contact. When they're stressed, as happens when an individual gets separated from the rest of the group, it blinks frantically to try and relocate its fellows.

Other animals have been equally creative in their use of light, employing it for both fair means and foul. One widely practised application of bioluminescence is camouflage. While it's true that there's very little light in the deep sea, the faint glow of the surface can serve to create a silhouette of a prey animal that's visible to predators lurking below. Camouflage, where animals blend into their background to prevent detection, is a familiar concept to us. The typical examples that spring to mind involve colour- or pattern-matching for concealment. In the deep sea, a similar principle is at work, but one that uses light to achieve its ends. The bellies of many open-water fish are covered in light-producing organs known as photophores, which emit a glow that emulates the intensity and colour of the light that filters down from the surface. The effect, known as counter-illumination, is to make those fish appear invisible from below. Exactly how they achieve this precise matching isn't known, yet fish have an innate understanding of how to pull off this incredible disappearing trick. Experiments using naïve, captive individuals that had never even seen the feat performed, let alone tried it themselves on account of lacking in their diet the raw materials needed to generate bioluminescence, revealed that, astonishingly, they get it right first time when they're fed with the planktonic animals that yield the necessary ingredients for the display.

It's been suggested that cookiecutter sharks turn this supposedly defensive measure on its head, using it instead as a ruse to attract the gullible. These small sharks are the chunk-biting scourges of the deep, excavating lumps of flesh from larger victims with their viciously sharp teeth. To perpetrate this fraud, the cookiecutters adapt their counter-illumination so that they appear like one of the prey species favoured by their victims. When the predator rushes in for the kill, the cookiecutter dexterously turns the tables and, instead of falling prey itself, manoeuvres to excise a neat hunk of meat, about the shape and size of half an apple, from the body of its dupe.

The mesmerising twinkling of blue light at the surface of the night-time ocean is the calling card of a group of phytoplankton known as dinoflagellates, which produce their bioluminescent exhibition when disturbed. The purpose behind this is to stymie their predators on the principle that my enemy's enemy is my friend. A hungry shrimp that sneaks into a phytoplankton patch wants not only to sate its appetite but also to keep a low profile. This latter aspiration is undone as soon as it trips the dinoflagellates' warning system, and the lights go on all around it. This draws the attention of nearby fish that are indifferent to the gustatory delights of the microscopic dinoflagellates but who are suckers for shrimp.

Not many deep-sea prey organisms live in the dense aggregations sometimes formed by dinoflagellates and consequently they must take a more individual approach to self-preservation. Some lanternfish emit a sudden burst of light from their tail to momentarily startle an onrushing predator, and squid are among many deep-sea animals that eject a cloud of shimmering bioluminescence that causes their attacker to misdirect its killing lunge. Worms, too, display illuminatory inventiveness. Freed by the buoyancy of water from the restrictions faced by their terrestrial kindred, plenty of marine worms have lifted themselves from the seabed and have adapted to a life in open water. In doing so, they have also shed the familiar worm shape. The bombardier worm looks like a pair of false eyelashes glued together, and it is small and vulnerable to all manner of threats. If challenged, however, the bombardier has a surprise in store for its persecutor: it throws globules of light out into the dark. The principle is the same as that used by fighter jets expelling chaff to distract incoming missiles; in this case, the predator is faced with the challenge of finding the worm among a confusing constellation of glowing balls. Alongside the bombardier is another worm that inspires pathos and ridicule in equal measure. The pigbutt worm is appropriately named, since it resembles a chickpea-sized, swimming bum that when chastised expels a fart of bioluminescence from the anatomically appropriate place.

Other animals take things a step further in their quest to evade death. The octopus squid, so named for its copyright-breaking eight arms, can wilfully shed the tip of one of its limbs to leave a predator short-changed. The discarded arm is equipped with glowing photophores and a tendency to wriggle appealingly, drawing the hunter's eye while the rest of the squid, an arm down but otherwise unscathed, jets into the safety of the shadows. And if the predator accepts the consolation of eating the lively, illuminated zombie arm, it potentially faces a new problem. The relocation of the arm to its stomach doesn't entirely staunch the glow, and now the predator, which relies on concealment for its own safety as well as for hunting, has a beacon in its belly that may be perilously visible for several hours. This isn't only true of octopus squid appendages. It applies to lots of other prey animals, and even those who've been entirely consumed can wreak this kind of post-mortem revenge. Naturally, a defence that is most effective after you've been eaten is arguably less than ideal, but if it causes a predator to think twice about attacking in the first place, it might just be an advantageous strategy.

As is common in nature, as one organism adopts a strategy to keep it ahead of the competition, its adversaries are driven to produce a counter-adaptation, and so an evolutionary arms race develops. One of the most interesting examples of this is the blackness of many deep-sea fish. And when I say black, I don't mean a half-hearted, everyday black. A material that's completely black looks that way because it reflects no light. Black paper might seem very black, yet it reflects about 10% of the light that hits it; even matt black paint returns around 4%. To be classed as ultra black, a material has to reflect less than 0.5% of incident light. Very few man-made things meet this criterion, with the exception of a special nano-engineered coating known as Vantablack that has a reflectance value of 0.05%. Incredibly, evolution has matched this in the deep; a delightfully named fish, the rough dreamer, has a reflectance value of 0.04%. It is 100 times blacker than matt black paint and is, in effect, completely invisible in the deep sea. Even transparent animals, of which there are many in the deep, are more detectable.

Why is intense blackness so important to deep-sea fish? For one thing, think of the fish that's just eaten an incandescent meal and finds its belly broadcasting its presence to any number of potential threats. To overcome this, fish like the bristlemouth have evolved ultra black linings to their innards, so that all the light they eat remains within. The defensive application of ultra black pigment in bristlemouths has been turned to a more predatory purpose in other species. The rough dreamer is an anglerfish, which hunts by use of subterfuge in the form of a bioluminescent lure, fashioned from the first rays of its dorsal fin. The lure arcs forwards to dangle temptingly in the water in front of its jaws. The ultra-black shading of the dreamer serves to avoid the occupational hazard of illuminating itself whenever it turns its lure on.

Lots of other deep-sea fish have adopted a similar form of trickery to entice their prey, but they're far from the only animals that do. The octopus *Stauroteuthis* is adorned with luminous suckers, while some chiroteuthid squid deploy glowing lures at the ends of their arms. Siphonophores, close relatives of jellies and corals, get in on the act too. *Erenna sirena*, named for the alluring sirens of

Greek mythology, beguiles victims into the embrace of its stinging tentacles by means of lures that dance among its deadly appendages. And what if you lack the means of producing light to tempt prey? No matter, you can borrow it from other organisms. Though yet to be confirmed, it has been suggested as a trick used by two giant hunters of the deep: sperm whales and megamouth sharks. The whales are thought to trigger a bioluminescent firework show by creating a disturbance in much the same way as we did during our ascent in *Idabel.* The whale simply opens its white-lined mouth, reflecting the light and amplifying it into a super-stimulus amid the dark to provoke the curiosity of its prey. Megamouths adopt a similar approach, instead making use of the white band that runs across their upper jaw like a chalk moustache to reflect the light and so attract its planktonic prey.

All of the previous examples have described localised bioluminescence, but there are instances that happen at much larger scales, some of which are visible from space. In one case, 100,000 square kilometres – roughly the equivalent in area of South Korea – of the Indian Ocean off the coast of Somalia was illuminated for ten days in early 2019. Another of the same size near Java the same year lasted for over two weeks. The organisms responsible are bacteria – countless trillions of them – and their goal, it's thought, is to try and find their way into a fish, or more specifically, into the animal's accommodating and nutrient-rich gut. Something similar is practised by the bacterial inhabitants of marine snow; by causing the snowflake to glow gently, they encourage fish to eat it, thereby giving them a passport to the microbial equivalent of Nirvana, a piscine intestine.

The glow generated by free-living, entrepreneurial bacteria is, however, relatively faint – the bioluminescence equivalent of mood lighting. In contrast, the springtime seas of Toyama Bay on the northern coast of the Japanese island of Honshu are host to a breathtaking natural spectacle. The waters sparkle with marine pyrotechnics, generated by one animal in particular, the firefly squid. These beautiful, thumb-sized cephalopods arrive in the surface waters of Toyama Bay in their billions from March to June, completing a journey that has brought them from the hundreds of metres below to spawn in the shallows. Stippled with a spectacular array of hundreds of photophores, each squid resembles a delicately beautiful constellation of electric blue stars. No one is entirely sure why they produce such an elaborate display; it may be for hunting or to avoid being hunted. Or it may be a signal that's central to the squid's amorous adventures in their deep, dark habitat. Despite their prodigious numbers, firefly squid are both chaste and choosy. The embodiment of monogamy, each individual selects one partner and mates just once. This complete, the males expire in a state of bliss, leaving the females alone to carry their cargo of fertilised eggs up to the distant shores. Here they spawn, before they, too, die.

* * *

An abiding impression from my trips in *Idabel* was how alien everything seemed. Even as a trained biologist, my familiarity with the organisms of coastal waters completely failed to prepare me for the sheer weirdness of what lives in the deep sea. Over millions of years in this cold, dark realm, deep-sea creatures have adopted all manner of extraordinary morphologies and behaviours that are a world away from the easily recognisable forms that we're used to seeing topside.

Perhaps disturbed by our presence in *Idabel*, a ghostly white sea cucumber does a most uncucumberlike thing: it propels itself from its station on the seafloor and, twisting its body like an energetic, spectral dancer, cavorts into the distance. Competent as this sea cucumber was at swimming, another echinoderm, vividly dubbed the headless chicken monster, has mastered the art. Its bulbous pink body is adorned at either end by diaphanous structures that call to mind a ballerina's tutu. It's these that the animal uses as fins to propel itself elegantly along. Tunicates, also known as sea squirts, are distant relatives of these cucumbers and are also here in the deep. While most are content to follow the tunicate blueprint of inoffensively filtering the water for morsels, one has adapted to the starving deep by being a little more proactive. One, the ghostfish, which is neither a ghost nor a fish, is essentially a giant mouth atop a stem-like body. Like its terrestrial equivalent, the Venus flytrap, it seizes onto any prey that ventures too close and slowly digests it alive.

At the edge of the *Lophelia* reef, a viperfish lurks like an assassin waiting for the unwary. This is a consummate deep-sea hunter, its long, tapering body fronted by nightmarish jaws bristling with outsized, inwardly curved and viciously sharp teeth. It may go weeks between meals, so this fearsome dentition is one means of ensuring that, once encountered, its prey has little chance of escape. An equally toothy cousin of the viperfish, the black dragonfish, has a further adaptation to boost its chances of success; it has no floor to its mouth, allowing it to snap close its jaws around its victim without the hinderance of water resistance.

The elastic stomachs of many deep-sea fish allow them to devour large prey, though few can match the capacity of the aptly named black swallower. In 2007, one of these was found floating dead at the surface of the Caribbean, having eaten a fish four times its own length. So distended was the swallower's belly that its victim could clearly be seen coiled up within behind a fine membrane of tautly stretched skin. Indeed, the stupendous feast was most likely what killed the swallower, since a meal that's too large will rot faster than it can be digested, poisoning the diner.

Not all deep-sea fish are swimming mousetraps. The tripodfish takes things easier, propping itself above the silty seabed on long, spoke-like fins. It doesn't seek the jackpot of large prey but instead sits, mouth agape, harvesting particles that drift on the faint current. The barreleye takes a more cunning approach to foraging. It's a strong candidate for the oddest-looking of all fish, looking overall like a Zeppelin with a dome-like transparent forehead that encases its large eyes. Why does a fish need what amounts to a visor? Recent research suggests that while the barreleye is a predator in its own right, it opportunistically filches food from the siphonophores with whom it shares its habitat. These rope-like colonial jellyfish may be anything up to 40 metres in length, and their strange woolly appearance emerges from thousands of stinging tendrils that cluster along their winding bodies. As the siphonophore entraps a bounty of shrimps, fish and plankton, it's a simple matter for the enterprising barreleye to parasitise its efforts, pecking off its prey and leaving the siphonophore empty-tentacled. The barreleye can't achieve this without running the gauntlet of the siphonophore's stinging cells, but its visor at least protects its delicate eyes as it does so.

As exhilarated as I feel to have experienced the deep, I realise that our 600-metre descent barely scratches the surface of the abyss. To put it into context, if you imagine the ocean's depth represented as a person's height, so that the surface is the top of the head and Challenger Deep is the soles of the feet, I've travelled only down to the eyebrows. The darkness of the realm beyond the submarine's portholes was almost complete, but not quite. Indeed, on a bright day, in the clearest waters, sophisticated equipment might register a few stray photons all the way down to 1 kilometre. To use my comparison of the ocean's depth in relationship to a standing human, this upper kilometre of water takes us down to the tip of the nose.

Beyond this lies a vast expanse of water, wreathed in permanent darkness, whose temperature seldom rises beyond a meagre 4°C.

Biomass decreases steadily with depth in the ocean, one reason for this being that there's less and less to eat the further you go down. The gentle fall of marine snow delivers some nutrients, but even that which isn't snapped up by animals in the upper layers is degraded by bacteria as it continues its slow drift towards the seabed. The dearth of food, in combination with the cold and pressure, places limits on biological activity. Nothing exemplifies this better than the story of the longest-lived baloney sandwiches in history.

The sandwiches in question were part of a packed lunch belonging to the crew of none other than the submersible DSV *Alvin*, the veteran of hydrothermal discovery and much else. Since its launch in the 1960s, *Alvin* has been attacked twice by large predators, faced jeopardy in the form of marine radioactive dumps and an unexploded hydrogen bomb, and even led the exploration of the wreck of the *Titanic*. Nothing, however, compares to the calamity in October 1968 when *Alvin* sank after a supporting cable snapped during launch manoeuvres. Fortunately, the crew managed to escape through the open hatch before the vessel disappeared below the waves and sank into 1,500 metres of water. It was the following September before *Alvin* could be recovered, and when it was, there was a surprise in store. The baloney sandwiches that the crew had left behind in their scramble to exit were still there, despite their exposure to seawater for almost a year. Indeed, they were in such good condition that researchers at Woods Hole felt confident enough to take a nibble of them, reporting that they still tasted good.

The phenomenon of the preserved sandwiches represents more than just a quirky anecdote; it provides important insights to life in the deep. The icy coldness of the water more than a kilometre down is one factor in slowing down the rate of decay, but more important is the pressure. Although seawater is suffused with bacteria, only a small fraction of those to be found in the deep sea are able to cope with the crushing conditions. The remainder simply lie dormant, a fact that slows down the process of decomposition to something like 1% of what it would be at the surface. When we're talking about a place where even bacteria struggle to operate, it tells you something of just how challenging life is for deep-sea organisms.

So how do animals survive in the bathypelagic and beyond? For one thing, they live life in the slow lane. They're dedicated energy conservers, with metabolisms that crawl along at about a tenth of the rate of surface animals. One upshot of this is the stupendous ages that some of these animals attain, and comparing the genomes of deep-water rockfish with those of close relatives in shallower habitats gives us an unrivalled opportunity to understand ageing. The cold and calorific restriction may play a part, but on their own, they aren't enough to explain the fact that some deep-sea fish can live for more than 200 years, a span 20 times longer than that of members of the same genus that live in the sunlit waters above. One of the main findings from a study carried out by Sree Kolora of the University of California, Berkeley, and colleagues, pointed at species differences in the immune system. This is especially interesting because the immune system is paradoxically both the hero and the villain of the body. It protects against pathogens, but at the same time, the kind of chronic activation of the immune system that becomes ever more typical as we age is responsible for inflammation of tissues and organs. This inflammation plays a critical role in triggering cancer, Alzheimer's disease and heart disease, among much else. One of the main differences in the genetics of the deep-sea Methuselahs versus their more ephemeral relatives relates to immune function, seeming to suggest that a little immune suppression might go a long way in promoting a long life. While you can't extrapolate directly from fish to people, it's nevertheless an exciting finding, and yet more evidence of the immense value of deep-sea research.

The slow pace of life in the deep extends beyond metabolism to encompass movement. That these animals are typically unhurried isn't just because they're conserving energy but reflects more generally the fact that they're not built for speed. Those that live in open water have delicate, gelatinous bodies and watery muscles to maintain neutral buoyancy, avoiding the need to expend effort on fighting gravity. The gooiness of specialist deep-sea animals represents something of a problem for biologists, because when they're hauled to the surface, something bizarre happens: they melt.

This is a by-product of their adaptation to the intense pressures that they're exposed to in the depths. It's a means of coping with the squeeze that would otherwise crush and deform both their cell membranes and proteins, preventing them from working. To counteract these forces, the cells of deep-sea organisms are equipped with high concentrations of chemicals called piezolytes. One of these, TMAO, is the same substance that assists sharks in osmoregulation. In deep-sea creatures, it plays another role, bracing cells against pressure and keeping proteins from warping disastrously out of shape. At the same time, their cell membranes are infused with specialised fats whose molecular configuration combats the crushing force while simultaneously relying on it: the membrane molecules only lock together in the right formation if there's enough pressure to bend them into shape. If the animal is raised to the surface by an eager researcher, the decrease in pressure causes its cells to fall apart and, hence, they seem to melt.

Despite the many challenges, the deep sea offers opportunities. Its open waters are populated by organisms that cruise around in near-infinite space, and which may never encounter a hard surface in their entire lives. As a result, many are delicate, almost ethereal creatures that take a bewildering variety of forms. Comb jellies, or ctenophores, are one of the most ancient animal groups. They're both simple and incredibly successful hunting machines, essentially mobile mouths that scull around using countless, tiny hair-like cilia that line their bodies like the oars of an ancient galley in search of prey.

Scyphozoans, more familiarly known as jellyfish, are also well represented in the deep. Like the comb jellies, they're predators, but unlike them, most true jellyfish rely on their stinging cells to subdue their victims. There are deep-sea crown jellies that look like disembodied eyes, fringed with dozens of thick, noodle-ish tentacles as a stand-in for lashes; psychedelic jellies whose central bell lies at the heart of a mass of stinging tendrils that project in every direction and in numbers that call to mind the seed head of a dandelion; and the most majestic of all, the giant phantom jelly, a deep sea behemoth with a bell the size of a yoga ball trailing thick, meaty arms, each as long as a telegraph pole.

Even the phantom jelly pales in comparison to what is, perhaps, the deep sea's most famous species, the giant squid, and its Southern Ocean kindred, the colossal squid. It's only in this century that either has been filmed, and still we know precious little about them. A juvenile giant squid, at 4 metres in length merely a third of its full adult size, appeared in Toyama Bay on Christmas Eve 2015, whereupon a fearless diver, Akinobu Kimura, jumped in and, in his own words, guided the animal back out to sea. There's no indication as to why the hulking, crab-stick-coloured creature appeared in the shallows, but it seemed to show little signs of aggression towards the diver, though it did give him a pat down with its arms. Perhaps few people would have done as Kimura did in approaching a giant squid, largely because these animals are such an unknown quantity.

We know little about them beyond that they're predators and far from uncommon in the deep. To date, they've only been filmed in their natural habitat a couple of times; even so, that's a couple of times more than we've filmed the colossal squid. Colossals have a much more limited distribution than giants and are larger still, reaching up to 14 metres long and weighing over half a tonne. The mystery that surrounds these behemoths is emblematic of our naivety with regard to the deep. One thing we can say, however, is that despite our tendency to create monster myths, these fascinating animals offer no threat to us.

Our propensity to demonise marine creatures is also exemplified in the branding of another squid, *Vampyroteuthis infernalis*, the 'vampire squid from hell'. While it may seem testy of me to quibble over a name, I'd argue that by applying names that vilify deep-sea creatures, such as the black seadevil, the hagfish, zombie worms and so on, we exaggerate the weirdness of the habitat and place fear in people's minds at a time when engagement is critical. The vampire squid is, in fact, an amiable creature that feeds by latching onto dolphins and drinking every last drop of their lifeblood. Only kidding. It eats nothing more contentious than marine snow, which it harvests by means of stands of sticky mucus that it casts out into the darkness like an ill-formed spider's web. It then collects the sum total of its efforts into an appetising sticky dumpling and transfers the item to its mouth. The small size of this creature – it reaches no more than 30 centimetres in length – means that it is at risk of being on the menu of other, larger animals. Accordingly, it has a number of defensive adaptations, from producing a barrage of bioluminescence to everting its canopy of webbed arms, the insides of which are lined with rows of spines. Arguably, however, its most valuable strategy is to occupy a mid-water habitat known as the oxygen minimum zone, an area of the ocean a few hundred metres below the surface where, as the name suggests, there's little oxygen to breathe. The oxygen minimum zone is sandwiched between the well-aerated surface waters, and the cold, dense, oxygenated currents that flow from the poles, and is the result of intense biological activity in the upper layers of the ocean. By confining itself to the minimum zone, where oxygen levels may be less than 10% of their values at the surface, the vampire squid finds safety in a place where its predators struggle to breathe.

Not only is the vampire squid not a vampire, it's also not a squid, though it is at least a relative. The same can't be said for the squid worm, which is a worm that looks distinctly like a squid, by virtue of a cluster of tentacles that sprout from its head and collect marine snow as it swims through the darkness. In common with the pigbutt and bombardier worms mentioned earlier, the squid worm is testament to the extraordinary flexibility of the group, using the simple worm body plan as a foundation and diversifying in the deep sea to produce a dazzling array of configurations. That the squid worm pays evolutionary homage to the eponymous cephalopods might give some sense of just how successful squid are in the world's oceans, yet it's the unheralded worms themselves who are among the masters of the ocean. There are some 17,000 different species of them in the sea, and they're found in every possible habitat, performing crucial ecological roles, we just don't pay them as much attention as we might. Then there are arrow worms, acorn worms, strange worms, horseshoe worms and penis worms, which aren't worms at all, strictly speaking.[7] More importantly, and despite living outside the limelight, the worms and worm-like animals are essential components of the marine

7 They're not true worms since they're not members of the phylum Annelida.

machine, mixing and refreshing sediments, recycling nutrients, and creating habitat for other animals.

If worms fail to capture our imaginations, that might be down to the fact that most of them spend a great deal of time buried in the mud of the seabed. Hidden from view they may be, but they can exist in astonishing densities. Each square metre of the ocean floor typically harbours thousands, or even tens of thousands of worms. With such wiggly riches on offer, it's little surprise that the ocean floor is a hotbed of predators, including fish, worm-eating snails and crabs. Even deep-diving whales descend to take their cut, leaving telltale troughs behind in the seabed as they gouge mouthfuls of worm-packed silt.

Joining the worm eaters on the sludgy surface of the ocean floor are a host of scavengers, some aesthetically appealing to us, others not so much. In the former category is an extraordinary and charming creature, the sea pig, a translucent pink, cartoonish creature that tiptoes along the seabed on a cluster of stubby legs.[8] Oftentimes, the sea pigs roam in herds, looking like a band of balloon animals, gleaning what they can from the slim pickings available. Sea pigs are relatively small – a typical one could sit comfortably on your hand – but scouring the seabed alongside them is an animal that, alongside the giant squid, is a poster child for what's known as abyssal gigantism. Although deep-sea isopods are marine relatives of the humble woodlouse, they're on an entirely different scale to their log-loving, terrestrial cousins. Some of these tank-like creatures can reach over half a metre long, and with clusters of hooked legs and large, angry, black eyes peering from beneath their dome of armour, they look like something from *Doctor Who*.

A friend and colleague of mine, Alex Wilson, studied these giant,

8 Sea pigs aren't pigs, obviously, but rather they're a kind of sea cucumber.

steroidal woodlice, retrieving them from the deep by capturing them in baited cages. Footage taken by cameras deployed with the cages show unsuccessful attacks made on the isopods by squid and even sixgill sharks, neither of which seemed to perturb the tough crustaceans. Nor did being hauled to the surface from a kilometre down, treatment that would rapidly kill most deep-sea animals, but which the isopods took in their stride. The only indication that they were ruffled by the experience at all came when Alex picked them up, an insult that instantly provoked the isopods into vomiting on him. Being covered in the regurgitated rotten mackerel trap bait isn't ideal, but Alex is made of stern stuff and continued with his assays, examining the behaviour and metabolism of the creatures. The gigantism that these deep-sea animals exemplify is seen in a host of other crustaceans, such as the Japanese spider crab with a leg span approaching 4 metres, as well as ostracods, shrimps, sea spiders, fish and, of course, the giant squid. The deeper you go, the larger some species get, the question is: why? There's no clear answer to this as yet. It's likely down to a mix of factors, from increasing oxygen levels beyond the minimum zone, the ability of larger animals to forage more effectively while avoiding being eaten themselves, and the increase in metabolic efficiency that comes with size.

Survival amid all the challenges of life in the deep sea has driven the evolution of an incredible suite of adaptations, from biochemical hacks to gigantism. But survival on its own is not enough; success in biological terms is defined by producing young and here, too, deep-sea creatures have had to be creative. The paucity of food means that animals are constrained by a tight energy budget, limiting their ability to produce large amounts of eggs or sperm. This is one major reason that deep-sea communities are slow to recover from overharvesting. It's a problem that some animals, such as deep-sea squid and possibly some fish, such as grenadiers and eelpouts, solve by reproducing only once in their lives. Other creatures spread the risk by breeding throughout their lives, but even then, they tend to produce comparatively small numbers of young, investing in parental care to protect their investment.

Another problem is the logistics of finding a partner in the expanse of the deep sea. Bioluminescence is valuable in this context, and the twinkling of deep lights in the darkness serve as a kind of sexual semaphore, bringing together lonely hearts. Even so, the range of these lights is relatively limited, and comes with the risk of alerting predators as well as lovers. Without a longer-range romantic radar, animals could find themselves in a state of enforced celibacy. Luckily for them, this is where the sense of smell comes in. Deep-sea animals have some of the most sensitive noses in the animal kingdom, capable of responding to infinitesimal concentrations of pheromones and following the dilute trail to its source.

Tiresome shuttling between distant lovers is not something that comes easily to an animal with the underwhelming athletic ability of a deep-sea anglerfish, especially since there may be only one available mate per cubic kilometre of water. It's a problem that anglerfish have famously solved by strict adherence to the vow 'until death do us part', since the male sacrifices his independence and fuses his comparatively tiny body with that of his beau in a permanent union, becoming essentially a testicle on the side of her body. As anyone who has ever undergone a transplant would know, there is an ever-present risk of rejection of the new organ because of immune incompatibility. The anglerfishes' fruitful fusion represents a massive challenge to the immune system of each partner, but one that they have managed to overcome, leading to them being promoted as a model study system for immunocompatibility and potentially leading to new insights for human medicine.

Long before the anglerfish arrived at their peculiar answer to long-distance relationships, indeed before any of the reproductive wizardry devised by the creatures here or elsewhere, an event occurred which remains the defining moment in the history of our planet. Perpetuating life through reproduction is one thing, but all of that life had to begin somewhere and there's tantalising evidence that that somewhere might very well have been here, in the deep sea.

* * *

Unravelling just how life began is one of the greatest challenges in science. Our current understanding invokes a process of increasing molecular complexity on a young planet more than four billion years ago, on the spontaneous assembly of diverse molecules into organised structures, and on the capacity of such structures to replicate themselves. Implausible as this once seemed, we have now garnered evidence for each of these steps in the laboratory, though they require exacting conditions. As a starting point, you need a rich source of chemical building blocks. You need water – the universal solvent – as a medium. And you need heat energy, lots of it, for the simple reason that the warmer things get, the more efficient chemical processes become.

Few places on Earth satisfy these prerequisites; fewer still possess the credentials of hydrothermal vents. The water that pours in scalding streams from the vents provides an intense energetic crucible and is charged with minerals, making it an elemental soup of pure potential. Among the tumble of chemicals are traces of organic substances. Sometimes referred to as prebiotics, these are the compounds that long ago formed the basis for the emergence of life, the moment that chemistry became biology.

If the conditions that exist at vents seem to allow the possibility of biological alchemy, there are other persuasive pointers, one of which comes from the striking similarities between what we know of the very first life forms and their modern descendants at hydrothermal vents. By some estimates, there are at least a trillion different species of microorganisms alive today. Within this mind-boggling diversity, however, there are whispers of a congruent past, a tiny proportion of the genetic and biochemical machinery that is universal to all. These shared features were inherited from a common ancestor, allowing us to build a picture of that ancestor, and where it lived. A compelling concatenation of clues tells us that sulphides, hydrogen and carbon dioxide were key to the workings of its metabolism, that it thrived in the absence of oxygen, and that it loved heat. In sum, this ancestor was strikingly similar to the microbes present at vents today – after all, why change a winning formula?

In 2017, another hint emerged, this time from a fragment of rock from the shores of Hudson Bay in northern Canada. What makes this rock so intriguing first of all is its age; dating methods suggest it may be over four billion years old, making it among the most ancient rocks on the surface of Earth. The other reason is that the rock is sprinkled with microscopic structures: tubes and filaments of haematite, nodules of graphite and rosettes of carbonate. These might theoretically have formed naturally by means of processes such as you might expect in a crystal garden, but there are other clues that suggest that these structures were formed by archaic microbes. In particular, the chemical make-up of the structures contains mineral by-products that are biological in origin, and the complexities of the patterns are unlike anything that chemistry on its own is likely to conjure. The arguments over this are ongoing; however, there is one other crucial piece of information. The rock itself was once part of the seabed, its make-up typical of that which you'd find – where else? – at a hydrothermal vent.

As the case for hydrothermal vents as life's fountainhead were taking shape at the end of the last century, one problem in particular was proving hard to solve. That problem was that alkaline, rather than acid, conditions would be far more favourable for the formation of amino acids and for early cells to be able to establish the energy gradients that are the basis of metabolism. Yet everything in the primordial vent scenario was acid, from the hydrothermal fluids that gushed from black smokers to the ancient sea itself, conditions that posed a substantial challenge to the idea of vents as the cradle of life. Then, in 2000, a deep-sea expedition in the mid-Atlantic stumbled across a completely new kind of vent field, unlike anything that had been discovered before. It was an undersea grotto of intricate, spectral white shapes like a petrified forest. In some places towering chimneys soared above the surrounds to heights of 60 metres, expelling plumes of water that, rather than emulating the heavy black of known vents, were a cloudy white.

The Lost City, as this new vent field came to be known, was distinct not only in appearance but in chemistry. It was rich with calcium, silicon and barium, as well as methane and hydrogen and, crucially, the fluids had a pH of between 9 and 11; this was the alkaline system that had been theorised but never found. The missing piece of the puzzle had been located. The case was further strengthened when, in 2019, a team of scientists at University College London and Birkbeck College, London, overturned one of the strongest remaining objections to alkaline hydrothermal vents as the wellspring of all existence. They demonstrated that protocells, the primitive precursors to life, could not only form in a warm, salty and alkaline environment but that these conditions actively favoured the process.

The evidence currently in our possession can't substantiate the vent origin theory on its own, but there's something else that just might. At a distance of 628 million kilometres from Earth, the moon Europa traces a lonely path in its orbit of Jupiter. Europa fulfils one basic criterion for life, which is that it has liquid water. So much water, in fact, that it is entirely covered by an ocean that may be as much as 150 kilometres deep. Recent examinations indicate that Europa is also volcanic, and that there are concentrations of carbon dioxide at certain regions, all of which suggests the possibility of venting far below on the Europan seabed. With so many pieces in place, it may be possible that the revolution that took place on Earth some four billion years ago has its counterpart in the outer solar system. To find out, we'd have to dispatch remote landers to Europa and then cut through fully 15 kilometres of ice to investigate the ocean beneath. Too difficult? NASA and the European Space Agency don't think so. They're gearing up to send technology to Europa equipped with nano-submarines to investigate, with the specific goal of finding life in an extraterrestrial ocean.

Wherever the search for life's origins takes us, we're rapidly gaining the appreciation of the deep sea as not merely the basement of marine life but its foundation. From the sunken currents that energise the oceans to the vent plumes that feed them, the deep sea plays a fundamental role not only in maintaining the health of the ocean but the planet in general. The interconnected nature of all the great marine biomes means that each exerts its effects on the others, an influence that indeed extends beyond the margins of the sea, to profoundly shape life on land.

2.
Hinterlands

Where salt water mingles with fresh

A LIGHT BREEZE SWAYS THE STEMS OF MARRAM grass covering the sand dunes away to my left. The breeze carries with it the salt tang of the North Sea lapping against distant sands on the horizon. I join a small footpath that traces a route along the banks of a river, the Great Eau, as it winds through a saltmarsh, and I head towards that horizon. It was low tide when I began my stroll, and on the far side of the channel a fishing boat is heeled over onto its beam, clear of the water and resting against the glistening mud. Next to it, a jetty, bereft of support, sags and points downwards from the bankside like the needle of a fuel gauge on empty. The scene speaks of abandonment and loss, as though the ebb tide stripped the vitality from the landscape as it retreated.

A change is coming, though. Almost imperceptibly, the current of the river has slowed. Ribbons of vegetation that fluttered and danced in the river's flow just a few minutes ago now hang limply, seeming to await new instruction. The stillness is an illusion; water is never truly at rest. The river has slowed as it approaches its final destination, as though hesitating in anticipation of meeting the sea, but the flowing tide will show no such apprehension as it rushes to reclaim the channel for its own. The appearance of ripples running upstream across the surface of the river foreshadows this imminent flux.

The tide begins to rise, slowly at first and then gathering pace as if trying to overwhelm the land's defences. The ribbons of vegetation that I saw streaming towards the sea in the river's flow a short time ago are now fluttering inland as the sea forces the river back. In the deepening channel, water now laps at the hull of the stranded boat; soon it'll be lifted from its slumber, buoyed on the flood tide. Crouching at the river's edge, I can see that the waters are layered, the clear freshwater resting on top of a wedge of heavier, sediment-filled brine encroaching from the North Sea – the currents have met but not yet mingled.

A little way upstream, a pair of wooden lock gates stand open, allowing the inland plains to drain at low tide. The gates are built into great earthworks, many of which were constructed over 300 years ago with pick and spade to safeguard these flood-prone lands from storm surges that would regularly inundate the fields and pastures beyond. Massive though the lock gates are, waterlogged and tressed with strands of algae, they're beautifully engineered to respond to the threat. As the channel fills, the pressure of tonnes of seawater comes to bear against the gates, which respond by swinging ponderously shut, forming a barrier to withstand the salty siege. The dull thud as they close reaches me downstream on the riverbank, confirming the changing balance of the waters.

This is my signal for a little alfresco swimming. As the river starts to back up on the landward side of the lock, it creates a perfect summer lido. Donning face mask, snorkel and fins, I slip into the water. It's crystal clear, with luxuriant stands of aquatic plants –

milfoil and hornwort – flanking a strip of deep water in the centre of the channel. As I push away from the bank, a huge school of silvery, juvenile roach swirls away from my outstretched arm, and three-spined sticklebacks hover nervously at the shallows and margins. Below them, squadrons of large perch, elegant in their livery of bold stripes and crimson fins, cruise in search of prey. Another predator, a pike, eschews the restlessness of the perch. It waits immobile, the irregular mottles on its flanks mimicking the caustics of refracted sunlight that dance on the riverbed below. Its head protrudes from the underwater forest, torpedo body poised, taut, waiting for the careless and the foolhardy.

An eel, thick as my wrist and long as my arm, makes its sinuous way downstream through the stands of weed. Unlike its cotenants in the river, this serpentine, olive-green fish began its life an ocean away from here, in deep Atlantic waters. Its size and direction of travel tell me that, soon, it'll retrace that journey. The eel, like so many animals, is a reminder of the intimate connection between salt and fresh, and between land and sea.

I'd like to stay longer – the underwater tableau is incredible, more than a match for any public aquarium – but an hour is enough to chill anyone. I swim to the bank and start to climb out when I notice a dog walker watching me from his vantage point on the bridge above the lock gates. He regards me suspiciously for a moment, before speaking up. 'Nothing in there,' he says.

I hesitate, unsure how to respond.

'No fish. All gone. You'll have to look somewhere else,' he snaps.

While I'm used to attracting strange looks for my choice of swimming spots, the man seems unusually hostile. As our exchange develops, he begins to thaw towards me, and I learn the reason for his initial abruptness. It seems that there have been some clandestine activities in the rivers hereabouts, activities that he'd assumed I was in some way connected to. Small, well-organised teams of commercial fishers from the continent who came in search of eels and, supposedly, cleared the river of its occupants in what he describes as a series of hit-and-run raids. In the following days, I ask a few other locals about this, and they seem to bear out his story. The motive behind these operations could be found in the price of glass eels, the near-transparent early life stage of this fish. Once superabundant, there had been an unprecedented fall in their numbers across Europe, and the scarcity drove a short-lived and entirely unsustainable fishy gold rush; at the time of my swim, about 15 years ago, glass eels were fetching an extraordinary £600 per kilogram. While that might seem a high price, it amounts to a miserable few pence for each of these transatlantic travellers.

The premium placed on eels is one of the reasons that the species has gone from abundant to critically endangered in little more than a generation. It's a worrying trajectory for an animal whose life history is one of the most extraordinary in all biology, as well as one of the most mysterious. Until a few years ago, from Iceland south to Morocco, as winter gave way to spring, the lower reaches of rivers would come alive with transparent slivers of energy, each as long as a cocktail stick and scarcely more substantial. In some places, their numbers were such that they would form a living cavalcade a metre wide, stretching unbroken for kilometres, each individual wriggling resolutely upstream alongside its fellows, the mass hugging the banks where the flow is weakest. The annual glut of glass eels provided a bonanza for a host of other animals. They formed a valuable part of the diet of people, too, who fried them in oil, steeped them in ale or baked them into eggy cakes.

To early naturalists, the sudden appearance of glass eels each spring represented a conundrum: where did they come from? Some speculated that they were generated from hairs shed by horses, others that they were the product of sea foam or sunlight, or arose spontaneously from the mud. Part of the reason that such propositions were favoured was a response to studies of fully grown eels that failed to find any means of reproduction. They seemed to have no ovaries or testicles, no eggs, no milt. Altogether, the enigma of the eels tormented many scientists, such as the German anatomist Max Schultze, who lamented as he lay dying in 1874 that all of the important matters of biology had been settled, 'except the eel question'. We might well sympathise with Schultze; the eel has been among the least willing of all fish to surrender its secrets. Many biologists attempted to crack the puzzle, but we owe most to an indefatigable Dane, Johannes Schmidt, who in the early part of the twentieth century spent almost 20 years of his life in pursuit of an answer. His quest took him out into the Atlantic, a giant oceanic haystack concealing eel-shaped needles. Schmidt reasoned that if he recorded the size of eel larvae at each of his sampling locations, he could build a kind of contour map of eel lengths. From this, he could then home in on the bullseye, the spawning grounds from which the larvae radiated.

It was painstaking and laborious work, as Schmidt himself admitted. His description of the highs and lows that he experienced in pursuit of his life's quest would resonate with anyone who has taken on a great challenge. On publishing his findings in the 1920s, he wrote of his experiences: 'Years of research rich in excitement and suspense: disappointment alternating with encouraging discoveries and periods of rapid progress with others during which the solution of the problem seemed wrapped in deeper darkness than ever.'

Schmidt consistently caught the smallest larvae in a region of the western Atlantic near Bermuda, leading him to deduce that the Sargasso Sea is the eel's breeding ground. It wasn't until very recently, a full century after Schmidt's sleuthing, that we finally gained incontrovertible proof of this, when we were able to follow adult eels fitted with electronic tags as they navigated across the ocean and converged on the area. It is only during the course of their journey that the adults ripen, developing the reproductive organs whose absence so puzzled early anatomists. The eels travel in deep water, in some cases down to 1 kilometre, guided through the darkness by an internal compass over a period of months, or even years, to rendezvous with other eels, to spawn and then to die.

In the comparative warmth of the Sargasso, countless millions of eggs incubate before releasing their contents, legions of so-called leptocephalus larvae less than a centimetre in length, into the void of the open ocean. There's no repose for the larvae, however; they're genetically programmed to recapitulate their parents' journeys, in reverse. Thousands of kilometres separate these hatchlings from their ultimate destination, but it is at least one that is assisted by the marine equivalent of a following wind. The Sargasso Sea forms part of the North Atlantic Gyre, a huge expanse of sea bounded by major ocean currents. It's one of these currents, the Gulf Stream, that helps to convey the eel larvae northeastwards towards their ancestral home.

At this stage in its life, an eel larva is a curious-looking creature, transparent and wafer-thin, and like a willow leaf in shape. Even with the assistance of the prevailing currents, it's an arduous and dangerous journey for such an insubstantial animal, and one where the chances of success for any single individual are vanishingly small. Nevertheless, some do make it, and the first of them start to appear off Portugal around November, around six months after hatching. Those heading further afield, either south into the Mediterranean or north to European coasts from France to Norway, take a little longer, reaching their goals in late winter or spring. Over the course of their journey, the survivors have grown from just a few millimetres to almost 10 centimetres, and once they reach the continental shelf, they begin a transformation in preparation for the next stage of their life. Their compressed body becomes cylindrical, though they maintain their secretive, see-through appearance.

Having crossed an ocean, the young eels are close to their journey's end, yet many more challenges await them. As their migration continues, they have to cope with the physical challenges of freshwater, as well as a new guild of predators, eager to mark the completion of the glass eels' epic trans-oceanic trek by eating them. Despite suffering huge losses, the surviving youngsters make their way relentlessly upstream and into the aquatic hinterlands of Europe and North Africa. Where obstacles block their way, they might even divert across land for short distances on rainy evenings, slithering over damp grass before rejoining the river. At other times, obstructions such as waterfalls are navigated by the diminutive eels' use of the water's surface tension, which allows them to cling to and then climb the barrier.

The translucency that has been their hallmark since they first emerged in the Sargasso now gives way to camouflaging pigments and they become elvers. By degrees, their wanderlust departs them, and they settle into a more sedentary life. Here they remain for most of their adult lives, a period that may last for anything up to 25 years, before a shift in their physiology sets in train a final transformation. The adult eels respond to a visceral biological imperative that spurs them to begin the pilgrimage that will take them back to the Sargasso, completing the circle of their existence in their final act of spawning.

Impressive as the European eel's life history is, it's not alone in staging such a migration. Around the globe, related eel species make similar journeys from oceanic gyres to the watercourses of America, Japan, South Africa or Australia. The eels' journey is the reversal of that of salmon, in that eels breed at sea and spend much of their development in rivers and streams, but both undertake long and hazardous migrations between distant points. The obvious question to ask is why do this? The energy needed to undertake such odysseys is vast, while a huge proportion of the travelling cohort is lost to predation. It seems an unwise venture, on the face of it. Evolution, however, is parsimonious. It doesn't select for behaviours that, on average, don't deliver a net benefit. It's up to us to work out where that advantage lies.

In salmon, at least, we have some answers. By laying their eggs in the clean, oxygenated headwaters of rivers, the vulnerable salmon eggs and fry develop away from the battery of predators that prowl the ocean. By migrating to the sea during their growth stage, salmon take advantage of the fact that in temperate regions, productivity, for which read food availability, is far greater in the ocean than in freshwater. The results are dramatic. A comparison of typical migratory sockeye salmon to populations of the same species that go through their entire cycle trapped in freshwater lakes reveals that, at maturity, the migrators are ten times heavier than their stay-at-home counterparts.

For eels, the picture is far less clear, and far more perplexing. The adults seem to drop the kids off right in harm's way in the Sargasso and force them to endure a stupendous journey whose entire route is lined with danger. Then, just as the youngsters' development really gets under way, they move into freshwater, where food is comparatively scarce. Though it may seem ridiculous, the eels may simply be following a blueprint that was etched out deep in their evolutionary history. The ancestors of today's eels were tropical creatures and, in contrast to the pattern that exists at higher latitudes, in the tropics there's typically more food to be found in freshwater than in the sea. So, when eels first established their pattern of spawning at sea but living in rivers, it's possible that they did so for similar reasons that cold water salmon go into saltwater today: they were following the food.[9] In the millions of years that have passed since, eels maintained their migratory tradition, first crossing the ocean and then following watercourses far inland. It's possible that, in moving to freshwater, they escape competitors and predators alike. Whatever the reason, that eels spent most of their lives in rivers and lakes was considered axiomatic; it was just what they did. However, nature doesn't always operate along such neat lines, especially in the face of a crisis.

The downturn in the fortunes of eels worldwide is shocking in its suddenness. In many places, the populations of these charismatic travellers have fallen by more than 99%. The gigantic transference of energy from the sea to the land that the arrival of trillions of glass eels each year represented is now all but gone. This isn't just a problem for the eels themselves; the many predators who relied on the annual glut to propel their own breeding efforts have also suffered. There is no shortage of explanations for the eels' woes, ranging from the destruction and pollution of freshwater habitats, damming of rivers and the installation of turbines that turn eels into smoothies, and the damaging effects of a shift in the Gulf Stream.

Then there's overfishing.

The only sensible response to a steep decline in the population of a food fish is to stop persecuting it. Sadly, that isn't what happened to eels. Their very scarcity made them more, rather than less, desirable. Although glass eels gained protection, as prices skyrocketed a thriving black market emerged to feed demand, largely from Asia. In recent times, European police agencies have intercepted dozens of shipments, one example being the detainment of

9 The eel's strategy of living in freshwater and migrating to the sea is known formally as catadromy, the reverse of the salmon's pattern, which is referred to as anadromy.

30 kilograms of glass eels at Lisbon Airport in 2023, with an estimated value of €200,000.[10]

Adding to this calamitous mix is an invasive parasitic nematode, most likely introduced to Europe in a shipment of live Japanese eels for the aquaculture industry in the early 1980s. The nematode quickly established and is now found in virtually every part of the European eel's range, in some places affecting over half of the local population. Its effects on eels are devastating, causing muscle inflammation and damaging their swim bladders, which together compromise their ability to migrate. The American eel and the Japanese eel are now listed as endangered, the European eel critically so.

If there's hope for eels, it might just reside in what was, until recently, an unrecognised adaptability. Not every larval eel that arrives at the coast makes the choice to enter the rivers. The punishing consequences of doing so, in the shape of high-tech poachers, artificial barriers and habitat loss, would tend to reward those who elect to avoid a journey inland. And, contrary to what we once believed, eels are capable of completing their life cycle without ever leaving the sea. The alarming population trends that we have are drawn almost exclusively from freshwater monitoring. I have the sense of clutching at straws as I write this, but it has been suggested that a proportion of the eel population has made an adaptive switch, eschewing freshwater entirely to remain at sea, where the ordeals imposed on them by humanity are felt less keenly. It's an indictment of our behaviour in recent decades that we are on the cusp of losing one of our most iconic river dwellers, but if the species can win through by sidestepping the inland waters, and us, then this represents the lesser of two evils.

* * *

10 The fortunate glass eels were released – alive – in the Tagus.

The story of life on Earth is one in which the oceans play a formative role, but it's also a story of transitions, of groundbreaking journeys made by ancient marine pioneers that paved the way for the conquest of freshwater and, eventually, the land. Eels and salmon are the exemplars of this journey, their lives connecting seas to the rivers of the continental interior, but what might seem a simple passage from one aquatic environment to another is far from straightforward. The defining characteristic of freshwater is, well, that it's fresh; it lacks salt, whereas the sea is full of the stuff. Where we might see a continuum of aquatic environments, bundling them all together as blue on our maps, to the overwhelming majority of aquatic animals, this chemical distinction defines the very boundaries of their existence. Their physiology is inextricably entwined with the chemistry of their surroundings, and any deviation from this will quickly kill them.

Those adventurers who sought to leave the ocean and expand into the saltless watercourses of the land faced a seemingly insurmountable barrier. It's testament to the extraordinary adaptability of life that they did so, and not just once. Even so, echoes of a marine past remain in the biochemistry of contemporary freshwater and land animals, bequeathed by those bygone trailblazers. Although evolution could rebalance the physiological equations necessary to support life beyond the sea, it couldn't reboot and start from scratch. The result is that all the descendants of those ancient animals remain, to some extent, salty.

Since both the most primitive single-celled organisms, as well as the more complex multicellular creatures that subsequently arose from them, had an oceanic origin, we might think of seawater almost as the amniotic fluid of life. Even the simplest organisms wall themselves off from the outside world by means of a cell membrane, and their saline interior is an homage to the seawater that surrounds them. Things get slightly more complicated when we consider animals. At the most basic level, all of them are comprised of cells that sit within the envelope of their body wall. Those cells are provisioned by a liquid solution known as extracellular fluid. For us, this is blood (or more correctly, the liquid portion of blood, known as plasma). Wonderfully well as blood serves us, however, lots of familiar marine invertebrates seem to regard it as little more than a highfalutin frippery.

Take sea stars, for example. They simply use seawater as a proxy for blood, filtering it to remove pathogens and any other organic unpleasantness, and then supporting their cells in exactly the same liquid that they live in. They're far from alone in this. Recently, on a vessel off Akaroa in New Zealand, I was invited by the captain to sample a salp, a creature closely related to the sea squirts found on the lower shore. This is the food of the future, he told me, with all the enthusiasm of the recently converted. I gingerly took the transparent, prawn-sized morsel from the captain's hand and, despite misgivings, popped it in my mouth. It tasted of essentially nothing but seawater, overlaid with the texture of slime, and I can't see salps coming to a menu near you any time soon. The reason that it tasted that way is because its extracellular fluid is seawater – these are sea creatures in every sense of the term.

Once you leave the extracellular fluid that bathes and nurtures the body's internal structures and go deeper, penetrating the cells that comprise those structures, the biochemical picture changes. The liquid inside the cells, where the magic of biochemistry takes place, has a different chemical make-up. The sodium and chloride ions that are so abundant on the outside are much rarer within the cell, while potassium ions, in particular, are much more concentrated on the inside. Also swirling around in the intracellular fluid is a rich assortment of organic molecules, such as sugars and amino acids, that the cell uses as part of its operations.

The crucial thing about all of this in the context of maintaining the physiological status quo inside the animal is that, although intracellular and extracellular fluids are made up of different mixes of chemicals, the overall concentration of the two is evenly matched. If this weren't the case, and one side was saltier than the other, water would spontaneously flood through the cell membrane to join the salty side in the process known as osmosis. You'd see this to dramatic effect if you were to take a cell from an animal and put it in pure freshwater; the concentrated mix of ions and organic molecules inside the cell would draw water in, causing the cell to expand uncontrollably and eventually pop. The reverse is also a problem. A cell placed in brine would shrivel as it donated its water to the salty environment beyond. The ability of animals to maintain the balance between intracellular versus extracellular chemistry is an energetically expensive, but vital, part of keeping their cellular metabolism working properly.

The cost of sustaining a bespoke internal chemistry is one reason that so many marine creatures simply use seawater for blood. This is something of a gamble, since it makes them reliant on the salty stability of the water around them, but this stability is precisely what the ocean provides across much of its domain. The picture changes near the coast, particularly where freshwater runs off the land and where rivers discharge. For us, freshwater is cooling, revitalising and essential. To lots of marine creatures, it's as toxic as any poison and near the coast, those that rely on reliably uniform levels of saltiness are potentially in trouble. If a sea star – or any of the myriad other marine invertebrates that have co-opted seawater as blood – were to get washed up onto the high shore and trapped in a low salinity environment, like a rock pool suffused with rainwater, then its entire biochemical house of cards will quickly collapse. Lacking the ability to regulate its osmotic state, freshwater will flood in, and the sea star will quickly die in a welter of exploding cells.

You could be forgiven for wondering why such animals don't simply avoid the perilously fickle conditions at the coast; the world's oceans offer a vast amount of living space, after all. What tempts them is the promise of abundant food at the margins of the sea; rich rewards await any critters flexible enough to cope. This, however, raises a question: how do marine organisms manage to expose themselves to low salinity without simply bursting like a flotilla of balloons alighting on a cactus farm? The most common, approach, used by inshore creatures such as mussels, barnacles and winkles is simply to close their shells, barricading themselves in when conditions worsen and sitting tight in their own private pool of seawater awaiting the relief of the returning tide. They're incredibly effective at this: put a winkle in freshwater for a fortnight and it'll stay safe and salty inside its shell for the duration; indeed, a greater risk is that it'll run out of food or oxygen first. Luckily for them in the normal course of events, this is a strategy that needs only to be employed for a few hours at low tide when they're at the mercy of freshwater.

Still, this kind of aquatic siege mentality isn't perfect; it means that animals have to shut down their operations and exist in a kind of suspended animation when salinity drops. In a world where adaptability is the key to success, ingenuity brings rewards. There have been few innovations that can match the evolutionary significance of osmoregulation. Employing some of this new-found cellular wizardry, animals began to take a more active approach to balancing their body's salt budget, becoming in a sense masters of their inner sea. By throwing off the osmotic shackles that had constrained their forebears, they were no longer in danger of suffering the terrible repercussions that exposure to fresher waters might bring. It was this that freed these trailblazers to expand their horizons, to travel beyond the sea and into the saltless watercourses of the land. This seminal event, first thought to have been achieved by primitive, shrimp-like creatures known as branchiopods around 500 million years ago, was one of the key moments in the history of life. In the succeeding years, they were followed into freshwater by others; the animals' conquest of the continental interior had begun.

* * *

Even as the first invertebrate voyagers were adapting and leaving the seas behind in one of life's great transitions, a completely new type of animal flickered into being in the primordial oceans of half a billion years ago. This new creature was a fish, although the earliest examples in the fossil record are very different from anything that you might recognise as such; certainly, there is nothing about them that gives any hint of a golden future. Among contemporary animals, the lancelet, though not itself a fish, is sometimes offered up as an example of how the ancestors of fish might have appeared. It's a finger-length creature that looks more than anything like a halfway house between a worm and an anchovy fillet. Indeed, eighteenth-century taxonomists found it so underwhelming that they initially classified it as a slug. The lancelet feeds by sucking water into a rudimentary mouth and relying on a sticky net of mucus to collect little particles. It can swim, but on the whole would rather not, preferring to nestle in sedimentary mud with only its head protruding. A complex beast it is not, yet running the length of its body is a rod-like structure, known as a notochord, the precursor to a defining characteristic of fish, amphibians, reptiles, birds and mammals, including you and me. That characteristic is a spine, and all animals that bear one are vertebrates.

A backbone was the key feature of a completely new bodily arrangement, the central pillar of an anatomical revolution: an internal skeleton. The skeleton acted to anchor muscles, and the spine itself served to protect the body's delicate nerve superhighway, the spinal cord, from damage. In evolutionary terms it was a great leap forward, and one that paved the way for the rise of one of the most successful animal lineages on Earth. The story of the fishes' rise from their unprepossessing beginnings has many twists and turns, and there were times at which their survival was far from assured. Nevertheless, it's said that history is written by the victors, and that's never truer than in the history of the vertebrates. The modern dominance of this group – our group – is one that was forged in the marginal waters between fresh and salt.

Like the animal groups that came before them, the vertebrates first arose in the ocean. The ancestors of hagfish, the wonderfully slimy carrion feeders of the deep, were among the first fish to emerge in the ancient seas. They don't have jaws, or even bones, which might seem to disbar them from the vertebrate club. But evolution is a curious process, and one that doesn't necessarily follow an inexorable path towards what we might think of as advanced features. New research suggests that the very earliest (and now extinct) hagfish did indeed have rudimentary vertebrae, a feature that was discarded in the course of their evolutionary history. Another feature of modern hagfish that seems to ally them more closely with invertebrates than with other fish is that their extracellular fluid – in their case, blood – is at the same salty concentration as the sea around them. Clearly, the evolution of hagfish didn't involve much of a departure from the established osmotic script that has long served plenty of marine invertebrates so well.

Hagfish, and their cousins the lampreys, are all that remains to us today from what was once a thriving and diverse group of ancient jawless fish. They could rasp away at algae, or they could filter feed in the manner of a lancelet, but they lacked a bite. At the same time, it's more of a challenge to breathe if you lack a jaw. If you watch one of the more familiar modern fish as it breathes, you'll often notice its mouth moving. Opening the jaw and expanding the mouth cavity draws water in, which can then be squeezed out through the gills. This process is far less efficient in jawless fish, and combined with the limits to their feeding style, placed something of a ceiling on their ambitions.

Next time you sit down for a meal, perhaps reflect for a moment on your jaw. Despite its familiarity, the human jaw is a truly extraordinary adaptation, inherited from fish ancestors that existed over 400 million years ago. As often occurs in evolution, the jaw emerged from a repurposing of another structure, in this case, one whose original function was to support the gills.[11] It was a revolutionary development, permitting a completely new and vastly more effective means of feeding. Prey could be sucked or snapped up, selectively and deliberately, and they could be bitten and crunched. A new, and incredibly effective, predator had arrived in the oceans, and from being one of the marine supporting cast, fish were about to take centre stage.

The Devonian period, which began 419 million years ago, saw an unprecedented, explosive proliferation of fish species. Modern sharks appeared in this period, establishing one of the most venerable of all organismal lineages; they've been on this planet longer than trees. That sharks were unable to completely dominate the Devonian seas is due to the emergence of another new group of fish that substituted a cartilage skeleton for one of bone. As we shall see, it was these enterprisingly named bony fish that came to inherit the waters of our planet.

Piecing together the events in the so-called Age of the Fishes is

11 The price that they paid for this was a loss of the first two pairs of gill arches; the new jawed fish had seven pairs of gill arches, compared to the nine sported by jawless fish.

challenging at this remove, but it's clear that it was a period of intense evolutionary activity. Some of the old guard, including almost all of the weird and wonderful jawless fish, and even some of the founders of the jawed fish lineage, gradually disappeared. In their place, a host of new-fangled fish appeared, flourished and conquered new territories in what was, in geological terms, little more than the blink of an eye. The descendants of these modern bony fish are divided into two main taxonomic camps: the ray-fins and the lobe-fins. Both remain with us today, but it is the ray-fins who dominate, making up the vast majority of fish species, and, indeed, more than half of all vertebrates. They range in size from the tiny *Pseudocypris*, fully grown when it reaches the width of your fingernail, to the oarfish, which at 11 metres is twice as long as a giraffe is tall. They exist in just about every aquatic habitat on Earth, from the snailfish that live in the deep cold of the abyssal seas, 8 kilometres below the surface, to the African killifish who complete their life cycle over the course of just a few weeks in the shallowest of temporary pools. In terms of species number, they are by far the most successful of all vertebrate groups, but they nearly didn't make it at all.

The Earth's history is punctuated by a series of catastrophic extinction events. The deadliest of these, the Permian-Triassic extinction, colloquially known as the 'Great Dying', occurred 252 million years ago when a period of intense volcanic activity in what is now Siberia triggered a sharp rise in atmospheric carbon dioxide and disastrous global warming. All of the planet's ecosystems suffered, but hardest hit of all were the oceans. Oxygen levels plummeted, the seas acidified and toxic methane belched from the seafloors. In the face of this onslaught, all but a comparative handful of marine species disappeared forever.[12]

12 In fact, an incredible 96% of all marine life went extinct in the Great Dying.

Those that persisted through these hellish conditions did so almost as a quirk of fate, surviving in habitable refuges while all around were wiped out. Some fish, including a group of sprat-sized ancient sharks, achieved sanctuary in the deep ocean, away from the chemical and thermal extremes near the surface. Others clung to life in freshwater, where their ability to breathe atmospheric air was likely essential to their survival in stagnant, oxygen-depleted waters. This latter group included early ray-finned fish, but rather than thinking of these as refugees from the oceanic apocalypse, the evidence suggests that their ancestors had moved into freshwater long before the extinction crisis, and that it was here that they first evolved. The fact that the oldest surviving ray-finned fish – veterans such as reedfish, bichirs, bowfins and paddlefish – are all river dwellers tends to point in the direction of a freshwater origin for this group.

It can take millions of years for Earth's ecosystems to bounce back from the trauma wrought by mass extinctions, but as the recovery progressed, the ray-fins were gradually able to repopulate the seas. It was a reversal of the journey that had brought the ancient ancestors of ray-fins into freshwater in the first place, and one that means most of today's familiar sea fishes have freshwater roots. Consequently, the large majority of marine fish are relative neophytes compared to their freshwater cousins, a fact that may help to explain why, even after all this time, there are still more fish species in freshwater than there are in the sea.

But what of the lobe-fins, the seeming poor relations to the exuberantly speciose ray-fins? Well, they persist but their heyday has passed, and only eight species of lungfish and coelacanths remain. Their legacy, however, is enormous. At the time when lobe-fins first appeared, placoderms dominated the oceans. These were truly extraordinary-looking fish, their heavily plated bodies calling to mind a biological equivalent of an armoured vehicle.

Some, like *Dunkleosteus*, were colossal, reaching 6 metres in length and possessing a hunger to match. Placoderms were predators who lacked teeth, but they made up for the shortfall by having a mouth fringed with large, viciously keen, self-sharpening plates of bone that served the same purpose. Perhaps it was the ever-present threat of being bitten in half by a rapacious placoderm that drove some of the lobe-fins into the comparative safety of coastal swamps. Or perhaps they were drawn to such habitats by opportunity in the shape of a smorgasbord of prey in a place where they faced little competition. Whatever the reason, it was here, in the shallow, marshy waters of the land that lobe-fins changed the course of history.

In the sludgy, oxygen-starved environment of lagoons, lobe-fins could propel themselves along in the shallows, punting along using their fleshy fins and periodically taking gasps of air at the surface. The primeval lobe-fins that lived in these habitats were thus equipped with the basic hardware – proto-limbs and proto-lungs – that would allow them to escape the confines of an entirely aquatic existence and to take to the land. Coelacanths and lungfish are the only lobe-fins to have survived the passage of time from the early Devonian period to the present day; however, these fish are not the only descendants of their line. Every amphibian, reptile, bird and mammal alive today is heir to this tradition, the seminal crossing of a biological frontier 375 million years ago.

* * *

I'm a hundred million generations, give or take, removed from my fish ancestors, yet despite the chasm of time that separates me from them, there are plenty of anatomical reminders of my heritage. Alongside my jaw, lungs and limbs, I can count even my sense of hearing as a bequest from those long-ago lobe-fins who first heaved themselves to a life above the waterline. I'm contemplating this as

I dry myself in the aftermath of my river swim while looking out over the saltmarsh from my vantage point above the lock. There's a quality of timelessness here, a sense that it has remained unchanged for aeons, and it's not hard to imagine that it was in a place something like this, where creeks and rivulets delineate tessellated islands of vegetation, that my forebears took those first, faltering steps.

On the seaward side of the lock below the tidal defences, I can see how the tendrils of the rising tide have snaked their way through the creeks to clothe the nakedness of their mudbanks. Above the tideline, a host of plants thrive in defiance of the conditions. The enemy here is the ubiquitous salt, which interferes with the plants' water and nutrient uptake. They deal with this by weeping salt from specialised glands back into the environment, by counterbalancing salt's effects using protective molecules such as betaine and proline, or by imprisoning the offending chemical in storage vessels within old leaves and then shedding those leaves along with their salt cargo. Among these hardy plants are succulents, such as samphire, whose fleshy stems carpet the lower ground like a grove of toy cacti. Scurvy grass, once taken on board ships to combat the dreaded disease it is named for, proliferates across the marsh, and sea aster and sea lavender attract butterflies and bees with their purplish summer flowers. A little higher up, the earth banks of the defences have been colonised by sea buckthorn, which ignites the slopes each autumn with an eruption of fiery orange berries.

From among the clouds overhead, a skylark maintains a running commentary. It's a beautiful, though poignant, song; one of the many gems of the natural world that my grandad revealed to me. He sparked my interest in animals when I was a child, and now I carry that torch without him. It's hard to remain downcast, however, in the midst of so much beauty. Little egrets, brilliant white against the marsh, patrol the water's edge on their corn stem legs, prospecting for fish, while a turnstone flicks a piece of dried bladderwrack over, seeking hidden morsels. Gulls alight and then stomp about flat-footed, paradoxically seeming both comic and imperious. A posse of redshanks skitter timorously across the soft ground, giving the impression of being close to nervous exhaustion. An old folk name for these anxious waders was 'watchers of the marsh', since their jittery, piping call seems to be issued at the movement of anything from a marsh harrier to a windblown leaf. Redshanks are skittish sentries, almost perpetually ready to take to the wing in fright. I don't wish to interrupt their fossicking, so I keep perfectly still, half concealed by the sandstone wall of the bridge, admiring the industriousness with which they probe the ooze with their long bills. Like many waders, the tip of this bill is richly supplied with nerve endings, enabling them to detect the presence of buried prey with surgical precision.

Unlike the ephemeral seabird colonies that form on the limestone cliffs in spring to the north of here, the saltmarsh and its tidal flats bustles with birds all year around. Enormous flocks of brent geese and pink-footed geese arrive in October and November from their breeding grounds in Greenland, Scandinavia and Siberia to gain respite from the Arctic winter. They're joined by a host of waders, including sanderlings, bar-tailed godwits, dunlin and knot, all of whom are similarly heading for warmer climes as ice and snow grip their boreal range. This part of the coast, from Spurn Point down to the Wash, is a place of plenty and consequently has huge importance as a waypoint on one of the world's most significant superhighways for migrating birds.

Before the tidal defences were built, the saltmarsh spread across thousands of square kilometres of this low-lying countryside, forming a gentle gradation between land and sea. The wetlands were nature's way of melding the two, creating a buffer zone, but to the people who lived here the marshes were no more than wasteland, useless for crops. Building the embankments allowed them to reclaim ground and to put it under the plough.[13] Now the marsh is squeezed between the coastal earthworks and the shore, permitted to exist only as a truncated ribbon along the coast, far less in extent than before. Houses and settlements have sprung up on the former saltmarsh over the centuries, but even now all of them live in the shadow of spring tides and storms that breach the defences just often enough to warn us that if the land has been reclaimed, it can only ever be a temporary arrangement, a loan.

The saltmarsh is a reminder that the sea's influence neither begins nor ends at the shore. It profoundly shapes the lives of terrestrial organisms who live near the coast. Even when we consider only aquatic organisms, there is often no clear point of demarcation between seawater and fresh. In some places, the sea infiltrates the land, travelling up the courses of rivers at high tide, whereas in others, freshwater penetrates far out to sea. The most dramatic example of this is when the Amazon River is in full spate and huge volumes of water spew from its mouth into the Atlantic. In calm conditions, when waves don't induce the waters to mix, ships passing to the northeast of the delta can recover water fresh enough to drink, floating on the denser salt, an extraordinary 400 kilometres from the coast.

In an engineered landscape such as the saltmarsh where I am now, however, there is an artificially sharp line between salt and fresh. The ebb is well under way, and the pressure that kept the river imprisoned behind the lock gates is weakening. A few moments later, the gates shudder, and the backed-up freshwater begins to trickle through, like a stream of attackers breaching a defile in a castle wall. Almost impatiently, the momentum of the

13 It's a similar pattern across the UK, where 85% of the saltmarsh has been lost in the last 200 years.

river's release pushes the gates ever wider, and a swirling torrent of freshwater floods into the estuary, chasing the retreating tide back out to sea. Watching from the bridge directly overhead, I see several tiny fish swept along, unable to resist the current. The destiny of these fish depends on their species, and specifically whether they are stenohaline, able to cope with only a narrow range of salinities, or euryhaline, capable of adapting to varying concentrations of salt. Sadly for them, the young roach are stenohaline, and I can only watch as their sudden exposure to seawater sabotages them. They become disorientated, darting in drunken parabolas then flipping over and revealing their beautiful silver flanks. The current carries them away, still twitching, but their fate is sealed. The fortunes of the other tiddlers, sticklebacks, is markedly different. They maintain their composure resolutely, despite the abrupt change in their circumstances, gathering at the water's surface and swimming in small shoals to the side of the channel where the current is slack.

Despite their apparent sangfroid, what's going on inside the sticklebacks' bodies is anything but calm. The fish's physiological systems are thrown into chemical chaos, something that relatively few fish species can survive. We have no comparable experience against which to measure this. So to understand it, we must return to the topic of osmoregulation.

While invertebrates were the first osmoregulators, it was fish who effectively mastered this physiological art and the vast majority of them maintain tight control over their internal solutes. Regardless of where they live, in the sea or in freshwater, the saltiness of fish blood is surprisingly similar at something like 30% of the salinity of seawater. This has the rather weird implication that, pretty much wherever they live, in this respect they're dramatically out of kilter with their environment.

Physiologically, this puts quite a bit of strain on fish. Since the blood of freshwater fish is saltier than their environment, water continually forces its way into their bodies, and salts are continually in danger of being lost. Their answer to this is to capture salts via specialised cells, known as ionocytes, in their gills from the outside water to balance their salt budget, and to piddle more or less continuously. Marine fish have the opposite problem. Their blood is less salty than their surrounds, so they must drink seawater constantly, hiving off the excess salt and actively pumping it back to the sea, again by means of their ionocytes, before using the newly desalinated water to stay hydrated. At the same time, they urinate sparingly, their bodies hoarding every precious drop of water.

The demands on fish to marshal their osmotic resources carefully, whether it's a goldfish's preoccupation with keeping water out and salt in, or a cod's need to do the reverse, are helped in part by the toughness of their skin. This skin is made up of multiple layers of robust proteins, such as collagen, designed to impede the free flow of water. It's so tough, in fact, that fish leather has historically been used for shoes, door hinges and even the sails of boats. Even with all of this waterproofing, however, fish remain leaky, so they have to use precious energy to power dedicated cellular machinery that keeps their internal chemistry in balance. Exactly how much they pay for this varies from species to species, but according to one estimate, it amounts to about 7.5% of the animal's overall energy budget. That might not seem much, but it's fuel that could otherwise go towards something more obviously valuable, like growth, or making baby fish. So why do they pay this price? The simple reason is that by keeping their blood between a third and a quarter as salty as seawater, they optimise the efficiency of all the biochemical processes that take place within their bodies.

Fishes' physiological processes comprise myriad biochemical pathways that come together to form a beautifully tuned organic machine, but a sudden and dramatic change in salinity, such as that experienced by the fish as the lock opened, represents a handful of grit being thrown into that machine. The osmolarity of their blood goes haywire, and its pH shifts. The foundations, upon which all of those neat biochemical pathways are built, crumble. Salt rushes into their bodies and binds to proteins, destabilising just about every bodily function. If you or I were to experience a fraction of this seismic shift in blood chemistry, death would be rapid, and that, sadly, is exactly what happens to the tiny roach.

Somehow, though, the equally diminutive sticklebacks soldier on. The pandemonium inside their bodies demands a response, and that response comes in many forms. They mobilise molecular emergency services, such as shock proteins, that swoop in to chaperone and protect other at-risk proteins. There's a rapid uptick in the production of enzymes that will allow them to reconfigure their osmoregulatory processes, while ion channels open or close as a step towards redressing the blood's chemical balance. This overhaul, however, is large and energetically expensive, and it'll be some time before stability returns. For now, the sticklebacks try to soften their osmotic crash landing by seeking out more favourable parts of their habitat, and that's one reason that they come to the top of the water behind the lock gates, exploiting the fact that freshwater floats on the saltier layers beneath.

Not far from where I'm standing on the bridge, there are pools in the upper regions of the saltmarsh that are inundated by the tide perhaps only a couple of times each year. Now, at the height of a dry summer and with no water to feed them, these pools evaporate down to brine that's twice as salty as the sea.[14] Even here, in this

14 I measured it at 71 parts per thousand, compared to 35ppt that is typical for seawater, but this is nothing compared to the fishy record holders, the black chinned tilapia, which can tolerate 130ppt.

harshest of aquatic environments, sticklebacks scoot around, apparently untroubled. Part of the reason for this is that change has been slow, and salt levels have been climbing languidly for weeks. This gradual transition between different water chemistries is far more manageable for fish than the sudden 'shock of the lock', but even so, the sticklebacks' ability to switch from fresh to salt water and back again is restricted to less than 10% of fish species.

This inflexibility means that estuaries are a no-go zone for the vast majority of fish, a chemical boundary that they simply cannot cross. In many cases, the distant ancestors of these avoidant fish would have been able to flirt with changes in salinity, but they've lost this capacity. This is something that's not often discussed when we talk about evolution; instead, we emphasise how natural selection conjures novel adaptations, but the flip side to this is that evolution is simultaneously a tidy housekeeper. The apparatus needed to maintain an adaptable approach to salinity has, in a sense, to justify its presence. If by sacrificing this, a fish becomes more efficient, then doing so gives that fish a competitive advantage, so long as it stays within its river, or at sea, and that's why the overwhelming majority do so.

For those animals equipped to cope with the shifting conditions, estuaries offer rich pickings. The extraordinary biological activity within estuaries – they're twice as productive as farmland and over ten times as productive as the open ocean – means that there's plenty of food to be had. A few metres downstream of where I'm standing, there's a disturbance at the water's surface. I'm expecting a cormorant to appear, surfacing from a dive, but instead a large, mottled head surfaces in mid-channel. A pair of rheumy eyes and long snout give it the expression of an elderly dog, albeit a very well fed one, but this is a grey seal bull, and he's arrived in anticipation of the tide in search of food.

He's perhaps not the most attractive of animals and this is reflected in his official taxonomy. The early naturalists who coined Latin names for species could sometimes be bland and uninspired with the monikers they forged, yet they seem to have reserved some inexplicable vitriol for the poor old grey seal. They named it *Halichoerus grypus*, or the 'hook-nosed sea pig', which, if not unfair, feels a little direct. Nonetheless, he's impressively large at over 2 metres long and 300 kilograms, testament perhaps to his winning foraging strategy.

Twice each day, the sea floods the lower reaches of rivers like the Great Eau, and the influx of saltwater brings droves of fish drawn by the prospect of a banquet as the water covers the estuarine mud. The slick cocoa-dark flanks of the channel may appear devoid of life to us, yet the fish know different. Concealed within the ooze are countless worms, crabs and molluscs, organisms who are stirred from their interment deep in the estuary's bed by an internal clock set by the tide, which drives them to emerge and feed on the flood. This is the bounty that draws the flotilla of fish and these in turn have brought the seal. Food begets food begets food.

After regarding me blearily for a moment, the seal gives a disdainful sniff and disappears down into the murk. The sediment-rich water permits little visibility, but this is no handicap to the seal whose exquisitely sensitive whiskers allow him to trace the movements of his fish prey with spectacular accuracy. The turbulence that fish leave in their wake as they cruise around on their own foraging missions surrenders their whereabouts to a hunting seal. Equipped with his whiskery sonar, the seal can detect even the slightest disturbances in the water, allowing him to distinguish trails of interest, caused by his prey, amid a swirl of other flows. Under perfectly calm conditions, the fish's invisible spoor of vortices might remain detectable for up to three minutes, but even here amid the turbulence, it's distinctive enough for the seal to follow the faint watery imprint left by the fish upstream to its source.

Many sharks, too, make use of estuaries, but there's one species in particular that is virtually unparalleled among its kind in its ability to capitalise on their opportunities, and that can even travel far into freshwater to improve its fortunes. It is a species whose remarkable ability to tolerate anything the aquatic environment can throw at it can perhaps best be illustrated over a game of golf.

In May 1996, southeastern Queensland was hit by some of the worst floods in memory. In just three days at the beginning of the month, six months' worth of rain fell over the area. At the same time, a low-pressure area over Moreton Bay produced catastrophic storm surges. The terrible combination of the two raised river levels by several metres, inundating huge areas of land. When eventually the waters began to recede, the focus of the local people was naturally on remedying the devastation caused to lives and property, and at first no one noticed some new arrivals in the lake at Brisbane's Carbrook Golf Club.

It didn't take long for this to change. Golfers strolling along the 15th fairway spotted, to their astonishment, the unmistakable triangular dorsal fin of a shark slicing the water surface of the club lake to their right. Closer encounters over time revealed that the fin belonged to a bull shark, a species with a fearsome reputation as a man-eater, and that there was more than one of them. In fact, there were at least six sharks in the lake, possibly more, which was all the more surprising since the lake is freshwater and aren't sharks, after all, sea creatures?

Across all shark species, bull sharks are among the best adapted to life in rivers. Indeed, it's part of their strategy to seek the shelter of estuaries and other low salinity habitats during the more vulnerable stages of their early life, where they are safe from the notice of larger sharks that might otherwise make a meal of them. Even as adults, bull sharks are known to inhabit freshwater. They've been seen hundreds of kilometres from the coast in the Mississippi, the Zambezi, the Amazon and the Ganges, among others. But even such determined travellers have usually been assumed to be enjoying a relatively brief sojourn in *agua dulce*, not least because they breed in coastal waters, and that's what makes the sharks of Carbrook Golf Club so compelling.

After some initial disquiet, the Australian golfers took the sharks to heart, celebrating their presence by naming a competition for them and even adopting the sharks for their club's logo. Nevertheless, the sharks were unwilling guests. Their size – some of them approached 3 metres long – and the fearsome reputation of the species seem to have discouraged ideas of capturing them and so, with no available exit to the sea, the sharks were trapped, getting by in the lake's muddy environs while putting the hazard in water hazard.

That bull sharks in particular are able to switch between fresh and saltwater so effectively distinguishes them from the majority of their kind. Most other sharks are considered osmoconformers, which is to say that the salinity inside their bodies closely matches that in the water they swim through. However, shark blood has only about half as much sodium and chloride as seawater, which means that they have to find something else to make up the difference. Unlike other animals who have little use for urea and flush it away in their urine, sharks use it as ballast in their osmotic balancing act, so that the saltiness of their blood matches seawater.

Although urea provides an inspired solution to sharks' osmotic challenges, there is a downside: urea is toxic, which is a major reason why most animals expel it, and why kidney failure is so dangerous to us. Consequently, it takes another compound to make the whole system work. That compound is trimethylamine oxide, or TMAO to its friends.[15] If urea is the workhorse of sharks' osmotic strategy, then TMAO is the fixer. Rather like the oil in a car's engine, TMAO maintains the smooth running of the bodies of many species of fish, safeguarding structural proteins from a host of problems. In sharks, TMAO protects the tissues from damage by urea and keeps the whole physiological show on the road.

Bull sharks, too, use the urea and TMAO strategy, but their ecological flexibility is underwritten by some exquisitely sophisticated adaptations, and they're supported in freshwater by the relentless activity of their kidneys, livers and gills. This frees up bull sharks to take advantage of the comparative safety of rivers and estuaries from when they're born, already the length of your arm, to when they make the move to the ocean at around three to five years of age. After that, as we've seen, bull sharks are more than capable of heading upriver to freshwater, but not usually for an extended visit.

The sharks of Carbrook Golf Club remained as hostages to fortune until another catastrophic weather event in 2013 put an end to their 17-year tenure in the lake. Perhaps the flooding that accompanied the storms that January allowed the sharks to cross into the nearby Logan River and from there to make their way downstream 14 kilometres to the sea; we'll never know. Nevertheless, the sharks' golf club residence led to greater public appreciation of these fascinating animals, while demonstrating the remarkable capacity of bull sharks to make their home wherever the currents take them.

---

15 If any further reason were needed not to eat sharks beyond the fact that doing so is disproportionately disastrous to marine ecosystems, the levels of TMAO in shark flesh pose a threat to human health. Ingesting it causes some rather challenging gastric issues, including explosive diarrhoea, and toys with the brain in a way that produces symptoms akin to severe inebriation. Being 'shark drunk', as it's termed, isn't much fun, however, and eating too much shark meat can even be fatal.

* * *

Just a dozen or so kilometres north of where I'm standing in the tranquillity of the saltmarsh lies the mouth of one of England's great estuarine rivers, the Humber, which drains the largest catchment in the country. It's the meeting point of a network of rivers that flow through the urban centres such as Leeds, Bradford, Sheffield and Nottingham, and its banks are home to the ports of Kingston upon Hull and Grimsby. These were the nineteenth-century heartlands of the Industrial Revolution, places whose soot and steel transfigured the Humber, making it one of the most polluted estuaries in the world, a sump into which drained the most pestilential of corrupted rivers.

The Aire, one of the Humber's tributaries, was killed over a century ago by the factories of Leeds and Bradford, poisoned by tanneries, slaughterhouses and textile mills. The effluent from dye-works would change the Aire's colour, lending the dead river a discordantly gaudy appearance, like lipstick on a corpse. As recently as 30 years ago, more than two thirds of its flow during summer was made up of the discharge from sewers. Bacteria stripped the oxygen from the water, and methane bubbled from the riverbed. It was an assault on the senses, a reservoir of filth, and a breeding ground for disease. I can remember walking along the banks of the Aire as a schoolchild in the 1980s and feeling a sense of unease when I came near it. Some days the river ran clear enough for me to see that no living thing stirred in its waters; it was a tragic simulacrum of a river and an affront to nature.

Even as the heavy industry of northern England fell into decline, the Humber remained a conduit for pollution. The old-style mills gave way to new kinds of manufacturing, and instead of contaminating the river with the signal stench of ammonia from the wool scouring, or the outlandish colours of discarded dye, the Aire began to carry novel substances that were both more powerful and more insidious. The term persistent organic pollutant, or POP, was coined as a catch-all for a diverse group of synthetic, chemically engineered compounds that are both long-lived and deadly. Some of the most notorious of these, known as polychlorinated biphenyls, or PCBs, are now among the most widespread and noxious chemicals in the world.

First synthesised in the latter stages of the nineteenth century, the heyday of PCB production stretched from the 1930s to the 1970s. Like so many innovations, they were initially welcomed as heralds of a bright new age, signifiers of mankind's brilliance. The unique properties of PCBs, particularly their durability – they're virtually immune to almost everything that degrades other substances – made them invaluable as coolants, lubricants, insulators and sealants, in each case providing near miraculous improvements to what had gone before. Their use mushroomed, until they were to be found in everything from electrical equipment to paints, and from engines to paper. In the early days, when enthusiasm for PCBs carried all before it, they were even recommended as an additive in chewing gum and in the manufacture of dentures.

It wasn't long, however, before the true nature of PCBs was revealed. Within a few years of mass production, reports began to emerge that this group of wonder chemicals had a darker side, one that damaged the health of anyone who came into contact with them. Workers at chemical production facilities as well as those using the chemicals, such as painters, and even submariners on vessels using PCBs in the hydraulic systems, developed unexplained sicknesses, and a terrible form of skin damage known as chloracne. Clinical trials documented systemic toxic effects of PCBs on the liver and suspicions began to arise that PCBs were carcinogenic.

By the early 1950s, little doubt remained about the malignance of PCBs and the moral case to cease production was unanswerable. Yet the companies invested in PCBs fought a rearguard action of cover-ups, obfuscations and denial. There was too much money to be made, and you must never upset the shareholders.

Worse was to come. PCBs rapidly infiltrated the human food supply via plants grown in fields treated with PCB-augmented pesticides and through the meat of grazing animals. They found their way into waterways and into the air itself, and from these, insinuated themselves into the tissues of almost every organism. No living plant or animal was unaffected, people included.

As the 1960s progressed, public awareness about the scale of the environmental catastrophe wrought by pollution was growing, fostered by powerful accounts such as Rachel Carson's seminal *Silent Spring* and by astonishing spectacles including the Cuyahoga River catching fire. Even so, the recalcitrance of chemical firms was breathtaking. They'd known for two decades by this point that they were in the business of peddling disease and death, yet still they continued to defend their practices. It took government interventions around the world to bring a halt to the commercial production of PCBs. By the time that the European Union joined the US, China and Japan's lead and proscribed their manufacture in 1987, it seemed that world production had all but ground to a halt.

The global consensus that had developed around the damage inflicted by PCBs drove further change, most notable of which is a landmark treaty, known as the Stockholm Convention on Persistent Organic Pollutants, which committed signatories to the eradication of 12 types of chemicals, PCBs among them, colloquially known as the dirty dozen. Since the treaty was signed in 2001, more than 20 further chemicals have been added to the list, with others currently being considered for inclusion. In an unusual display of international unity, the convention has been ratified by almost every country in the world.[16]

The problem hasn't gone away, however, and the Stockholm Convention represents a beachhead rather than a victory. Official production of the chemicals listed in the convention may have ceased, but huge stocks of them remain in existence. Up to 2 million tonnes of PCBs alone were manufactured before the ban, around a third of which has already escaped into the environment. Most of the rest was consigned to landfill and without proper safeguards this will leach out to augment the contaminants already in nature. Moreover, it comes as a nasty surprise to learn that we're still producing PCBs, despite the prohibition. The difference is that PCBs used to be manufactured intentionally, whereas now they're a by-product of other chemical processes, such as in the making of paints, pigments and industrial solvents. Although this unintended production typically means that there are only trace amounts of PCBs in any final product, at the scale of the industries involved, it collectively amounts to thousands of tonnes each year.

While the Stockholm Convention enshrined our collective attitudes to organic pollutants, it is forced to rely on individual governments across the world to police it and some are more diligent than others. At the same time, the substances in question are highly complex and incredibly diverse. Some modest re-engineering of a banned chemical by a manufacturer can yield a new product, one that has similar properties and is just as malignant in the environment, but which is not on the Stockholm blacklist. Consequently, regulators constantly find themselves playing a deadly game of whack-a-mole with new and emerging contaminant

16 *Almost* every country in the world… But not the US, Israel or Malaysia.

threats. And once these POPs are in the environment, they linger. PCBs and similar substances stick around for decades to centuries; another group, known portentously as 'forever chemicals', have estimated half-lives of well over a millennium.[17]

The ultimate destination for all of these chemicals is the sea. Every day, rivers like the Humber deliver a toxic cargo of contaminants to marine life. From the North Sea beyond the saltmarsh to my home in Sydney, where the ocean sparkles like blue champagne in the wake of the city's emblematic green and gold ferries, nowhere is untouched.[18] The result is that POPs like PCBs, dioxins and forever chemicals are now ubiquitous in virtually every species of commercial fish. The worst affected tend to be those that live in close proximity to the seabed, where the concentrations of pollutants are greatest. Flatfish at one of the North Sea's most important fishing areas, Dogger Bank, suffer an astonishing prevalence of abnormalities, including cancer. The effects of exposure to contaminants start to show very early in the fishes' lives, and by the time they've reached maturity, up to 80% of the flatfish at the locale develop liver tumours.

As well as being exposed to pollutants directly from the environment, animals also take in these substances through their diet. When a fish eats a contaminated shrimp, it gains the shrimp's chemical baggage along with its calories. These toxins insinuate themselves into the fish's body and with each additional tainted

---

17 'Forever chemical' is a colloquial term for PFAS: per- and polyfluoroalkyl substances of which there are several thousand different kinds.

18 Indeed, the deceptively clear waters of Sydney Harbour are home to some of the highest concentrations of deadly organic pollutants in the world. Commercial fishing is banned within the harbour indefinitely, and recreational fishers are advised to strictly limit how much of their dioxin-loaded catch they eat. The ferries that pass the site of the former chemical works in the middle harbour must slow to a crawl to avoid stirring the dioxins from their muddy resting place into the water column.

meal, they accumulate. When in time the fish, too, falls prey, it passes on the poisonous legacy of a lifetime's foraging. So the toxins ratchet up with each ascending level of the food chain, their number and concentrations intensifying in the tissues and organs of animals in a process known as biomagnification. Right at the top of this food chain, the effects can be stark.

Numbers of the UK's only resident pod of killer whales have fallen from around 20 in the 1980s to just two today. There have been no live births in the pod for over 20 years, and since both the remaining individuals are male, there is no possibility of recovery. The culprits are organic pollutants, which are not only highly carcinogenic, but also weaken the whales' immune system and render them infertile. A 'safe' amount of PCBs in marine mammal blubber, if there is such a thing, is reckoned to be around 9mg/kg. When Lulu, the last female in the UK-resident killer whale pod, died in 2016, her post-mortem revealed that she was carrying 950mg/kg of PCBs in her blubber.

With such an appalling chemical burden, it's scarcely surprising that Lulu never produced a calf. Sad though this is, in some ways it's a blessing because organic pollutants have an awful habit of crossing between the generations. Gestating females inadvertently transmit some of their accumulated PCBs to their developing foetuses through the placenta. And when the calves or pups are born, their mothers sustain them with milk that's laced with PCBs. It's a heartbreaking thought that this most intimate act of maternal nurturing has become the deadly conduit of our pollution. The youngsters' noxious inheritance provides the worst possible start to life. The rapid development of infant body and brain take place amid a swirl of the most harmful substances known, and at a time when they're most vulnerable to them.

If organic pollutants were the only toxins that channelled their way to the sea via our inland waterways, it'd be bad enough, but in estuaries like the Humber, pesticides rub shoulders with heavy metals, microplastics jostle with herbicides, and oil mingles with organophosphates.[19] The effluent of 15 million people also makes its way to the Humber. Within this baneful flow, residues of the drugs that we take – legal and otherwise – are increasingly coming under the spotlight as pollutants. We might imagine that when we swallow a pill for a headache, say, or rub a gel on the skin to ease the pain of arthritis, the medications would be absorbed by the body as they do their work in alleviating symptoms. The truth, unfortunately, is that the drugs leave us in the traditional manner, and when we pee them out, they go into the sewers and from there to the environment, where they remain bioactive. Collectively, we excrete vast amounts of pharmaceuticals and narcotics, and these produce profoundly unsettling effects in aquatic organisms.[20]

Nature once offered us a solution to all of this, but in typical fashion, we squandered it. Long ago, the Humber housed millions of oysters, stretching in a single unbroken reef for 7 kilometres across the estuary. The purifying work done by oysters has led them to be described as the sea's kidneys. Each oyster feeds by slurping up the equivalent of a bathful of water every day and as a by-product of this relentless sucking, they clean the water, removing faecal matter and bacteria, as well as nitrogen and phosphates that would otherwise fuel toxic algal blooms. Their shells even absorb some of the worst heavy metals, such as cadmium, lead and copper, though

19 Most people imagine that spills from tankers and drilling platforms as the main sources of oil in the marine environment, but run-off from our roads and streets contributes around 20 times as much each year.

20 Residues from the contraceptive pill feminise male fish to the point that they are rendered infertile, antibiotics flooding into the environment accelerate the process of bacterial resistance, and even painkillers, such as ibuprofen, induce high levels of oxidative stress that sabotages the workings of organisms' cells.

it has to be said that this part of the process does the molluscs no favours. By degrees, we exterminated the Humber's oysters through overharvesting and dredging, while the conditions that we imposed on them fostered disease. Our assault on the oysters was such that, by the early 1900s, they were gone.

The same pattern was repeated across the globe, notably throughout Europe, in Australia and the shores of North America. In our ignorance, we gouged almost 90% of the world's wild oysters from the sea, in many cases selling them as food at a fraction of the value they provide to their environment. As we began to flood our rivers and estuaries with the dreadful by-products of our civilisation, we compounded the problem by destroying the very animals that could have helped to remediate the problem.

The juxtaposition of the biological exuberance of the saltmarsh and the tainted environment that lies just beyond the horizon is startling. It's almost as if they belong to two separate worlds, one pristine and the other defiled. Appearances can be deceptive, however. The saltmarsh isn't pristine, and its inhabitants aren't immune. The Humber has long broadcast its burden of chemicals into the sea, where they may linger unseen in the water column or settle into the bottom sediment. Entire communities of organisms are steeped in a noxious medley of pollutants. Along large parts of our coastlines, every animal, from the largest to the smallest, carries in its tissues a cocktail of contaminants. Many die, but even those that manage to survive the chemical onslaught face a terrible array of consequences. Some chemicals antagonise hormones, the body's signalling system, others render them infertile. Their cells and organs are compromised, their DNA is damaged, tumours develop, and their immune function is derailed.

The chemical warfare that we have waged upon our oceans in the past two centuries will take many human lifetimes to resolve.

Oysters are just one of countless species of marine creatures that have been driven from their habitats. Others remain, displaying stupendous resilience. Atlantic tomcod in New York's Hudson River have evolved resistance to the high levels of PCBs to be found there, while the killifish of the nearby New York-New Jersey harbour are now thousands of times more resistant to the toxins that circulate in the waters there. Incredibly, these fish aren't simply surviving but thriving in the face of these hostile conditions.

This does not provide us with carte blanche to forego our responsibilities on the basis that nature will find a way. Relatively few animals have these fishes' evolutionary capacity to adapt so rapidly, and those that can are typically short-lived species; it's not so easy for cetaceans, for instance, or indeed for us. Tomcod and killifish pay a price for their toxic tolerance, growing more slowly and being less fecund. They may also be less adept at escaping their predators, which makes it all the more likely that they will pass on a huge payload of poison to whatever might eat them. And they may even be less well equipped to live in pristine waters as a consequence of their acclimatisation to pollution. The endurance of tomcod and killifish, poster animals of the Anthropocene, is incredible, but the fact that they have to dramatically reconfigure their genetic architecture to survive in the conditions that we have created is, to me, more a cause of sadness than celebration.

In the two centuries since the dawn of the Industrial Revolution, estuaries have come to be the places where the tides breathe poison. By corrupting them, we've disseminated civilisation's contamination to the wider ocean. But times are changing, and among all of this gloom, there is good news. Since the Stockholm Convention came into force in 2004, monitoring projects from places such as the Arctic and the North Sea show a fall in the levels of organic pollutants in the environment, as well as in the bodies of marine organisms. In some cases, the decrease is dramatic, such as a hundredfold drop in POP concentrations in prawns, underscoring the value of the measures taken.

This slow recovery of the marine environment has been bolstered by restoration projects. Plans are under way to return half a million oysters to the Humber. If these pioneers find conditions to their liking, there's every chance that they will proliferate and that their descendants will one day rebuild a thriving reef. This would then provide refuge for crabs, barnacles and juvenile fish among much else. It's for this reason that oysters are regarded as ecosystem engineers: they create habitat for myriad other creatures. In a reversal of the well-worn phrase, for around 500 animal species across the globe, the oyster is their world. Other filter-feeding animals, such as mussels and sea squirts, would be drawn in by the presence of oysters, and would bolster their efforts, collectively purifying the water. And as the reef grows, it will fulfil other roles, enriching the environment and at the same time taming wave action and protecting the shore.

The Humber restoration project follows on the heels of groundbreaking projects in Chesapeake Bay, in Maryland, the Solent, on the south coast of England, and across various sites in Australia. At each, the stage for the oysters is often set by laying a bed of oyster shell, before adding oyster seed, or spat, to settle on their bespoke foundations. In some cases, the seed oysters have been selectively bred for their resistance to disease, which alongside our own depredations, have hampered the ability of the animals to recover naturally. It will take time for these projects to develop to their full potential, but the fledgling Solent project has already recorded a remarkable success, with almost a hundred other species already living among the oysters and benefitting from the protection they offer. There are no guarantees of success, of course, but by working with nature, rather than against it, and by harnessing the incredible ability of nature to bounce back, we can restore precious habitat in a way that repays the initial investment many times over.

Oysters, small as they are, make a heavyweight contribution to the renewal of damaged estuaries and coastlines, but a full solution has many facets. It's not a job we can simply leave to the molluscs, and nor should we. Better water treatment facilities have played a critical role in the restoration of the rivers and estuaries, yet more needs to be done. The weak link in any sewage solution is that the facilities can easily become overwhelmed during heavy rain. With urbanisation comes growing densities of people and the disappearance of an ever-increasing amount of land beneath paving and tarmac. Instead of storm water soaking into the ground, it hits the hard apron, flooding first the drains and, ultimately, the treatment plants. Since these plants have finite capacity, something has to give; a slew of untreated, chemical-infused sewage flows out of storm drains and into the natural environment.[21]

This is an entirely solvable problem, but it's one that requires investment and therein lies the problem. A politician calculating effects on the polls of preventing gallons of effluent sailing away down into the sea versus the more visible impact of providing more hospital or school equipment is likely to make the most eye-catching choice. Nonetheless, in some jurisdictions, legislation has compelled them to act. We have powerful new technologies to treat our waste, to strip it of contaminants and nutrients, and though we've scarcely fallen over ourselves in our haste to implement these, recent transformations seen in the Humber, as well as in the Thames, the Rhine and the Hudson, have been encouraging.

21 Worldwide, almost half of all sewage goes into the environment entirely untreated.

And what of the dead river whose banks I walked as a schoolboy? The Aire has been the beneficiary of change, inspired by a committed group of volunteers. Mobilising the public and local businesses, they brought pressure on the culprits behind the river's demise. The efforts of this group – Eye on the Aire – have borne fruit in the most dramatic manner imaginable. The lifeless river of my childhood is no longer. The Aire is being reclaimed by nature, its once barren margins have been planted with trees, and in summer are animated by dragonflies. Fish have returned, bringing with them kingfishers, herons and otters. In recent times, there have even been sightings of the most emblematic of all river fish, two centuries since they seemed to have left forever. Whisper it, but it seems that the salmon have returned to the Aire.

* * *

The grizzly stops in mid-stride and lifts its great muzzle to the wind. Then, with an appalling suddenness, it snaps its head around to look across the river to where we stand watching. A moment later, the ponderous and somehow benign creature that ambled down the far bank is transformed. It now radiates intensity and purpose as if the sight of humans has flipped some switch in its carnivoran brain. Leaving the shore, it steps into the shallows, eyes fixed on us, alert and intent. Then, quickening its pace, it begins to close on us with awful speed, sheets of water arcing from each galloping footfall. My idea that the boiling river rapids would provide some barrier is revealed as a ridiculous conceit; there is no refuge to be had. The immediacy of the danger meant that there is no possibility of outpacing the bear. Instead, my companions and I instinctively close ranks, the unspoken idea to try and present a more intimidating target. Yet ultimately, it isn't our collective defiance that saves us, but a fish.

The grizzly had approached to about 10 metres of us when a commotion at the water's surface distracted it. Perhaps disturbed by the bear, a salmon torpedoed across the mammal's path, its domed back breaking free of the river like a surfacing submarine. The bear lunged, but missed its quicksilver target, then stood gazing down at the point where the fish had been. With our pursuer's attention distracted, we edged cautiously away to higher ground. I kept an eye on the bear as we retreated, but it showed no sign of wanting to re-engage with us. Its focus was now on the river and its submerged wealth of spawn-heavy salmon.

The weird, dissociated sensation I'd had as the bear charged had now dissipated, replaced by a delayed adrenaline rush, and I was jittery by the time we reached the top of a granite escarpment, cut into the bank by the river's bend. I chided myself furiously for being so cavalier, putting myself in the vicinity of an apex predator who wanted me for a chew toy. In my defence, there was little in my early life in northern England that might have prepared me for something like this; the worst that nature could fling at me there might be a discombobulated stoat. But my self-recriminations were nothing compared to the reactions of some locals when I related the story to them later. Just take a shot in the air, that'll stop 'em. I don't own a gun. What? Wow, OK, well then you could just blow your whistle. Didn't have a whistle either. For crissakes, did ya *wanna* get eaten?

Still, my telling-off was in the future, and right now the high ground commanded a panoramic view of the river's reach and an armada of torpedo-shaped shadows beneath the surface, toiling upstream against the churning flow of the rapids. In late summer throughout the northern climes, millions of salmon cease their oceanic wanderings to return to the freshwater rivers that were their nurseries. They have travelled for years, and for thousands of kilometres in the great swirl of the North Pacific's Alaskan Gyre, gorging themselves on the sea's bounty until they approach, or even exceed, a metre in length. As the summer advances, they are stirred by a new urging. A choir of cellular pings, conducted by hormones, announce a bodily ripening, impelling the salmon to return to freshwater.

Guided initially by the Earth's magnetic field, and ultimately by the siren smell of their natal river, with its unique chemical fingerprint, they funnel inland in a quest for the headwaters that once served as their cradle, and which are now destined to be both spawning ground and grave. The final journey is one of incredible hardship for many salmon. They ate their last meals when still at sea, and now as they travel upriver, the demands of negotiating the unrelenting current and the many obstacles takes its toll. At the same time, their bodies have diverted resources towards the production of eggs or sperm, and away from hitherto critical functions like bodily repair. With their immune systems in decline, many of the salmon begin to decompose even before they reach the headwaters, their flesh pock-marked by ulcers, or even coming away in strips. The fearsome, rotting appearance of some has inspired the description 'zombie fish', but while the wreckage of once sleek bodies might call to mind the living dead, they retain the unyielding, obdurate determination to complete the arc of their existence, to spawn a new generation before, exhausted, they succumb.

It's this reliable, essential glut of salmon arriving that has drawn a host of their predators, including the bears. Over the few short weeks of the salmon run each year, brown bears, and their less fearsome black cousins, feast relentlessly, cramming in as much energy as they can to support them through the lean winter ahead. This so-called hyperphagia is thought to be driven by leptin, or rather a lack of it. Leptin is the hormone responsible for governing appetite, making bears (and us) feel full, but at this time of year, the bears' bodies either produce less leptin or become desensitised to it. With the hormonal shackles off, the bears gorge themselves almost to inanition, and then, still unsatisfied, keep right on eating. The bears don't have limitlessly capacious stomachs, however, so they focus on the most calorific parts of the salmon, especially the brain and other internal organs, the eggs and roe, and the skin. The parts of a salmon that we most cherish, the succulent fillets, are often discarded. With portion control gone out of the window, and with a relentless focus on the richest, fattiest parts of the salmon, the bears might cram on as much as an extra kilogram or two of fat every day. By the time the salmon run is over, the now rotund bears are likely to have the fuel needed to see them through a six-month slumber, beginning when they enter their dens in late autumn to hibernate.

From a vantage point high above the river, I can see the grizzly standing in mid-channel, head down, its nose almost touching the water, waiting for one among the seemingly endless procession of salmon to stray too close. Framed by the stands of spruces that form a guard of honour for the rushing white water, the bear was the centrepoint of an iconic Alaskan panorama. Crows and jays squabbled among the trees, while overhead, bald eagles circled against a mountainous horizon. The one thing that all these organisms have in common, as well as the countless others that go unseen in my field of view, is a dependence on the annual bounty of salmon.

The Pacific salmon that grace these waters today began life perhaps just metres from where they will spawn and die. Years before, these mighty fish emerged from their eggs as minute simulacra of their adult selves, each perhaps a gram in weight. Once they reach the ocean, they grow rapidly amid a rich soup of plankton, krill and smaller fish. The result is that by the time they return, the survivors are several thousands of times heavier and just about every one of the extra grams that they pack on comes from the sea.[22] Tens of billions of salmon fry begin their lives in Alaska's rivers each year, and although the overwhelming majority of those never make it home, their abundance is such that perhaps 250 million adults find their way back to those rivers annually. Their upward migration represents a huge flow of energy, running counter to the river, and in a reversal of our typical assumption that rivers carry nutrients to the sea, and not the other way around.

Each adult salmon amounts to something like 25,000 kilojoules of food energy – enough to feed an adult human for three days. Furthermore, they represent a rich source of essential nutrients. Nitrogen, for instance, is crucial to all living things. It forms the chemical backbone of the proteins that are, in turn, the building blocks of all organismal bodies. And despite being surrounded by nitrogen as a gas – it comprises something like four fifths of every breath we inhale – it's unavailable to animals and most plants in this form. Instead, animals must gain it through their diet. Plants, meanwhile, absorb it through their root systems, but while nitrogen is one of the most common elements on Earth, it's relatively rare in the Earth's crust. Consequently, finding sufficient nitrogen in the soil represents a perennial challenge to them. Here in Alaska, that's where salmon, and bears, come in.

Shortly after our run-in with the grizzly, we were hiking along a trail at the top of a thickly wooded ridge overlooking a steep ravine when a black bear suddenly materialised right in front of my friend, Mike. It remains a mystery how a 200-kilogram animal could scale such a gradient through dense undergrowth without alerting us, yet here it was, virtually face to face with Mike whose mouth hung open

22 This transformation is a phenomenon common to many aquatic animals; the equivalent in human terms would be if we were born the size of an undernourished bumblebee.

in shock. The bear's ability to adopt an appropriate expression was rendered more challenging by the fact that it carried crossways in its mouth a colossal salmon; nevertheless, each held the other's gaze for a moment in mutual astonishment. It was the bear who moved first. Snorting its chagrin, it diverted around us and trotted into the undergrowth with its bounty. It's this tendency of bears to take the salmon that they catch away from the river that's key. The reasons that the bears do this is straightforward enough: a salmon can't escape once on land, and by withdrawing to a safe place, there's less risk that a larger, more powerful bear can steal the prize.

As we've already seen, however, the bears' ability to catch these superabundant fish outstrips their capacity to eat them. In fact, they typically discard much of their booty, in some cases even leaving an entire fish that disappoints in some way. The result is a windfall for other animals, from tiny opportunistic insects to birds and smaller mammals. Hundreds of different species come to the bears' feast, and little goes to waste.

Liberated from the salmon, molecules of nitrogen now begin journeys of their own, travelling far and wide into the web of riparian life. The flies that hatch from eggs laid on salmon cadavers carry in their bodies the marine nitrogen brought here by the salmon, and when they fall prey to spiders, or frogs, those same nitrogen molecules are reallocated and reincorporated over again. For the plants in the vicinity, the chemicals that leach into the soil from spent carcasses are the natural equivalent of the artificial fertilisers used in intensive agriculture. Grasses, shrubs and trees greedily absorb the windfall to the extent that up to three quarters of the nitrogen found in these plants comes from the sea, delivered by salmon. One vital use made by the plants of this nitrogen is in the production of chlorophyll, and so the salmon's legacy can be seen in the green of the leaves, the capture of the sun's energy and its transformation into sugars at the foundations of the food chain.

The marine nitrogen brought in by the salmon can be seen throughout Alaska, and indeed right down the west coast of the US and Canada. That nitrogen has a distinctive chemical identity, an isotope known as 15N, which means that we can trace and measure it. From this we know that it's concentrated around the salmon rivers, naturally enough, but evidence of it can even be detected in the Rocky Mountains, hundreds of kilometres from the sea. The contribution of the salmon to healthy ecosystems on the land, both in scope and scale, is enormous. In return, those ecosystems act to maintain healthy spawning rivers as the vegetation filters the rainwater, provides shade, and regulates extremes in water flow. The intimate interdependence of the salmon and the habitats through which they navigate to spawn isn't an exception; in fact, it represents the norm in natural systems.

We often fail to account for the broader implications of our actions, either through ignorance or indifference. Tweaking one part of the web of ecological interactions has ramifications that are not always immediately obvious. By overfishing salmon, by logging and by polluting spawning rivers, we take a machete to this web, haphazardly severing its supporting strands one by one, until it collapses.

Right along the west coast of North America, wild salmon numbers have shrunk to a tiny fraction of what they once were. There are all manner of explanations for this, including the competition that they face from hatchery-released fish, overfishing and the struggles of salmon to find sufficient food in a warming ocean. Moreover, the wild adult salmon that make it to the spawning ground are getting smaller, likely because they're returning at a younger age. All of this means that the great nitrogen conveyor belt that they comprise is far weaker than it once was and the effects of this echo through the ecosystems that depend on them. The same pattern can be seen in other salmon species elsewhere. In my lifetime, the population of wild Atlantic salmon has dropped by two thirds, while the number of eggs that a female has to lay to be confident of just one of her progeny surviving for a single year has doubled.

Despite the fact that some populations are contracting at an alarming rate, as things stand none of the seven major species of salmon are at imminent risk of extinction. They are, in many ways, one of the luckier groups of fish. Their iconic status as both a food fish and as bellwethers of healthy rivers, alongside their huge economic value to both commercial and recreational fishing, mean that salmon are among the most closely managed fish in the world. For instance, the Pacific Coastal Salmon Recovery Fund was set up in 2000, with the remit to restore salmon habitat on the US west coast. It achieves this by supporting projects that restore damaged and depauperate salmon rivers to a healthy state, planting bankside trees, installing fish passages past dams and enforcing tight controls on pollution. Their efforts are paying off in many cases, with growing numbers of adult salmon returning to spawn, yet despite this, concerns remain.

Protecting and revitalising freshwater salmon habitats represents a huge challenge, but once the fish head out to sea, they're off our monitors and largely beyond our help. Aquaculture installations in estuaries can attract sea lice in enormous numbers, which can infest healthy wild salmon as they pass by on their migrations. Meanwhile, fishing operations exact a staggering cost. Trawling for food fish such as Alaskan pollock in the Gulf of Alaska and the Bering Sea is huge business, bringing in well over a billion dollars per year. The problem with trawling is that it's a crude and untargeted means of fishing, and as iceboxes overflow with pollock, other species that are in the wrong place at the wrong time get swept up in the colossal nets. In 2023 alone, hundreds of thousands of salmon, as well as tonnes of the squid that they feed on, were captured by industrial trawlers in the region. This bycatch is very often just dumped, representing a criminal waste of life and a massive drain on populations, and it should be borne in mind that the quantities reported are likely only a fraction of the true numbers.

In 2023, a stark statistic emerged from one of the most iconic rivers in the Pacific Northwest, the mighty Yukon. As of mid-June, not a single chum or chinook salmon had passed the monitoring stations in the lower river, against a historic average in the tens of thousands. The loss isn't just ecological, or even just commercial in terms of the loss of sport fishing revenues, it's something that has a deep cultural significance. For many of the First Nations people who have lived along the Yukon and its tributaries for thousands of years, salmon are part not only of their lifestyle but also their identity. In a tradition that stretches back into history, the beginning of the salmon run has been marked with ceremony, with fires, drumming and dancing to mark the passage of the first fish. That is now gone. In its place, some have tried to maintain some element of the past by buying frozen salmon from elsewhere, but this is a sad pastiche of the vivid tradition that has been lost.

The decline of the salmon can't entirely be laid at the feet of the profligate trawlers, however, and indeed this is a point made repeatedly by National Oceanic and Atmospheric Administration (NOAA), the arm of the US government tasked with, among other things, managing fisheries. They claim, not incorrectly, that global warming is hitting salmon hard, changing the abundance and distribution of both the salmon's prey and their predators at sea. It's difficult, if not impossible, to partition the causes of the salmon collapse accurately between overfishing and climate change. Nevertheless, the refusal by NOAA to put a ceiling on the numbers of chum salmon that can be caught incidentally by wealthy industrial trawlers as bycatch is perplexing. The trawlers could, in theory, capture ten times as many chum as they already do, without hindrance. The trawlers themselves don't gain either, but the cost of instituting measures that might reduce bycatch deters them into inaction. The toothless and seemingly supine attitude of NOAA to the salmon crisis is even more surprising when you consider that it is they who oversee the much-trumpeted conservation efforts of the aforementioned Pacific Coastal Salmon Recovery Fund. This bafflingly counter-intuitive approach is an all-too-common motif in our endeavours to protect our natural resources.

The loss of wild salmon from Pacific rivers contrasts with their return to what were once some of the most polluted waterways in Europe, navigating the Humber to reach the Aire. The one does not balance out the other; instead they highlight the dramatic consequences of our actions. If we make smart decisions, we can revive nature. My generation, and that of my children, can be the fulcrum for a change in humanity's attitude to the ocean. We have everything to fight for, and we have the capacity to make a difference. Of all the environments that contribute to the health of the seas, the rivers and estuaries of the hinterlands are perhaps the most important. If we can get these right, the rest will surely follow.

It's late afternoon now on the marsh, and the waders cast long shadows as they follow the ebbing tide to access virgin mud. This margin of land and sea, where salt shades into fresh, is a magical and ever-changing place, but there are shadows here that are not born of the sun. As I watch, one of the redshanks latches onto a ragworm, pulling the reluctant victim from its muddy retreat, and hurriedly swallows it. I can only hope that, in striking gold, it hasn't also struck mercury or cadmium and that, if it survives to raise new generations, those offspring will prosper in a cleaner world than the one in which their parent prospects.

3.
A Kaleidoscope of Shells

The biological wonderland of the seashore

NORTH LANDING WAS MY CHILDHOOD'S SUMMER playground, a place that seemed as far removed as it was possible to be from the close-packed streets and houses of the suburban neighbourhood where I grew up. Visits to this cove, in the coastal Yorkshire village of Flamborough, awakened me to the magic of the seashore. At each foray, I was spellbound; each fresh discovery charging me with fresh wonder. My enchantment stemmed in part from its twice daily reveal, as though each ebb tide were the prelude to a gala performance, and an invitation for me to explore. It's a show that's run for as long as Earth has had seas, but I've never lost this sense of enchantment in the years since for the simple reason that it's never quite the same twice.

The shore is a place of endless flux, a shifting frontier between terrestrial and marine. The sea is the arbiter of this relationship, the rhythmic constancy of the tides contrasting with the capriciousness of its moods. Waves that lap against the land's margins like liquid caresses one day might pound with exhilarating fury the next.

In the turbulent, frenzied maelstrom of a storm, steepling breakers thunder against exposed rock. Displaced rubble is scooped up by the waters to be co-opted as artillery to bombard the shore, gouging and fracturing the cliffs, and quarrying pockets in the bedrock where water collects at low tide in rockpools. The evidence of the sea's many incursions can be seen in crooked coasts, rounded cobbles and, more than anything, in the sand that represents the ultimate result of ceaseless attrition, the milling of boulders to grain.

Change happens almost imperceptibly much of the time, so that the seeming stability, year on year, lends a false impression of permanence. But the shore wears and ages just as we do, and nothing lasts forever. Another of Flamborough's coves, South Breil, was home to two chalk pillars, known as the King and Queen Rocks. Running along the cliff path to the lighthouse as a boy, I'd pause to pay my respects to these twin stacks, each towering tens of metres above the waves, festooned with seabirds and imperiously defying the North Sea. Yet the summer came when I reached South Breil to find that the King had gone. All that remained was a shallow platform breaking just clear of the water and surrounded by an armada of paddling kittiwakes and razorbills, some of whom perhaps fledged years before among the King Rock's erstwhile crags and ledges. I later learned that a winter storm had brought the King down and swept him from his dais. Time and tide will take the Queen, too, one day. Her future, like ours, is a leasehold.

Just as the surf sculpts the shore, it scours and winnows the community of animals that live here. This, however, is not the only privation that these creatures must face. The relief provided by the ebb withdrawal of pounding breakers is tempered by the fact that they now find themselves stranded above the waterline in what, for marine organisms, is one of the most inhospitable environments imaginable. Our familiarity with the shore makes it easy to overlook just how physically difficult life on a rocky shore is at low tide. Compared to those animals that enjoy the comparative stability of permanent submersion, the amphibious creatures of the intertidal have to cope with massive fluctuations in temperature, the risk of drying out, and the inability to feed, or even to breathe freely until the tide returns.

The intertidal zone describes the part of the shore that lies between the lowest extent of low tide and the highest of high tide and one group of animals has adapted so successfully to this habitat that they've come to dominate it. These are molluscs, the largest single phylum of animals in the sea, making up around a quarter of all marine species. Gathered here are periwinkles, whelks, clams, mussels, oysters and chitons, a constellation of different forms. Collectively, their success is built on a molluscan hallmark: the shell.

To construct their shells, molluscs produce proteins, known as conchiolins, and a mix of other molecules that together make a framework for the formation and growth of calcium crystals. It's this composite material, an interwoven matrix of proteins and minerals, that gives the otherwise brittle calcium carbonate its strength, in much the same way that fibreglass, with its complex structure, is more resilient than its chief constituent, glass. When pressure is brought to bear on a mollusc shell, the layers within the mollusc shell shift, confining damage to pits in the shell's surface and, unless the force is overwhelming, preventing fractures from radiating.

Using some adroit biochemical tinkering, molluscs are also able to finesse the construction process, encouraging different types of calcium carbonate crystals to assemble at particular parts of their shell. The smooth and beautiful mother of pearl lining to the inside of their shells is sculpted by one form of crystal, aragonite, while the stronger and rougher bulk of the structure is built with another, calcite. As the animal grows, it adds neatly to the hem of its shell so that it remains sheathed in its mineral armour, fortified against the world beyond. And if the shell is damaged, it can make repairs in much the same way that our skin knits together after injury.

The simple expedient of encasing themselves within a keep of blended calcium provides molluscs with an answer to the challenges posed by harsh physical conditions. Just about every biochemical process relies on water, so keeping wet is critical. As well as protecting the animal against damage, the shell also retains a reservoir of precious seawater during low tide. It's from this reservoir that the animal gains its oxygen, acting as a kind of inverse scuba tank that supports these marine animals during their emersion. Their one weakness in this regard, of course, is the aperture to their shell; bivalves like mussels solve this problem by clamming shut, while periwinkles have the mollusc equivalent of a closeable front door, a bony plate beloved of pendant makers known as an operculum. Even so, the snails are far from blasé about the risk of desiccation and as the tide ebbs, they retreat into moist, shady spots on the shore and huddle together to enjoy the supportive humidity provided by a close press of molluscan bodies.

The ability of molluscs to make water canteens of their shells is a major reason that they have such a commanding presence in the intertidal. They're not, however, the only organisms here, and all face similar challenges at low tide. Fast-moving animals such as crabs withdraw into crevices or rockpools. Their crustacean relatives, barnacles, emulate molluscs by sealing themselves off against the elements behind a calcium redoubt. Anemones lack the shelly wherewithal to do the same but retreat into tight blobs and laminate themselves in waterproof mucus to remain hydrated. Even so, the most impressive adaptation to the strictures of low tide belongs not to an animal, but to the high- and mid-shore seaweeds. Wrack can survive losing 90% of its water; laver, which typically grows higher on the shore, can safely forfeit an incredible 95%, turning into a crisp, mummified version of its fleshy submerged state.

If drying out weren't bad enough, the low tide isolates rocky shore creatures from the stable thermal conditions of the ocean and, in winter, puts them at risk of being frozen. Happily, they're equipped for the challenge: plenty of mussels, winkles and barnacles are able to survive subfreezing temperatures, and some can even bounce back from more than half of their tissue being frozen solid. Summer brings a different problem; hours of direct sunlight threaten to turn shells into bespoke crock pots and boil their inhabitants alive. To combat this, limpets on sun-exposed shores grow taller shells, like traffic cones, as a means to disperse excess heat on the wind. Others actively engage in cooling behaviour. Some stand on the snail equivalent of a tiptoe to lift their shell away from the scorching bedrock and into a refreshing breeze. There are even periwinkles that gain relief from the heat by channelling their inner acrobat, climbing on each other's backs to form towers that might be five snails high.

Respite from the extremes of weather is provided by the returning flood, but this brings with it a new danger: that posed by the swirling, battering power of the waves, which threaten to dislodge the animals and sweep them to oblivion. Those creatures that have the advantage of mobility scurry to safety and hold on tight, the remainder are compelled to be stoic. It's not enough for the shells of limpets, mussels and barnacles to shelter them and to retain a water supply; now they have to be sleek. Shore animals rely on streamlined shells to thwart the sea's energy, parting the waves and allowing the surges to sweep past them. Mussels and barnacles anchor themselves against the surf's chaotic energy by respectively lashing and cementing themselves in place. For grazers like limpets, which need to move to feed, a little flexibility is needed. Limpets rely on their muscular foot to cling tight to the rock, but their grip is powerfully augmented by the multipurpose slime that oozes from the sole of that foot. That limpets have become a byword for tenacious sticking power is due in large part to this glue. A single large specimen attached to the underside of a rock shelf could, in theory, support the weight of an average-sized person dangling from it.[23]

As the burgeoning flood tide carries the surf further up the shore, the immediate danger of being washed away recedes. This, however, is no time to relax. Feeding isn't possible for most of the intertidal creatures at low tide, and now they need to make up for lost time. Such is the value of every minute that they don't simply react to the stimulus of the incoming tide, they anticipate it. Their lives are entrained by an internal clock and even before they're reclaimed by the rising waters, physiologically speaking, they switch on. The metabolic slumber that sustains them through the austerity of emersion gives way to a cellular quickening as they awake in readiness for the tide. At the earliest possible moment, filter feeders break their fast by sucking in draughts of fresh seawater, shrimps emerge from hiding, and barnacles part their craggy shutters to scour the currents for morsels. Now back in their element, limpets reconfigure their mucus from glue into lubricant and set off to glide across their stony savannah, rasping away with their file-like radular tongues at a near invisible film of algae. Like the limpet, many of the cast of the intertidal take the opportunity to move about their inundated stage. Even anemones, if they find their environs unsatisfactory, are capable of shuffling along at the stately pace of around a millimetre a minute to a better spot, away from competitors and into the path of a food-laden current. It might be

23 The largest limpets can resist up to 80 kilograms of pulling force.

imagined that having survived the hardships of the last few hours, high tide would be a haven for these creatures. Alas for them, the rising waters bring with them a posse of predators.

* * *

Some years ago, I was involved in a marine field ecology course at Dale Fort, on the stunning Pembrokeshire coast of South Wales. What had been built as a defensive clifftop fortification to protect the deepwater harbour of Milford Haven was now given over to the study of sealife, the guns replaced by aquariums that allowed close examination of the local fauna. Among the wayfarers enjoying a temporary sabbatical from the wild were a trio of shore crabs, each of whom had a carapace about the size of a playing card. The crabs would glower from within their aquarium at anyone who passed, but they distinguished themselves chiefly by a general lack of activity. This all changed, however, when the tide began to come in. The crabs might have been exiled from the sea, but they're equipped with the same kind of internal clock as other intertidal animals. The moment that this intrinsic alarm went off inside their crustacean brains, they'd start to rampage around the aquarium in feverish excitement. So deeply ingrained was this that they did it day after day, without any sense that they needed the tide itself to provoke them into action. The reason for this is simple: the high tide is hunting time for the crabs, and the rise of the flood is their dinner bell.

Shore crabs can give you a painful nip, but this fades into insignificance by comparison with that delivered by their larger relative, the brown crab, one of the many animals that I first encountered as a small boy in Flamborough. North Landing was home to a small collection of the wooden fishing boats peculiar to that part of the coast, known as cobles: wide-beamed, clinker-built and open-topped, their gunwales painted a vivid orange. Whenever I heard the thrum of *Summer Rose* or *Spring Flower*'s motors announcing their return from a trip, I'd run to the shore, excited to see what treasure they'd hauled from the sea. They didn't disappoint. Sometimes the boats would disgorge a posse of charter fishers carrying their trophies of iridescent mackerel, glistening haddock, or fascinating peculiarities like flatfish, with their strange, asymmetric faces. Most days though, the fishers were after lobsters and crabs, and the cobles' decks as they turned into the bay were stacked with the mesh pots recovered from the seabed.

Craning my head to see over the side of the coble as it was dragged slowly from the sea by a rust-ridden tractor, I came face to face with a tray of brown crabs, each a hulk of menace the size of a lunchbox. I stared at these other-worldly creatures, fascinated by their heavy terracotta armour, their black, expressionless eyes and foaming mouths, and, of course, their threatening pincers. Getting a little too close in my curiosity, a grizzly old fisherman with a windburned face and a woolly cap warned me to stay away from their black-tipped claws. 'Steady, lad,' he said with a chuckle. 'They'll 'ave your thumb off!'

Crabs are demolitionists, their powerful claws perfectly adapted to the task of overcoming the shelly defences of molluscs. Although crabs have two claws, they're unmatched; one claw is heftier than the other. This is the crusher, its job to break the resistance of fortified prey. Decent-sized brown crabs like these can exert about 20 kilograms of force, which, if not quite enough to sever an unwary digit, is more than sufficient to ruin your day. Once the crusher has done its work in shattering the bivalve's mineral fortifications, the other claw, more svelte and dexterous than its partner, and known as the cutter, gets to work snipping up the meal within before delicately transferring morsels to its waiting mouth.

The larger crabs hunt where their favourite prey, mussels, cluster on the low shore. Mussels, unlike their periwinkle or limpet cousins, have a pair of symmetrical shells, connected by a tough, leathery hinge. The delicate animal within is protected from every angle, ensconced in its wrap-around shelter. When submerged, the mussels cautiously ease their paired shells apart a few millimetres, allowing them just enough space to suck water in from outside and to collect titbits of food. The moment that danger threatens, however, they clamp their shells together, sealing themselves off from the world beyond.

Confronted by a predator as capable as a crab, however, mussels are at a considerable disadvantage. In crab hotspots they can bolster their defences, laying down additional layers of minerals and thickening their shells, but this comes at a cost. Energy can only be spent once, so reinforcing a shell might mean fewer offspring. What's more, even the thickest shells are unlikely to withstand the brute force that a hefty and determined crab can unleash. Crabs do have one significant weakness, though; their tank-like bodies are ungainly, and they struggle to operate effectively in the turbulent surf of exposed rocky shores. In such places, mussels mass in great beds, protected both by the restless waters and, to an extent, each other. Indeed, mussels seem to recognise the assurance of clustering with their own kind, since those that live and grow in these crowded conditions make less of an investment in thick shells.

To consolidate their position, mussels lash themselves to the rock of their unforgiving habitat with strands of proteins akin to the silken threads produced by spiders. The mooring ropes that each produces to hold it in place are known to biologists as byssal threads, while chefs refer to this tangle as the beard. Regardless of terminology, they're incredibly tough – roughly on a par with the tensile strength of a human tendon. But if mussels have an answer to the power of the surf, and indeed use it to gain respite from crabs, they still need to be inventive when it comes to other threats.

Dogwhelks are fellow molluscs, carnivorous snails that offer the mussels a very different challenge than that posed by crabs. All snails have a tongue-like structure known as a radula, which works like a kind of biological bandsaw. For the majority of snails, committed herbivores, the radula serves to rasp away at plant matter. Whelks, unlike their grazing relations, are carnivores, and their radula has been adapted into a drill to pierce shells. Sat atop their mussel victim, the whelk brings its hardware to bear and trepans a perfectly round hole in the shell. This done, it injects a cocktail of digestive enzymes into the cavity, dissolving the flesh within and allowing the whelk to extract the mollusc in the form of mussel chowder.

For a creature such as a mussel, the approach of a predator as ingenious as a whelk might seem nightmarish. Firmly attached to the rocky substrate, mussels can't make a break for it, so their fate seems to rest on chance alone, each hoping not to be the one to suffer a soupy fate. That, however, isn't quite true – mussels do not go gently into that good night. Instead, they deploy one of the most extraordinary anti-predator strategies in the entire animal kingdom. As a whelk cruises menacingly in search of a victim, mussels do something so odd that it seems like lunacy. Using the same fibres that they use to fasten themselves to the rock, the bivalves anchor themselves to their enemy. Sometimes multiple mussels get in on the act, collaborating to weave a cat's cradle of byssal threads to more effectively ensnare their predator. There's no escape from this improvised trap. Locked in a deadly embrace, the whelk can't manoeuvre sufficiently to apply its drill. Its fate is to starve slowly to death amid a banquet of potential prey.

Sea stars, too, converge on the mussel beds. Like the crabs, their strategy for breaching the mussel's defence relies on strength.

Rather than trying to crush the mollusc, however, they put their adhesive expertise to use in an attempt to prise the mussel open. While diving, I've watched as sea stars fold themselves around their intended victim like a hand closing on a priceless keepsake. The tube feet that line the underside of the sea star attach themselves to the smooth contours of the shell and begin to pull. What follows is a trial of strength, the might of the predator vying against that of its prey. The sea star is patient, and the leverage it exerts is enormous. If it can open the mussel by the merest fraction – less than a millimetre may be enough – the game is up. A squirt of digestive enzymes through the cleft between the mollusc's paired shells begins to dissolve its tiring flesh and brings its resistance to an end.

By clustering in groups, the mussels have a chance of stymieing these attacks, since to pry open their prey, the sea stars need to wrap themselves around as much of the mollusc's body as possible. The need for neighbours is something that mussels seem to recognise at some level. Although mussels aren't the most mobile of animals, by a process of breaking and remaking byssal threads, and heaving themselves along by means of their fleshy foot, a mussel can warp and manoeuvre like a ship in harbour. In this way, an isolated individual can gradually edge closer to its companions.

* * *

Groves of blue-black mussels mark the furthest extent of low tide, their shells darkening the water's edge. Around them, sea squirts stand forlorn, like the trunks of a miniature petrified forest. The concentration of life is often greatest here at the very limits of the land, a place where the sea seems to surrender only grudgingly, and at that for just an hour or so at a time. As the tide ebbs at North Landing, it reveals, by degrees, narrow rock platforms at the foot of the cliffs on either side of the bay. To my younger self, all of the intertidal was Eden, but the place where the mussels huddled held the shortest lease. I'd skip across a field of slippery boulders to the end of the platform and watch as waves swirled and stirred the blades of leathery kelps around the twisted iron remains of a wreck that, half a century after the ship foundered, was now draped in the living architecture of the sea.

The deep water that lay beyond the limit of the rocks lent the frisson of danger, as did the risk of getting too wrapped up in my fossicking and being cut off by the returning flood. I retreated as the tide turned and began to reclaim ground. As I did so, I noticed that the cast of creatures changed the further I went up the shore. The mussels gave way to ever more limpets, which mingled with winkles and topshells, as oysters crimped the shore's smooth contours into wicked ridges. In place of kelp, wracks and other seaweeds lay prone against the dry, stony shelf. Further still, leaving the cadence of the waves behind me, the rich diversity of shells dwindled until only barnacles remained, the abundance of their limestone cones creating an ashen strip just below the high tideline. The rocks beyond this were glazed with a faint rime of salt, but even here there was life. Lichens, conglomerations of fungal, algal and bacterial partners, invigorated the stone and coloured the sea's margin in sable or with bursts of sulphur yellow defiance.

Every 12 hours or so, the sea reclaims and submerges the shore of North Landing, but I returned whenever the tides allowed me to explore, absorbing everything that I could about the creatures of the cove.[24] In time, I came not only to know them but to know where they could be found, so that if I'd been transported down

24 Not all shores are subject to the typical semidiurnal pattern of two roughly equal high tides and two roughly equal low tides over the span of just over 24 hours. Some experience diurnal tides, that have just one high tide and one low, and others experience mixed semidiurnal tides that have two highs and two lows of different magnitude.

to the rocks in a thick fog and could only use the animals at my feet for reference, I'd know exactly where on the shore I stood. In my naivety, I thought that what I'd learned was unique to North Landing. I later realised that this is true of most coasts: a pattern forms from bands of distinct organisms that collect at different parts of tidal shores. This patterning of life into domains within the intertidal emerges from a confluence of powerful forces that govern both existence and coexistence.

The ribbon of barnacles that decorates the upper shore at so many coasts seems to hint at a habitat preference on the part of the crustaceans. Yet as any boat owner knows, barnacles readily muster at any available space and they're seldom to be found without company; acorn barnacles cram together to such an extent that as many as 100,000 of them can collect in every square metre. What keeps the barnacles from coating the entire intertidal like white buckshot is a combination of predators and competitors. Hunters are concentrated on the lower shore and here the gimlet mouths of whelks and pulverising claws of crabs make an easy meal of stray barnacles. Those that survive this risk being outcompeted. Barnacles are small creatures and can easily be overgrown and crowded out by larger, more robust creatures like mussels, or even bulldozed out of the way by foraging limpets.

Making a living on the upper shore is torturously challenging for marine organisms. Barnacles, however, possess a sort of steely crustacean determination and it's this that saves them from being jettisoned entirely from the intertidal. The opportunity for upper shore barnacles to feed presents itself only for an hour or so each day, at the limit of high tide. Meanwhile, the water that they most store inside their stony home gradually becomes depleted of oxygen and, since all animals must excrete waste, the animals are trapped in a steadily increasing interior puddle of their own concentrated piddle. It's the ability of barnacles to withstand this and other hardships, essentially to be more resilient than their predators and competitive superiors on the seaward side, that allows them to persist in their stronghold at the top of the shore.

Similar forces dictate the position of all the organisms on the shore. How high they can go depends on their tenacity in the face of the physical challenges of exposure; how low they can go depends on their ability to cope with the rocky shore rat race of predation and competition. Most species find a sweet spot to suit them between these two countervailing pressures and settle there.[25] There is, however, another vitally important factor in determining where on a shore a species is likely to be found, and it's one that emerges from choices made by the animals themselves in their first few days of life.

As the year progresses on the shore, tiny barnacles, like petrified mustard seeds, establish among the ranks of comparatively Vesuvian elders. Beds of mussels are bolstered by an influx of tiny freshers, and, in late spring, minuscule crabs can be seen pittering about the damp shore, each a perfect yet diaphanous replica of their hulking, armoured elders. These, and the young of many of the intertidal's cast of creatures, seem to appear, small but fully formed, as if by magic. Though we don't need to invoke sorcery to explain it, the arrival of new cohorts is a fascinating and important facet of life at the shore. Unlike most members of the animal kingdom, creatures like barnacles, mussels and limpets do the majority of their wandering very early in their lives, when they live a peripatetic existence out in the open sea. And when they do eventually return to settle at the shore, they have firm ideas of where might make a suitable home.

---

25 Some, of course, adopt a flexible approach, but even then you can see a pattern. The size of crabs tends to increase the further down the shore you go. This is because vulnerable youngsters linger at the high shore, out of reach of the unsentimental claws of their larger brethren, gradually migrating down towards the sea as they age and grow.

The hardships of intertidal life mean that the shore is no place for delicate youngsters. Perhaps wisely, a large majority of rocky shore parents loose their prodigy onto the ebbing tide to drift into the sea beyond. The larvae of mussels, barnacles, crabs and limpets mingle in the plankton, living a footloose, floating existence as specks of life in the expanse of the open ocean. At this tender age, they're almost unrecognisable from the more familiar forms that they will assume in later life. Apprentice mussels look like flying saucers, and the crab larva bristles with defensive spines with a tail that it curls protectively around its head, miming a 9. The barnacle, meanwhile, is even odder in appearance, resembling a winged Zeppelin. This microscopic marine fancy dress party doesn't last long, however. As each grows and develops, the impulse to head shoreward grows stronger. Many make their move in the days leading up to the peak spring high tide, riding flood currents that deliver them inshore.

The precise choice of landing site is of less importance for the crab, whose future is one of itinerant scuttling, than it is for the others. Barnacles and mussels have to be discerning; they're choosing a home for the rest of their lives and if they get it wrong, their tenure on the shore will be a short one. The larvae factor an array of different information into their decision, noting appealing odours that emerge from some microbial biofilms as well as being repelled by the smell of danger that arise from predators. Other desiderata for the home hunters include the shelter offered by a given microhabitat, and the smoothness of the rock. But what really draws these animals is the presence of their own kind. The larvae gain encouragement from the knowledge that others of their own kind have prospered here, meaning that their chances of success in life will be much greater.

For barnacles there's another consideration: sex. Most barnacles are hermaphrodites, but they still need a partner if they harbour hopes of becoming a parent and that means cosying up to other barnacles. Once a barnacle larva finds a place that ticks all of its boxes, it attaches its head to the substrate, using a glue whose properties make it the envy of the natural world, and begins to secrete the calcium plates that, in time, will give it the characteristic volcano-like appearance of its adult form. When it's eventually ready to make babies, it unfurls its secret weapon: the largest penis, relative to body size, in the animal kingdom. This appendage is eight times longer than its own body and which Darwin described breathlessly as 'wonderfully developed'. Were humans to be similarly equipped, we'd be encumbered with 15-metre members. Such a disproportionate appendage in the barnacle is necessitated by its inability to move.[26] If it wants to mate, then there's only one thing for it: its penis has to visit the neighbours.

The net result of the decisions taken by the larvae are that established patterns of species distribution tend to be reinforced. Where adults are, juveniles will come. The larvae aren't infallible, though. Some with the invertebrate equivalent of a faulty radar will settle beyond the protective embrace of their kindred. Initiatives like this are likely to end in failure or, more specifically, in the stomach of a predator. Nonetheless, if fate plays them a kind hand, there's just the whisper of a chance that these exiles might survive and, ultimately, establish a new locale for their tribe to gather. In this way, waves of larval recruits coming in from their open sea odyssey not only buttress the positions taken by previous generations of settlers, but they also have the possibility of reshaping the ecological patterns of the shore.

* * *

26 Animals like mussels are less inclined to engage in this kind of sexual tomfoolery and simply broadcast eggs and sperm into the water.

From a not-quite-comfortable seat on the craggy sandstone of Sydney's Pacific Coast, I'm looking out to sea. It's late afternoon and the sun has mellowed to gold, lacing the ocean with warm copper and cobalt. The tide is turning, and just beyond the line of breakers, a multitude of tiny larvae are perhaps steeling themselves for an assault on the shore, to fulfil their destiny by joining the ranks of the rockbound. I wish them well, but if I'm to finish my exploration of the intertidal, I'll need to get moving before the flood swallows my living laboratory.

My main focus today is on the rockpools that strew the shore and the pen-lid-sized gobies that inhabit them, and which I'm studying. Pools like these are one of the chief delights of shores the world over, each a sea in microcosm and an oasis amid the unforgiving rocks. For me, as for so many others, these miniature wonderlands of life provided my first window into the marine. They carry the promise of new and exciting encounters with an ever-changing cast of marine animals, showcased within arenas of stone.

The gobies are eluding me today, but among the lush, emerald leaves of sea lettuce, transparent prawns roam, their whereabouts disclosed by a pair of dark eyes that seem to float, disembodied among the foliage. Pressing my face close to the water surface, I take advantage of a shrimp's frozen caution to study it close to. I'm entranced by the exquisite precision of its gossamer form. Crystalline mouthparts twitch as an infinitesimal heart pulses in the midst of its glassy body. Beyond the lettuce, winkles and topshells slide along, creating a Jackson Pollock pattern of paths through a sprinkling of sand on the rockpool's floor. Then one shell mobilises in a way that seems unmollusc-like, its motions angular and speedy, rather than smooth and slow. I twitch and the shell stops dead, a giveaway that this is no snail but a different creature entirely, disguised in the garb of a gastropod. Hermit crabs are one of my favourite creatures of the shore, their ubiquity and extraordinary lifestyle making them irresistible.

One of the first things that I did when I arrived in Sydney was to install an aquarium in my office so that I could study the local marine fauna on the days when I was unable to make at least a quick detour to the seashore. A rotating cast of different animals graced that aquarium, each on a temporary secondment to the city before being released back to the wild. Cushion stars oozed across the glass while barnacles wafted at the water with net-like, orange cirri, in the gesture that I still whimsically imagine as waving. Anemones lent a carmine glow of opulence to the scene. Since the anemones were at risk of being fleeced at dinnertime by the more dextrous prawns that roamed the tank, I hand-fed them. Grasping their meal of defrosted shrimp in a gelatinous embrace, they folded inwards, their crown of tentacles disappearing into their bodies' interior, taking their prize beyond the reach of the prawns.

The action in my ad hoc rockpool, however, was provided by a trio of hermit crabs, who roved, stomped and clambered ceaselessly throughout their watery world. In the weeks after I installed the aquarium, I began to hear strange noises; trains of loud clicks started to become the soundtrack to my office. I tried to localise the source over the course of several days to no avail. I looked everywhere in the room, and in the street outside, but it wasn't until I happened to be looking at the aquarium at the precise moment that the clicks broke out that I realised that, incredibly, it came from the hermit crabs. More exactly, it came from two of them. Crab number one grasped the other's shell then rapped its own shell against it, in rapid, staccato fashion. The clashes caused an impressive racket, despite the fact that neither crab's shell was larger than a glass marble. If it was loud to me, I can only imagine how it felt to crab number two, but this is entirely the point of the first crab's actions. The rapping is an aggressive measure; a form of crustacean intimidation whose goal is to make the recipient surrender its shell.

A hermit crab's shell is both its comfort and its chief source of angst. Like their cousins, the 'true' crabs, the foremost parts of hermit crabs – their heads, legs and claws – are protected by a tough exoskeleton. Unlike their relatives, however, the hermit crabs have a delicate and seldom-seen corkscrew-shaped abdomen. It wasn't always this way. The ancestors of hermit crabs were better protected than their modern equivalents and had bodies that looked somewhat like lobsters. It's a challenge to piece together the process by which the hermits took to their unique lifestyle more than 150 million years ago, but it seems likely that what started as an opportunistic use of empty shells as refuges developed gradually into a reliance. The evolution of a softer abdomen allowed the crabs to thread and curl around snugly into the interior swirl of a shell and maintain a tight hold. The downside of this adaptation, of course, is that the hermits have horribly vulnerable hindquarters, making them dependent on their mobile homes. Another problem they face is that top-quality shells are in short supply. As a result, hermits face fierce competition for their bespoke armour, and when another hermit takes a shine to the shell being worn by another, fights break out.

A dispute between two hermit crabs over the ownership of a shell might lack the gory fascination of, say, two bull elephant seals clashing heads and lacerating one another with their teeth. But when an aggressor seizes hold of an occupied shell and performs its percussive assault, it does so with all the ferocious intent of the seal. The defender, meanwhile, withdraws into its shell, yet still gains a very clear idea of what it's up against. Larger, stronger hermits deliver louder, more powerful raps, and it's often better for an outmatched defender to surrender than to suffer injury at the claws of its aggressor. If the defender makes the decision to give up, it relaxes its grip on the interior of its shell and allows itself to be dragged out. Thereafter, hermits don't succumb to the temptations of spite. Having decided the issue, what follows is a model of decorum. The victor eases itself into its new abode, and the evicted hermit takes the consolation of its conqueror's vacated home, before both go their own separate ways with no further tussling.

Such is the dearth of suitable shells that hermits usually end up making do with a home that's lacking in some way or other. First and foremost, this shell – essentially a portable burrow – has to be the right size to provide a close fit; too small and they can't withdraw to the safety of its interior, too large and their inability to grip its interior means they can easily be extracted. Beyond that, the hermits are discerning about the species of donor snail. Smaller hermits favour periwinkle shells, but as they grow, they branch out to consider the shells of a wider range of gastropods. They tend to become conservative about this as they age, however, and once they decide on a type of shell, they tend to stay loyal to that particular one, like car owners sticking with the same marque.

Beyond their careful selection of shell type, hermits also give careful consideration to a suite of other qualities. Shells that are in good condition are desirable, as are lighter models – no crab wants to have to lug around their equivalent of an overpacked rucksack. Even colour is important, and in northern European waters, hermits with a yen for periwinkle shells have a choice to make. Periwinkles that specialise in feeding on the upper side of a seaweed frond tend to be dark brown in colour, the better to blend in when seen from above. Those that occupy the underside of the fronds are typically yellow, camouflaging themselves against downwelling light when viewed from below. This gives the hermits something to think about when the snails eventually bequeath their shells. Generally, hermits plump for darker shells, allowing them to slink around secretly in the gloomier recesses of their habitat, but if they find themselves standing out against a light background, then they'll reconsider their choice and make a switch.

With so many considerations to balance, it's perhaps little surprise that the hermits are especially tenacious in defence when they find the ideal home. Indeed, the lengths that they go to in order to possess the perfect shell in the first place illustrates key principles in ecology as well as in economics. Take, for instance, the vacancy chains that develop when collections of crabs move shells, which have much in common with sequences of events in human society, such as buying a house or switching jobs. The crabs' ability to use nuance and make decisions under constraint, sometimes rejecting what seems to be a perfect shell for another that meets its criteria more broadly, has echoes in our own behaviour as consumers. The sophistication of their actions gives the lie to the old canard 'small brains equal small minds' and has made the humble hermits into model species in the scientific study of animal competition.

* * *

Hermit crabs are emblematic creatures on the rocky shore and have played their part in inspiring the nascent interest of countless budding naturalists. And while hermit crabs have made important contributions to science, the intertidal more generally has also provided a natural test bed for the development of some of the most influential theories in experimental biology, especially those concerned with the question of how species coexist. Despite the privations endured by the creatures here, the rocky shore has many compensations, not least of which is a profusion of food. Algae prosper due to the rich supply of nutrients in the swirling, oxygenated waters and form the basis of a thriving food chain. Consequently, rocky shores support abundant life, representing a dazzling variety of species. Wonderful as this diversity is, it presents a central problem in ecology, especially in light of an influential series of experiments on a pair of small and unassuming species of beetles.

In the late 1940s, Thomas Park, a professor in the University of Chicago's zoology department, ran a series of painstaking and persuasive experiments in his laboratory. In them, he tested the ability of the red flour beetle and its close relative, the magnificently named confused flour beetle, to live in harmony. Park's approach was straightforward: he placed the two species in sealed test tubes along with a good supply of food and waited to see what happened. Under virtually all of the conditions that Park imposed on his beetles, one of the species was wiped out by the other.[27] What Park's results suggested was that if two species contest the same resource, it's inevitable that one will eventually be extirpated by the other. Yet the fact remains that in the world beyond the boundaries of a test tube, species – even similar ones – do coexist. In the unsentimental arena of nature, what is it that prevents the strong from driving out the weak?

A partial answer to this is that each species exists in its own niche, an idea that had been introduced by the pioneering British ecologist Charles Elton in the 1920s, who compared the niche of an organism to the professions occupied by working people. Apt as this is, nowadays the concept has been expanded to encompass the totality of a species' way of life. When ecologists talk about niches now, they're referring to where an organism lives, what it eats and what eats it, and indeed how it interacts with all parts of its environment, both living and non-living. From this comes the idea that to coexist, species must differentiate their niche from that of their competitors.

27 The outcome was determined by factors such as temperature and humidity. Under cold, dry conditions, the confused beetle won out, while in warm, humid test tubes, the red flour beetle persisted at the expense of the other.

On the rocky shore, niche separation is exemplified by the barnacles' flight to the high shore in the face of competition from below. Even here, however, niches have an awful habit of overlapping. There is no definitive territorial border between 'barnacle country' and the rest. And in their lofty stronghold near the strandline, barnacles do not live in glorious isolation but exist, shuck by shuck, alongside numerous other shelly species. Indeed, multiple, closely related species of barnacle coexist here, feeding on the same things and occupying the exact same niche. Clearly, niche separation on its own can't explain the paradox of diversity.

Step forward a charismatic American researcher, Robert Paine, a giant both in ecology and, at 2 metres in height, stature. At the time that Paine was beginning his research career in the early 1960s, ecology was fizzing with radical, transformative ideas. Rather than continuing with the historic focus on small-scale interactions between species, ecologists were raising their sights in an attempt to work out how entire ecosystems functioned. Answering this is reliant on working out not only how myriad different species interact with one another and with their environment, but also how the dynamics of those interactions change over time. It also involves getting to grips with feedbacks and second-order effects as they ripple through the interconnected parts of the ecosystem. It was a daunting undertaking, but one that had already begun to bear fruit in the form of groundbreaking new perspectives. Yet the discipline that Robert Paine entered was almost exclusively pursued indoors, using arguments founded on mathematical equations and theoretical models. Valuable though these are, there's no substitute for getting out and testing hypotheses in nature. This is where Paine's combination of relentless intellectual curiosity and his deep connection to nature came in. He was first and foremost a field ecologist, one who was happy to get his hands dirty in the pursuit of knowledge, running simple yet elegant experiments that he modestly described as 'kick-it-and-see' ecology.

Robert Paine's long association with intertidal research began serendipitously, on a student field trip that he supervised to Makah Bay on the spectacular coast of Washington state. The bay faces due west into the immensity of the Pacific Ocean, its wave-carved rocks host to a burgeoning community of animals. Many of these are vividly coloured; sea lemons slalom around bright green anemones, which cluster alongside resplendent marigold-hued tunicates and crimson sea urchins. And what the neighbours of these species lacked in chromatic glory, they made up for in sheer abundance. The rocks were strafed with barnacles, limpets, chitons and mussels. One animal in particular caught Paine's eye, a huge sea star known as *Pisaster*, which measured almost half a metre across, and whose different morphs were attired in royal purple, deep terracotta or warm orange. Aside from their impressive colours, Paine noted *Pisaster*'s predatory mettle; these were opportunist hunters that took on a wide range of prey. Limpets, barnacles and snails all fell victim to the sea star, but its favourite were the mussels that coated the lower shore.

At the time that Paine first encountered *Pisaster* in Makah Bay, his academic contemporaries were proposing novel theories about the importance of predators in shaping the dynamics of ecosystems. These flew in the face of the conventional wisdom within the discipline, which centred on the straightforward idea that resources were the main factor in limiting animal populations at each position in the food web. In other words, the system is controlled from the bottom up; the quantity of algae on the rocky shore determines the number of herbivores, which ultimately constrains the size of the predator population. The new ideas, while not denying the importance of resources, flipped this belief on its head and proposed that control of the ecosystem was from the top down. At the heart of this was the simple observation that the world is green, and not just slightly green, but overwhelmingly so. The only way of explaining the huge amounts of plant biomass, proponents of the top-down ideas reasoned, was that predators played a key role in limiting herbivore populations, in turn allowing plants to flourish. Keen to test this idea, Paine embarked on his kick-it-and-see philosophy. He levered the giant sea stars off the rocky shore one by one and flung them, like meat frisbees, as far as he could out into the bay.

Paine returned to his rocky shore plot every couple of weeks to repeat his *Pisaster* evictions. At the same time, he kept his eye on a similarly sized patch nearby, where he left the sea stars in peace to provide a control for his experiment. The results were dramatic. Just three months after the experiment had begun, a noticeable shift was obvious in the *Pisaster*-free plot. Barnacles were the first to take advantage, settling in huge numbers to coat the shore. The following summer, a year since Paine started the experiment, things had taken a different turn. Mussels had monopolised the available space and crowded out the barnacles. They now carpeted the rock, painting the shore black with their shells. And so it continued for the remainder of the experiment, a once diverse community of organisms transformed into a mussel monoculture.

Paine repeated his experiment further up the coast from Makah Bay and then later on the other side of the Pacific, in New Zealand. Wherever the sea stars were removed, an avalanche of mussels descended and took up the available space. One by one, their competitors fled, unable to contend with the relentless march of the mussels. What Paine had clearly demonstrated was that the sea stars, the top predators on these rocky shores, played a disproportionate role in regulating the ecosystem. By controlling mussels, sea stars created the space for other species to coexist and rather than destroying diversity, the predator promoted it. Paine later coined a term to describe the role of sea stars in this ecosystem, using an apt analogy that he borrowed from architecture. A keystone is the wedge-shaped stone at the top of an arch that holds the entire structure together and prevents it from failing. The sea stars of the rocky shore were the ecological keystones of their system.

Paine's research, however, went far beyond simply identifying the implications of apex predators for their prey organisms. By keeping mussels in check, *Pisaster* allowed space for herbivores like limpets, periwinkles and chitons to graze, and in doing so exerted an indirect effect on the ability of algae to flourish. At the same time, the huge mussel beds that developed in *Pisaster*'s absence provided a miniature labyrinth of space among their shells. Shrimps, marine worms and small fish, such as gobies, could live in safety among this newly created protective matrix. The actions of one predatory species, in other words, had huge ramifications for other animals at different levels of the food chain, creating both winners and losers. Years later, Paine dubbed this a trophic cascade, a phrase that neatly captures the idea of flow-on effects. More broadly, it also emphasises an idea which was barely acknowledged by ecologists at the beginning of Paine's career, that the organisms that coexist within an ecosystem are intimately and inextricably connected to one another.

On the rocky shore, as elsewhere, each species affects its neighbouring species at least to some extent. This is true even among those that don't interact directly, neither competing nor feeding on those neighbours. For instance, take the oystercatchers that, in frequent defiance of their name, assume the challenge of eating limpets. A deft but powerful lateral jab at the hem of the limpet's shell dislodges it and leaves the snail defenceless to the bird's probing beak. With space at a premium on rocky shores, the limpet's loss is another's gain, as barnacle larvae swoop in to settle and take its place. Similarly, and despite the fact that a periwinkle may have little to say to a mussel, since they forage in different ways and neither is partial to eating the other, the presence of mussels affects periwinkles and vice versa in various ways. For example, mussel beds change the foraging landscape for winkles, while the grazing done by periwinkles maintains patches of algae-free rock, allowing an opportunity for juvenile mussels to settle. In this way, coexistence breeds interdependence. As John Donne didn't quite say, no mollusc is an island.

The diversity that we see on the rocky shore emerges from this interdependence, from the influence wrought by each species on its cotenants within an ecosystem. Direct competition and predation are clearly important, but no action in nature occurs in a vacuum. Chain reactions of second-order effects propagate through a community of species to regulate their ecosystem. This creates a series of ecological checks and balances that actively, albeit blindly, encourage and maintain diversity. This diversity is further promoted by the natural disturbances that are intrinsic to the ocean's margins. A winter storm tears through a rocky shore community, scouring organisms from their moorings. Opportunity arises in the wake of the destruction, as the openings created represent a chance for other creatures to claim the newly vacated space. Physiological tolerance, too, plays a part, offering a means for the hardiest of marine species to escape the intensity of competition on the lower shore, allowing them to persist by venturing to the furthest limits of the sea's dominion.

And so rocky shore ecosystems evolve, each species waxing and waning in response to a complex and shifting mix of pressures. An organism that gains the ascendancy at one time is brought low at another through a flux of resources, threats and sheer random chance. Taking a broad view, the patterns of abundance and diversity fluctuate, sometimes by increments, sometimes by lurches, but often in cycles. This self-correcting quality of ecosystems makes them robust to natural change, but they're seldom equipped to cope with the intensity and destructiveness of our actions. Insightful as Paine's research was, we're now as a species undertaking what amounts to a global stress test on ecosystem resilience. With overharvesting, pollution and climate change, we're challenging ecosystems to an extent that has no precedence. And even as the baleful results come in, we're fixedly refusing to respond to them. Paine's work taught us that the species in the ecosystem are interconnected. The part that we too often forget is that this includes us, and the way we're going about things at the moment makes us more disaster than *Pisaster*.

* * *

As naturalists make a beeline for the rocky shore, most seaside visitors enjoy nothing more than to stretch out their towels on the sand. One obvious explanation for this parting of ways, of course, is the distribution of marine life across the different parts of a shore. The natural wonderland among the boulders and pools of the rock platform offers a cornucopia of life; the sandy beach, meanwhile, seems desert-like in every way. If the study of biology has taught us anything, however, it's that life conquers all. There is no place to hide for animals on the sandy shore, and the open spaces leave them vulnerable to birds and other mobile predators, such as crabs. For that reason, the creatures that live here excavate themselves to the relative safety of the sediment. Buried beneath the sands, a subterranean community exists, lying in wait for the returning tide. Most of this interred throng are small creatures, visible only under the microscope, but if this is the rule, it exists to be broken.

On Mona Vale beach in north Sydney, a man waits for a

wave to retreat and then patiently trails a gauze bag filled with an unspeakable mash of decaying fish across the wet sand, first one way and then another, as if training an invisible dog to the leash. At first glance, his behaviour might seem to indicate a troubled mind, but this is a fisherman and he's prospecting for bait hidden deep within the sand. Finally, his patience is rewarded. Piqued by the enticing aromas released by the bag's contents, a creature has risen from deep below to investigate. The slight disturbance it creates on the surface as it ascends is its undoing. The fisherman swoops and begins carefully to pull. A thin, brown ribbon emerges, festooned with pairs of thick, leg-like bristles along its length. It's a polychaete, a worm resembling a colossal centipede. The worm is the length of the man's leg when fully extracted, though some of these so-called 'hairy Marys' reach 3 metres in length, with terrifying mouthparts to match. Having dropped the giant into his collecting bag, the fisherman returns to his patient work. Over the course of just half an hour, he caught more than a dozen such worms, making a nonsense of the idea that the sands are barren.

If the thought of giant polychaetes adds a certain frisson to building sandcastles, then take heart in the knowledge that much of the sand's concealed bestiary lives lower down the shore. On beaches around the world, the telltale casts of lugworms – another species of polychaete – lie like unspooled, sand-sculpted coils of string on the surface of damp sand. Lugworms are the intertidal equivalent of the earthworms on land, scarfing quantities of grit and extracting what goodness they can find in and among. Unlike its terrestrial counterpart, however, the lugworm works smarter rather than harder. It builds a U-shaped burrow with its tail end pointing upwards and its head below a column of saturated sand that rises vertically above it. Every mouthful of sand that the worm sucks in, gravity replaces as more sand descends to fills the space. After that, it's a simple matter for the worm to digest out particles of organic material and microorganisms and to make room for more by expelling its artistically sandy turds at the surface.

Much else lives down here in the darkness beneath the sediment. Bivalves, such as cockles, live just out of sight beneath the uppermost layers of sand. They sustain themselves by sucking water in through a snorkel-like siphon tube that protrudes slightly above the surface of the sand. Having filtered out algae and any other rewarding matter, they return the cleansed seawater to the surface via a second siphon. A canary-yellow muscular foot protrudes beneath the cockle's shell, ready to pull its owner deeper into the seabed if danger should threaten. Alongside the molluscs and worms, shrimps inter themselves along with sea stars and flattened urchin species known as sea biscuits, or sand dollars, whose disc-shaped bodies allow them to slide easily under a covering blanket of sand to escape detection by predators. They're assisted in their endeavours by their strange habitat of ingesting, and then storing, particles of metal among the sand, the extra weight stabilising them in turbulent surf.

But there's another animal in the northern European intertidal that has less need of caution, even though at first glance it would seem easy prey. The bootlace worm is an anatomically undistinguished creature – it's essentially a slender, chocolate-brown tube – but for one important detail: it's the longest animal on Earth. In 1864, a bootlace worm washed up on a beach in St Andrews that measured 55 metres, almost twice the length of a blue whale. Its size isn't what keeps it out of trouble, however. That distinction goes to a powerful toxin that oozes from its skin along the stupendous length of its body. It's a chemical defence of such potency that the merest hint of it is enough to cancel a crab; the bootlace worm may be little more than a long thin piece of string, but you wouldn't want to tangle with it.

Staying out of harm's way by mobilising chemical defences or by hiding only provides an answer to one of life's conundrums. Foraging effectively often means moving about. This is easy enough for nimble creatures like shrimps, but others aren't quite so fleet. So rather than heave themselves along under their own power in search of the perfect feeding grounds, plenty of sandy shore animals instead catch the tide. Some species of clam, such as the coquina that lives of the Eastern Seaboard of the US, are referred to as swash riders. In effect, they surf the waves in order to move up and down the beach along with the ocean's ebb and flow. They achieve this by listening out for the sounds of large waves, which produce deep infrasonic rumbles as they gather and break. When it hears the approach of a particularly large wave, the coquina jumps out of its sandy hiding place and gains a thrilling ride up the beach.

Rather than embrace the drama of the coquina's travel plans, another intertidal animal, the laver spire shell, eases its way up the beach on the incoming tide. When their internal clock tells them the time is right, these snails surrender their underground bed and burrow up through sediment to reach the great outdoors. Once there, each produces a snotty pillow of mucus that they use to raft upshore on the flood. This ultimately deposits them at the high tide mark, where they gorge themselves on a biofilm of algae, away from most of their competitors. Having dined to their satisfaction, the mass of laver spire shells follows the retreating waters of the ebb tide like the baggage train of an army to return to their favoured burrowing sites.

A laver spire shell is a tiny creature, only a couple of millimetres across, and while they can reach densities of more than a quarter of a million for every square metre of the shore, they typically pass notice. The same couldn't be said for a chunky, finger-length snail from South Africa that emerges from the sands to engage in some ghoulish beachcombing. The plough shell is a large whelk, possessed of a surprisingly large and meaty foot, which it uses for surfing the waves in search of carrion. Ploughs have an incredibly keen sense of smell, which they use to locate the dead and the dying. When they perceive the alluring scent of putrefaction, they emerge from the sand in enormous numbers, riding the waves to converge on stricken fish, crabs, and even jellies.

* * *

Most animals to be found on, or under, sandy beaches are year-round residents, but the strand occasionally receives visitors from the sea beyond. Turtles are one such, of course; another is one of the strangest and most fascinating creatures in nature. My first encounter with one of these took place quite by accident as I took a stroll on Maryland's Assateague Island, following the waveline across sands flecked with mica and quartz that sparkled in the rays of the setting sun. Contemplatively, I placed my feet alongside the hoofprints left by three chestnut horses, whose ancestors were brought here by Spanish colonists centuries ago. The horses stood at the foot of the dunes to my left, watching me with the caution of the freeborn, their dark manes catching the onshore breeze. Atlantic surf washed up the gentle slope of the seaward beach and rolled across a dark, squat shape. Intrigued, I approached to investigate and then noticed another, and another. There were half a dozen more scattered across the shallows, the waters of each retreating wave cresting over their rounded profiles. As I grew closer, the shapes took form and resolved, incredibly, into animals: horseshoe crabs. The horses have walked this beach for perhaps 500 years, but these surf animals have the older claim; they've been coming to the ocean's margins at places like this for hundreds of millions of years.

I picked one up, fascinated to take a closer look at a living fossil. It was about the size of a dinner plate, metallic olive in colour, with a long spike of a tail trailing behind. The helmet-like, armoured dome of its back bears faint ridges, beneath which a pair of large eyes looked out at a world that has changed beyond recognition across the enormous span of the horseshoes' heritage. A thicket of legs, each tipped with a pincer, descended from within the dome and began to explore the unaccustomed space. This was my signal to return the creature. I placed it back on the wet sand where it shrugged off my intervention and made its way stiffly through the waves.

The horseshoe crab is a reminder of the incredible venerability of marine life, a creature that, like the coelacanth and the nautilus, belongs to a different time. Horseshoe crabs are older by far than all but a handful of modern animals, somehow managing to cross the narrow bridgeheads of three of the five mass extinctions to persist, more or less unchanged, to this day. Even the name horseshoe crab is an anachronistic fudge, since they're more closely related to spiders than crabs and, in any case, appeared at least 100 million years before their crustacean namesakes. I'm lost in wonder watching these Permian time travellers, heaving themselves along this American coast just as, long ago, they contended with the waves of the Tethys Sea as they scaled the shores of Pangaea.

The horseshoes are here today not for the past, but for the future. As I watch in the fading light, the females climb higher up the beach to the high tide line. Each is accompanied by her swain, who holds tightly to his beloved's hindquarters with a pair of modified front legs. For animals like horseshoe crabs, which aren't lithe at the best of times, travelling in lock step like this – sometimes for several days – is quite an effort. Finally, however, the pair heave themselves to the brink of the land, allowing the female to begin digging a shallow depression in the wet sand. This done, she deposits a clutch of eggs, and the male reciprocates in kind by releasing sperm. Regrettably for the attached male, the wantonness of the couple's behaviour has attracted the attention of other males.

They're strangers to decorum, these so-called satellite males. They rush in to join the party, butting their shells aggressively against the pair. Satellite males are typically older, and in poorer condition, than the young, virile male who's attached to the female. Fastening yourself to a female is tiring and interferes with feeding, so at a certain point in life it seems that the more seasoned males give up on monogamy and instead become horny old chancers. Not that they seem tired now as they strive and barge to get at the trove of eggs and make their own milty contributions. Ultimately, chance, along with a fair amount of sperm competition, will decide who is to be the father.

Of the many peculiar aspects of the horseshoe crab's anatomy, the strangest thing isn't their light-sensitive tail, their possession of at least nine eyes, widely distributed across their shell, or the positioning of their mouth, which lies in the crotch of the third of their five pairs of legs. Instead, it's their baby-blue blood, and in particular the specialised cells that it contains. The colour is a result of haemocyanin, the horseshoe crab's equivalent of our haemoglobin, which employs copper instead of iron to bind oxygen. The cells that circulate within the blood, however, have some unique properties that stem from the primordial lineage of these animals. Amoebocytes, as these cells are known, are the horseshoe's equivalent of our white blood cells, and they're spectacularly sensitive to pathogenic bacteria. It's a sensitivity that, once discovered in the 1950s, we began to exploit for our own use. Nowadays, the horseshoe's amoebocytes form the basis of a simple, yet powerful, medical assay that tests for the presence of sepsis-causing bacteria in vaccines, medicines and implants. If you've ever had a vaccine, you owe a debt of gratitude, if not more, to a horseshoe crab.

Horseshoes are collected in their hundreds of thousands each year to be clamped to laboratory benches and to have a steel needle inserted into their body, near the heart. There the crab remains, its exotic blood draining until it fills an alarmingly large bottle. The intent then is to return the animals to the sea, but at least a quarter of them don't make it, and those that do are compromised to the point that they don't breed in the aftermath of the procedure.

From an anthropocentric perspective, the cost to the horseshoes is outweighed by the benefits that we accrue from their blood, and besides, many might point out they're only 'crabs' after all. But here's the kicker: a tested and proven synthetic alternative exists. The problem is that it's more expensive. Consequently, there's been a great deal of feet dragging over the adoption of the synthetic test, and there are few places where a ban on the bloodletting is mooted. In the meantime, populations of the horseshoe crab are in steep decline. This is an animal that has survived three mass extinctions, yet it now faces an even worse threat in the shape of humanity. It's a poor, and unnecessary, reward for an animal that has saved millions of human lives.

* * *

At the other side of the Pacific, on the shores of K'gari, off the coast of Queensland, an army has gathered. This isn't a military offensive, rather it's a congregation of thousands of soldier crabs, the rustling sound made by countless armoured legs in motion audible above the swash of the waves. The crabs are packed tightly together and parade in loose formation as if directed by some invisible sergeant-major across the low shore of this, the world's largest sand island. Each foot soldier of this crab army sports a uniform of pale, porcelain blue across its ping-pong ball of a body, and warm, cream-coloured, spidery limbs. Their unusually rounded torso gives them the appearance of eggs on legs.

Despite their martial name, soldier crabs are almost completely defenceless, their minuscule pincers carrying little threat. The crabs' inability to mount a convincing defence is exploited by ibis and kingfishers above water, and by toadfish that snatch the crabs if they go beyond the shallows. Even carnivorous moon snails feel sufficiently emboldened by the soldier crabs' lack of weaponry to attack them. The snails first coat their victim in an envelope of mucus, then drag the stricken crab down into the sand to devour it at leisure. Confronted by a threat, the soldier crabs rely either on the rather passive protection offered by safety in numbers, or, alternatively, their ability to merge into the sand with impressive speed. As danger approaches, each corkscrews itself into the waterlogged substrate, getting deeper with every turn of its blue body until it vanishes from sight. It only takes a second or two for them to achieve this.

Working with soldier crabs is a claggy business, since their preferred shores tend to be muddy. Nevertheless, being shoeless and covered in a mire of sludge was, for me, a small price to pay for studying these extraordinary creatures. I wanted to get to grips with their collective behaviour, to understand how they self-organise into their crowded marching formations and to work out how this closing of ranks benefitted them. Each day, I followed the tide as it retreated, like a bipedal version of the laver spire shell I described earlier, armed with a camera and tripod. The crabs are only visible for part of the tidal cycle, making their appearance once the sea releases its hold on their foraging grounds. Around halfway down the shore, I knew that the soldiers were waiting in their underground barracks, ready to emerge and feed at the low waterline. I set up my equipment and waited, watched as I did so by confused holidaymakers and, occasionally, dingoes. The first sign that the crabs are about to make their appearance comes in the form of a hummock that rises above the surface of the flats. Moments later, a blue back breaks through, and a crab rises to its feet. After a little wiping down and sprucing, occasionally incorporating a somersault to knock off the excess mud, the animal begins to collect clawfuls of silt and to sift from this the meagre rations lodged in and among. All around, more hummocks rise and yield more crabs. Then, at some unknown cue, the crabs begin a troop to the waterline.

One of the odd things about soldier crabs is their nonconformity when it comes to walking. Most crabs scuttle sideways, which is usually ascribed to a mix of them having wide, flat bodies and legs articulated to bend from side to side rather than front to back. Soldier crabs, with their globular bodies and more flexible joints, walk forwards, which somehow makes them seem more purposeful than their relatives. When they're on the move, the soldiers sometimes scuttle single-mindedly along as if late for an appointment, while at other times they pause after virtually every stride to scoop up and process a mouthful of gunk. This muck-munching came to a halt when one of their many predators loomed, however. Their behaviour changed, and though they kept moving, they closed ranks so that the army became almost liquid, its margins shifting and flowing, each crab in the embattled mass attempting to avoid the peril associated with the edge of their group. Despite their manoeuvrings, there were always casualties, with crabs taking to the air, locked in a beak, or being dragged off to sea by an opportunistic fish. After each attack, the army trekked on; there was no latitude for either anxiety or sentimentality when there was food to be gleaned.

Occasionally, the crabs' march would cause them to stray towards me, but even if I kept still, they'd veer off, cautious of an incongruous vertical object in the largely two-dimensional sandy tidelands. When occasionally one did get too close, it was a simple matter to scoop it up for a closer look. Its spindly body felt insubstantial and brittle in my hands, just a slight prickle of pointed feet on my palm. Once released, the crab hared after its disappearing colleagues, rejoining the ranks seconds later.

The coordination that the animals show within their armies has much in common with that seen in shoals of fish, herds of mammals, and even crowds of people. Each individual responds to the movements of its near neighbours. Amplified to the group level, this makes for beautiful patterns of self-organisation. There is no drill instructor; instead the crabs decide collectively where to move, each continually evaluating and updating its own actions in light of those of its fellows. The net effect of this is to make the army behave as a single, coordinated, cohesive unit. Even more impressively, the group as a whole follows a pattern known as a Lévy walk as they forage, which involves lots of short distance moves around the shore, interspersed with occasional longer journeys. By doing this, the army optimises its ability to find rich – but hard to find – patches of their food. The crabs may not know it, but in doing so, they're operating close to mathematical perfection as foragers.

As the tide turned and the sea began to recoup its lost territory, the army's manoeuvres faltered. Crabs broke away piecemeal singly or in small groups and headed for the mid-shore. Once there, they spun themselves in their unique way to the obscurity of the muddy sand, the descent of the whole like the mirror image of bubbles rising in a drink. A few minutes later the army was gone; the beach stilled and the rising waters claimed the scarified sand tracks of a crab army on the march.

The garrison of the crustacean horde on K'gari's shoreline is flanked by two of the world's most important plants. To the seaward side, exposed only at low tide, lie verdant beds of seagrass. To landward, massed stands of mangroves create a maritime forest of surpassing beauty. Though very different, both plants have a critical role in coastal marine habitats, and, predictably, both have made way for what we call progress. Lately, however, we've begun to gain an appreciation of their value.

Seagrasses are a unique group of around 60 species of plants that grow in shallow seas and estuaries across the world, from the shores of Iceland and Greenland through the tropics all the way to the southern tip of New Zealand. While there's a tendency to lump them together with seaweeds, these are utterly distinct. Seagrasses are true plants that, as their name suggests, look much like broad-leaved grasses. Where conditions are to their liking, they grow in profusion, creating spectacular marine meadows rich in animal life. Some of these meadows can be truly immense, and astonishingly long-lived. The seagrass meadows of the Bahamas are huge in extent; the largest is the size of Portugal. In Australia's Shark Bay, researchers found a single, individual plant that's 4,500 years old. It has spread outwards across 45 centuries via a network of rhizomes running beneath the sediment until it is now the size of a large city.

Runners aren't the only way that seagrasses propagate, however. The clearest reminder that these are true plants, rather than seaweeds, is that they flower. Admittedly, this isn't the kind of floral exuberance that attracts people in the way that the cherry blossoms do in Japan, or the tulips in the Netherlands' Keukenhof. Seagrasses are more demure than their terrestrial counterparts, and their flowers are delicate and pale. Moreover, they open at night when the water swarms with tiny invertebrates. Rather than using nectar, seagrasses entice larval crabs, shrimps and worms by producing a sticky bundle of rewarding mucus. Inevitably, some of this, and the pollen it contains, adheres to the diners, which fulfil the plant's purpose by transferring the pollen from a male to a female flower like miniature swimming bees.

Here in K'gari, the seagrass beds are neither so large, nor so ancient, as those in the Caribbean or on the west coast of Australia, yet their value is incalculable. Dugongs graze these meadows on the high tide, like blubbery aquatic cattle. Green turtles browse them too, indeed they get their name indirectly from the plant, as their habit of gorging themselves on luscious ribbons of seagrass lends an emerald tint to their body fat. Fish from seahorses to sharks live among the marine pastures, which serve as a critical nursery for them and for hundreds of other species. Worldwide, seagrass meadows support the juveniles of a fifth of the world's commercially exploited fish. Aside from the diversity that they foster, seagrasses stabilise the seabed and absorb excess nutrients that might otherwise fuel toxic algal blooms. But the one particular quality of seagrasses that has come under the spotlight most in recent times is their phenomenal ability to suck up and store carbon. By some estimates, seagrasses, while covering only a fraction of 1% of the world's seabed, store 10% of the ocean's carbon. The extraordinary and disproportionate capacity of seagrasses in this regard is further exemplified by the startling statistic that a meadow of them absorbs carbon at a rate 35 times greater than a similar area of tropical rainforest.

Regrettably for seagrasses, and the animals that live among them, the fondness of seagrasses for shallow waters puts them squarely in our path. They're dredged to make way for development, gouged by anchors and moorings, trampled by feet, poisoned by our toxic run-off and shaded out by the algal blooms that our fertilisers foster. In consequence, seagrasses are contracting at an alarming rate. Worldwide, according to the Seagrass Restoration Network, we're losing wild seagrasses at a rate of two football pitches an hour. This isn't just a problem for the future; it's one that's hitting us now. When we destroy a seagrass bed, we don't just lose the carbon-taming capability of the plant in the here and now, we liberate the carbon that has been locked away by the plants in the sediments beneath, sending it spiralling into the atmosphere to augment the concentrations of greenhouse gases already there. Small though the coverage of seagrass beds is across the globe, they're currently holding back a reservoir of 20 billion tonnes of carbon dioxide.

Thankfully, there is some good news on this front. Armed with a new appreciation of the value of seagrasses, recovery efforts are under way in some countries and are already garnering impressive results. Seagrass seeds, known for their appearance in parts of the Mediterranean as 'the olives of the sea', are collected either for planting right away, or to be germinated in nurseries. These, alongside fragments of mature plants that wash up in the aftermath of storms, are forming the basis for the rewilding of coastal waters off the UK, the US, Australia, the Netherlands and Sweden. A *million* seeds planted in Pembrokeshire in 2019 to 2020 by Project Seagrass, a UK charity, are thriving and beginning to establish a meadow. Similar successes are being reported in Australia and New Zealand. It remains a challenge to scale this up and to expand these efforts into new territories, but the very visible outcomes of these initiatives in the shape of new seagrass beds offer a beacon of hope.

The rising tide has submerged seagrass and slumbering soldier crabs alike, and forces me to climb the shore. Before long, waves will lap at the roots of the mangrove trees that fringe the bay. As I watch, the waves collect a spear of vegetation from its repose on the sands and send it spinning into the churn of the next wave. The spear is a propagule, a seedling mangrove that took shape and grew on its parent tree before ripening and falling to the sands. The next wave is a large one. It carries the propagule up onto the high shore between two larger trees, where it catches and remains. Left undisturbed, the miniature mangrove will waste no time in anchoring itself to its new home with lateral roots. It'll then bend over the course of weeks into a hook shape from its base and use its strengthening roots to lever itself into an upright position. This new stance means that its leaves, when they sprout, will be borne above the level of the water, allowing the seedling access to precious sun and air even as its base is inundated at high tide. By working with the sea as a disperser, mangroves re-enact the march of the crabs to radiate along the coastal margins. It's a march that takes place at a different timescale to that of the trees' crustacean neighbours, though its value is evidenced nonetheless in an unbroken line of mangroves that stretches as far as the eye can see.

Alongside the newly deposited seedling, the whorls and ridges of roots protruding from the flats at the foot of the adult trees trap seawater in a woody reconstruction of a rockpool. Within the confines of one of these, a sudden splash betrays the presence of an animal. A closer look reveals a wonderful creature, a fish that has evolved to cope with the rigours of a tidal environment to a greater extent perhaps than any other: a mudskipper. Its glistening head protrudes from the water and features a pair of large, googly eyes that sit on the very top of its head, giving it a frog-like mien. In common with frogs, the mudskipper is amphibious, motoring along the damp flats between pools using its pectoral fins as makeshift legs, or occasionally by lunging forwards using its tail. I keep perfectly still until the mudskipper relaxes and begins fossicking around among the muddy crevices of its world. Occasionally, it dips its head into the water, possibly to rewet its gills. As much as mudskippers are adapted to land life, their need for moisture to lubricate their gills and skin for breathing ties them to the ocean.

There's another reason that the mudskipper has need of water, and it's to do with feeding. Many fish make use of suction feeding to capture their food. To do this, they rapidly and quite dramatically expand their oral cavity, creating a powerful though short-lived vacuum that causes water and – if they've got the timing right – prey to rush into their mouths. The problem for mudskippers is that what works well underwater doesn't work at all on land, which has forced them into a dramatic rethink. Their solution is astonishing, and again has parallels with frogs. The amphibians use their long, sticky tongues almost like an extra limb to grab prey. The mudskipper isn't equipped with anything so lavish, so instead, as the authors of a 2015 study describe it, they've conjured a hydrodynamic tongue. First the mudskipper squeezes the water out of its mouth until it contacts the prey animal, then, milliseconds later, it sucks the water back in, complete with the prey. The whole process happens so fast – less than a tenth of a second in total – that we can't easily perceive it without the use of high-speed cameras. Using water to construct a functioning tongue represents a truly extraordinary adaptation, and one that makes skippers the masters of their muddy realm.

A thousand kilometres to the south of K'gari, I continued my exploration of mangroves, this time on the shores of Sydney Harbour. The forests offer a gratifying contrast to the modernity of the city, though it's one that was almost lost. In the two centuries from European colonisation to the introduction of legislation to conserve mangroves in 1994, it's estimated that almost two thirds of these marginal forests were destroyed. Thirty years on, Sydney's mangroves are recovering, but there's a long way to go before they ascend to their former glory. Five of the world's 70 or so species of mangrove can be found here, although the most common by far is the grey mangrove, whose glossy, dark-green leaves create a thick canopy. The leaves are flecked with pearls of pure salt, exuded by the trees as part of their ongoing struggle to cope in the harsh, saline conditions of their habitat. Between the trees, on the glistening mud of the shore, a forest of short, woody stems pokes from the waterlogged ground. These are pneumatophores, or peg roots, that act as snorkels, their purpose to draw air down into the tangled mass of roots hidden in the anoxic sediment below. Among this miniature forest within a forest, a host of miniature crabs scuttle about busily, like pedestrians in a crowded city centre.

It's a glorious experience for any naturalist to enter a mangrove forest, a kind of mud safari in an environment like no other. The air is humid, with a salt tang that lingers in the windlessness of the close-packed trees. A loud pop makes me jump, and then another, and another, as though I were surrounded by celebrants opening champagne. I'm told that the animal responsible is a pistol shrimp, which generates the sound with a snap of its disproportionately enormous claws. They're seldom seen, however, preferring the safety of seclusion while maintaining a fractious commentary to deter would-be visitors to their burrows. This in-your-face pugnaciousness is partly because real estate is hard to come by in the hustle and bustle of the mangroves, and partly because there are plenty of predators ready to make a meal of an injudicious crustacean. For this reason, if I'm to see the mangrove animals, I'll have to keep very still.

Minutes go by, and a mosquito is toying with my ear. I let it have a beakful of blood because the curtain has just been raised on an action-packed crab drama playing out in front of me and I'm reluctant to disturb them. There's a diverse cast of different species, each of which displays a distinct colour. There's the purplish semaphore crab, the marsh crab, with its jade carapace and bright-orange claws, and the mottled shore crabs, whose deep terracotta back is spotted with cream. All of them are small – matchbox-sized – and they're frantically busy. The males, more vibrant than the females, wave their claws rhythmically, beckoning, threatening and gesticulating. Their audience, the bustling congregation that surrounds them, respond in kind. There are countless crab conversations going on, mostly through the medium of this hectic sign language, a forest of brandished claws silently shouting at all and sundry. What's being communicated depends on whether the conversationalists are rivals or potential mates, and whether any one crab is an aggressor, asserting itself, or a supplicant, just trying to get by in the melee. As if this wasn't enough, they're doing all of this while eating, scooping tiny portions of mud and chewing through it with their strange, upright mouths to extract microorganisms, or dragging mangrove leaves into the larder of their burrow.

No crab, not even the most dominant, strays far from its bolthole. Indeed, each seems fastened by an invisible tether to the mouth of its burrow. The reason is that for a crab to lose its bearings when danger approaches, as it often does, means almost certain death. Sure enough, when I flick at the sister of the first mosquito, the entire muddy stage dissolves into a bedlam of activity, before resolving into a still, crab-free scene less than a second later. It's musical burrows, the crab equivalent of the children's party game, where on the signal, every last individual scoots underground. The crabs have an unerring sense of where their own private tunnel can be found, a facility which, as someone who almost invariably loses his car in multistorey car parks, I can't help but admire. Only very rarely do some scoot into the wrong burrow. That can be a deadly mistake, because the homeowner will inevitably evict them to face whatever's out there alone. It takes a few minutes before the ensemble cautiously returns to their surface to resume the hawking and jostling of this bustling crab bazaar. I leave before nightfall brings the mosquitoes in earnest, but already the crabs are retreating to rest underground, sealing their burrow entrance behind them with individually tailored mud front doors.

Worldwide, mangroves not only provide habitat for a rich assemblage of animals just like these, but they also yield enormous benefits for us. Exactly like seagrasses, they're a vitally important carbon sink, sucking in carbon greedily at a rate that far exceeds similar, terrestrial forests, and locking it away both within their living structures and in the mud below. So long as the mangroves persist, the carbon will be imprisoned indefinitely. And, also in common with seagrasses, the tangle of vegetation at the base of the mangroves acts as a highly structured and safe environment for marine animals, especially vulnerable juveniles. Lower shore mangroves, whose lower portions are immersed much of the time, are thus vital fish nurseries. In Indonesia, for example, more than half of the country's commercially important fish stocks are dependent on the protection of these trees during early life. And since the effects of pollution are so often concentrated at the margins of the sea, it's also worth noting that mangrove wetlands are known to play a vital role in mitigating these, acting like natural filters and purifying the waters.

The mangroves' ecological heavy lifting is continuous, but so gradual we hardly notice it is happening at all. Yet there is another role played by mangroves that is immediately obvious. Mangroves are wave tamers, the labyrinthine interior of the forest deflects and distorts, cutting each large wave into dozens of chaotic smaller ones and dissipating their energy as it does so. They reduce wave heights by anything up to 95% and so act as a critical buffer that is far more effective than any concrete sea wall. They stabilise foreshores, soak up the power of storms and protect the hinterlands. The role of mangroves in mitigating the effects of the tsunami that struck Southeast Asia on 26 December 2004, provides a stark illustration of this. Across the disaster zone, from Simeulue Island near the epicentre, to the Sundarbans of West Bengal, wherever there were healthy mangrove forests, the death toll was a small fraction of what it might otherwise have been.

Even if we're immune to the intrinsic beauty of mangroves, we should at least give thought to the services that they provide – the carbon capture, the sea defences, the remediation of toxicity. Yet, across the globe, in virtually every one of the hundred countries that host mangroves, they are being cut down and ripped up at an appalling rate. In Vietnam, for instance, more than half of the original mangrove cover has been destroyed to make way for aquaculture, in particular for shrimp farming. Meanwhile, in Indonesia, home to more mangroves than any other country, hectares of forest are also being cleared for the production of charcoal, which is then exported to Europe, Japan and China, where in a regrettable elision it might well end up fuelling a barbecue to cook cultivated shrimps. At the current rate of clearance, mangroves are expected to be entirely gone from Borneo by the end of this century. In other places, mangroves are cleared for coastal development, to provide acceptable seafronts for wealthy tourists, or to expand oil palm plantations. Throughout their range, mangroves are being muscled out as we vandalise coasts in characteristically mindless fashion.

The economics of this are absurd. Calculations of ecosystem services ascribe a financial value to the benefits that we derive from nature. In the case of mangroves, assessments in different parts of the world typically set that value at between $15,000 and $100,000 per hectare per year. According to the United Nations, harvesting of mangrove wood for charcoal yields only a little over $1,000 per hectare, and since this is a destructive practice, the pay-off is a one-off. Any commercial enterprise that sold off assets on this basis would go bankrupt in no time, yet here we are, blithe in our irrationality.

My perspective on this, however, is that of a comfortable Westerner. The viewpoint of the people on the front line, those who live in places like west Borneo, for instance, is rather different. Here, in communities where charcoal production is prevalent, people often have little alternative; there are few other jobs to be had. The mangroves might be producing $15,000 per hectare per year, but they won't see a cent of that. They can't eat ecosystem services. For a year of exhausting work, cutting mangroves and making charcoal, they might take home as little as $1,000. So, while their collective actions cost the planet an amount vastly in excess of the value they extract by making charcoal, it's useless to point the finger. The destruction of the mangroves is not the fault of those who make a subsistence living from charcoal, it's a much bigger flaw that resides in the inability or unwillingness of commerce to account for the loss of these ecosystem services. We demand cheap seafood and chafe at paying more than a pittance for charcoal, yet we forget that the difference between the price of these things and their value is a cost that must be paid by nature.

* * *

Aquaculture – the farming of marine species – represents one of the main drivers of mangrove deforestation throughout the tropics. It's a trend that began long ago but picked up pace in the 1960s and '70s. Since those times, according to some estimates, we've lost around half of the world's mangroves. Satellite images of coasts around the world show remnants of forests shoved aside by farms in a landscape scarred by industrial shrimp farms. Asia supplies the vast majority of the world's shrimp, and in Vietnam, Thailand, Indonesia and the Philippines, enormous swathes of mangroves have been cleared. Similar trends have been seen in India, Sri Lanka and Bangladesh, in Central and South America, in Africa and the Middle East.

The most obvious effect of the explosion in aquaculture for most Westerners is that shrimps and prawns are cheaper than they've ever been, and they're available in every supermarket. But each bag of frozen prawns from a farm where mangroves once grew is freighted with habitat loss and accelerated coastal erosion. And since the trees are no longer in place to hold closed their vaults of greenhouse gases, a dinner of farmed prawns also comes with a great belch of liberated carbon dioxide on the side.

There's pollution, too, a multifaceted problem and something that the aquaculture industry has long struggled with. There's typically little to no barrier between these farms and the marine environment. Something that's immediately apparent to any visitor in the vicinity of a scaled-up shrimp farm is the effect of the run-off from intensive farming, the uneaten food and the waste products that pour nitrogen into surrounding waters. You can smell it in the air, but for the marine organisms downstream of this sewer, the impacts are greater than mere odour. The run-off promotes toxic algal and bacterial blooms that strip the sea of oxygen and kill marine animals. Coastal zones that were once characterised by their clarity become turbid, loaded with sediments that starve seagrasses and corals alike of light. The contamination isn't purely biological sludge, either. The concentration of shrimps into densely packed pools creates a perfect breeding ground for pathogens and parasites that then spill over to infest and decimate wild populations.

Chemicals, too, go hand in hand with industrial aquaculture. Shrimp farms in the Philippines, for instance, routinely use more than 100 different substances in their operations. These variously sterilise the ponds prior to the introduction of each new batch, disinfect the incoming water supply, kill pests and fertilise the stock tanks to trigger algal growth for larval shrimps to eat. Oh, and let's not forget the unregulated antibiotics that are pumped into the ponds on many farms as a prophylactic to keep crustacean yields high. The vast majority of these chemicals go straight out into the environment where they sabotage local marine ecosystems and, inevitably, drive anti-microbial resistance in bacteria. A recent analysis found a high prevalence of antibiotic-resistant bacteria in both raw and cooked farmed prawns, posing the risk of resistant strains of human pathogens entering the food chain.

Regardless of its problems, aquaculture is huge business, worth globally more than $300 billion per year. It's grown rapidly from essentially nothing before the 1970s to the point that, in 2022, aquaculture production surpassed wild fisheries by weight for the first time in history. We now consume more fish and shellfish from farms than we take out of the ocean. We're also eating far more seafood than ever before. The average human eats more than 20 kilograms of fish and shellfish a year nowadays; back in the 1960s, that figure was less than 10 kilograms. According to the Food and Agriculture Organization of the United Nations (FAO), it's likely to grow by another third by 2050, and that growth will have to come through aquaculture. The FAO sees this as an unqualified boon. Seafood, it says, is an excellent source of high-quality protein that's rich in vitamins, minerals and omega-3 fatty acids, and has the potential to 'contribute to food security, nutrition and poverty reduction'. The question is, can aquaculture feed the world without destroying it?

On the face of it, aquaculture has enormous potential for good. It provides a reliable food supply while, at least theoretically, taking the pressure off wild stocks and offering a solution to overfishing. It already provides employment for millions of people and allows the possibility for some of the world's poorest nations to develop economically. Additionally, fish farms take up less space than traditional livestock and arable farms, and the feed conversion ratio, a measure comparing food input to yield, is much better. In other words, it takes less food to grow each kilo of fish than it does to raise pigs, or chickens, or cattle by the same amount. The reason for this is simple: warm-blooded animals, like birds and mammals, burn a huge amount of energy just to maintain a steady body temperature, energy that cold-blooded creatures such as fish and shellfish can instead put towards growth.

There's no question that the aquaculture industry has promise, and that many farms are doing an impressive job when it comes to sustainability. There's even the potential in some aquaculture sectors to deliver a double whammy, benefitting the environment at the same time as growing seafood. Wild-farmed shellfish, for instance, can provide ecosystem services as they grow, since by filter feeding they improve water quality and create habitat for other species as they do so. These animals, including oysters, clams and mussels, are also self-sufficient – they collect their own food naturally – obviating the need to shovel pellets into their midst to support them. With inputs close to zero, the pollution that arises from shellfish farms is minimal. That's not to say there's nothing to be concerned about. Shellfish farms can damage delicate habitats like seagrass meadows and compete with natural populations for food, and they can attract and harbour disease. There are also legitimate questions about where the farms get their stock, especially if supplies of juveniles are taken from the wild, and about the use of chemicals to clean their equipment. Overall, however, and especially when compared to other forms of aquaculture, properly managed shellfish farms can be sustainable and ecofriendly.

The brave new world of aquaculture, however, has a darker side, one that has been exposed recently in a series of shocking reports and images from the salmon industry across Europe. Anaemic fish coated in blood-sucking lice, their flanks riddled with ulcerated sores, swimming listlessly in overcrowded pens. Others with missing eyes, torn fins or trailing strips of necrotic flesh. Slicks of corpses floating on the surface of fouled water or piled in their hundreds in bins for disposal. On the seabed beneath rafts of salmon feedlots, each a circular cage, 50 metres in diameter and home for up to 200,000 fish, thick layers of anoxic sludge accumulate. This sludge is a terrible amalgam of decomposing food, excrement and chemical treatments that poison the local environment. This is the counterpoint to the FAO's enthusiastic advocacy of aquaculture and to the wholesome notion many have of an ethical industry raising healthy fish in sparkling waters.

Salmon farming in its modern form has been with us for around 50 years, when it began in Norway as an answer to collapsing wild salmon populations. Nowadays, it's become an enormous industry in the country, producing well over a million tonnes of salmon a year. According to Dyrevernalliansen, the Norwegian animal protection alliance, in 2023 there were an astonishing 849 million salmon in captivity in Norway alone, a figure that is 300 times greater than the entire world's population of wild Atlantic salmon. It's also thousands of times greater than the remnant populations of free-living salmon in that country, which has declined dramatically as the numbers in captivity have skyrocketed. In 2024, 33 of Norway's rivers were closed to salmon fishing because there simply weren't the numbers to sustain it. If salmon farming was devised as a means to conserve wild stocks, it's done a conspicuously bad job of it.

By concentrating huge numbers of salmon in coastal waters, the farms act as reservoirs of pathogens as well as parasites, and in particular fish lice. The lice are crustaceans that attach to the flanks of fish and gorge themselves on their host's blood and protective mucus. In addition to the direct costs that the lice impose, the wounds they inflict attract secondarily infections, all of which is terrible news for the fish. Dealing with lice is an enormous problem for fish farmers because the treatments that are the most effective at killing the parasites are devastating for the environment, and, in any case, the lice have developed resistance to some of these treatments. Alternatives such as thermal delousing – putting the fish in a hot bath to destroy the lice – or adding other fish species, such as lumpsuckers or wrasse, to salmon pens to eat the lice are both associated with their own problems, including welfare issues for the salmon, labour intensity for the handlers, and high mortality among the cleaner fish.

So, lice are a problem that isn't going to go away while salmon farming is done at sea. As wild populations migrate past their captive counterparts, there's a risk of cross-infection between the two groups. Perhaps the worst of this occurs when wild juvenile salmon, known as smolts, leave their natal rivers for the first time and enter the sea, where they are embraced by a welcoming committee of lice that swirl around the salmon pens. When as few as six lice may be enough to kill a smolt, and even a lighter burden induces chronic stress in the fish and depletes its energy reserves, this represents a serious problem. There is considerable debate, however, on the question of just how much damage lice from farmed fish inflict on wild stocks. Environmentalists point to the precipitous decline of wild salmon, which mirrors the increase in farmed salmon in coastal waters. Aquaculturists, in response, point out that correlation is not causation, and claim that wild salmon mortality as a result of increased lice burdens from farms is overstated.

Lice aren't the only problem, however. Each year, hundreds of thousands of farmed salmon escape their pens and mix with their wild equivalents. At face value, this might be seen as a boost to flagging natural populations, but there's a catch. Farmed salmon have been artificially selected for generations to make them grow faster and to make them more resistant to disease, among other things. Genetics, though, is complicated. It's enormously difficult, often impossible, to select for one trait in isolation. Consequently, the faster pace of life induced by these breeding efforts also seems to make farmed salmon more aggressive, less averse to danger and comparatively ill-suited to life in the wild. In populations that have experienced the greatest influxes from farmed fish, interbreeding compromises the wild stock. Adults mature at a younger age and produce fewer, smaller eggs, and the resulting juveniles show lower rates of survival, potentially because they're more vulnerable to predators. Rather than augmenting wild populations, therefore, escaped salmon may be hastening their demise.

Salmon is a premium product: it's good to eat, synonymous with opulence and in high demand. Unfortunately, salmon are predators. Carnivorous fish need meat to eat, and much of that comes from wild-caught fish. It's at this point that the logic of preserving wild stocks by switching our emphasis to fish farming starts to falter. Worldwide each year, 17 million tonnes of fish, a fifth of all the wild fish landed, are caught to supply the aquaculture industry. To put that into context, 17 million tonnes is about twice the weight of all the wild birds on the planet. So, fish farming overall uses far more fish than it produces. According to a recent report in the journal *Science Advances*, it may take as much as 6 kilograms of wild caught fish to produce a single kilogram of salmon.

Controversy surrounds the exact calculations of this 'fish in–fish out' ratio. Where the new report put the upper limit of the ratio at 6 kilograms in for 1 kilogram out, previously, more aquaculture-friendly estimates put the figure for salmon farming at around 2:1. Why the difference? For one thing, there's a serious lack of transparency about exactly what goes into feeds; reporting is done largely on a voluntary basis. For another, the lower values that are preferred by the aquaculture industry seem to rely on a little creative accounting. A slew of different fishmeal and fish oil sources used in the production of aquaculture feed are simply overlooked in some assessments. But even if we were to take the industry's word on the lower figures, the sourcing of much of the fish that are caught to supply protein for aquaculture is ethically questionable. Factory ships hoover up fish from the coasts of West Africa and South America, putting subsistence fishers in those regions out of business and taking food off the plates of local people.

In an effort to make the industry more sustainable, feed manufacturers have switched increasingly to plant proteins, easing the pressure somewhat on wild fish stocks. But if increasing the demand for crops such as soy means that more land is turned over to agriculture, then the problem is shifted rather than solved. As things stand, there's a limit to how much plant material can be substituted into the mix for species like salmon. Central to this is salmon's requirement for oils, in particular long-chain omega-3, in their diet, the best source of which has historically been in fish. New technologies, such as genetic modification, already enable plants to produce these oils, which could one day entirely replace fish in the diet of farmed salmon. Objections to GM foods among the public, however, make this an unpalatable option for fish farmers.

The problems facing aquaculture aren't easily solved. In a bid to address the problems of pollution, disease and escapes, the Canadian government recently announced plans to transition salmon farming from open pens at sea onto land. In the decades to come, Canadian salmon will be raised in recirculated, temperature-controlled and filtered water, and because they'll be cut off from the sea, they'll be free from the lice that plague their counterparts in cages. Unsurprisingly, perhaps, there are problems with this. The energy required to sustain a high-tech onshore fish farm means higher costs and a higher carbon footprint, and regardless of where they live, for now the salmon still require wild fish in their diet.

Other choices exist, such as carp and tilapia, omnivorous species that can be sustained on a diet containing much lower levels of protein from wild-caught fish. Consequently, the fish in–fish out ratio for both of these is way below one; indeed, for herbivorous carp it approaches zero. Freshwater aquaculture also offers the possibility of using the waste nitrogen produced by the fish as a fertiliser for growing hydroponic crops. In this win-win situation, the fish feed the plants, which in turn cleanse the water for the fish. Carp and tilapia, however, are not viewed as premium products in the way that salmon is, at least in the West. There's an irony to this because a farmed salmon is a shadow of its sleek, wild counterpart. While the free-living animals migrate over thousands of kilometres, those on the farms are crammed together with comparatively limited scope for movement. The enforced lassitude of the animals means that they grow flabby; their underworked muscles are a pastiche of those of their wild ancestors, and their bodies contain four times as much fat. Even the pinkness of the farmed fish's flesh is a cosplay, induced by colouring agents such as astaxanthin.

Regardless of the quality of its product, aquaculture is now a huge business and, all too often, commercial considerations take the driving seat; the bottom line becomes the headline. A combination of slick messaging, lobbying and belligerence are employed to deflect criticism, prevent scrutiny and keep the cash rolling in. Claims made by business about the sustainability of marine fish farming have to be taken with, what else, a pinch of salt. Though it's essential that governments and NGOs maintain pressure, individual consumers have a vital role to play in this. Purchasing power comes with a moral responsibility to make ethical, informed choices and to buy sustainable seafood, sourced from producers who are doing things the right way.

Any of us browsing a menu, or pushing a trolley past a supermarket display, is susceptible to being seduced by the thought of some meaty prawns or the delicate juiciness of a salmon fillet. They're popular because they're delicious, and the growing disconnect that many feel between the natural world and our own encourages the idea that our actions don't have consequences. The seminal studies of rocky shore diversity, however, give the lie to this. We know that living systems exist in a complex, dynamic and sometimes delicate balance. We also know that if one species becomes too dominant, ecosystems have a way of correcting, to the detriment of that species. Our future is inextricably intertwined with that of the ocean and its inhabitants, and if we fail to take the lessons provided by the tribulations of intertidal animals like sea stars, mussels and barnacles, it will be so much the worse for us.

4.
Where Shadows Drift

The seasonal riches of temperate seas

HOLDING MY MASK TIGHT AGAINST MY FACE, I TAKE a single stride from an oyster-encrusted rock overhanging the bay. In that moment, I leave what the poet Philip Larkin calls the 'miniature gaiety of the seaside', with its noise and bustle, and plunge into the Pacific. In place of the excited chattering of children haring around picnic rugs and the traffic driving along the seafront, is a gently swaying forest of kelp. The static crackle of snapping shrimps hidden beyond my sight supplants the hubbub of humanity. For me, the sea surface isn't just the boundary between air and water, it's the point at which I pass from the cares of everyday life to the serenity of the world beneath the waves. The water soothes me; my heart slows and my quotidian anxieties fade. I'm reborn in the ocean.

The feeling of peace that comes with being in, or near, the ocean is what the author Wallace Nichols calls a 'blue mind' state. Those who live near the coast report being happier and less stressed than those who don't. And even those who simply visit benefit from a kind of detox from the fractious overstimulation of modern life, a fact that Nichols ascribes in part to our evolutionary history, when our survival depended on access to water. Whatever the truth of that, it certainly works for me. Just the sight of the ocean is enough to lift my mood, and joining with it feels like a homecoming.

A blue mind is apparently not for everyone. I remember my dad visiting me in Sydney a few years ago and his very apparent discomfort at the idea of joining me for a swim in the sea. Perhaps I shouldn't have mentioned an incident that occurred just a year before his arrival. In 2009, an Australian Navy diver, Paul de Gelder, was attacked by a shark as he worked beneath the docks in Woolloomooloo Bay, within sight of the famous opera house that stands sentry over the middle harbour. De Gelder lost an arm and a leg in what he later drily described as 'a bad day at work', but in a sense, he wasn't the only casualty of the attack. The vanishingly rare incidence of shark encounters in the decades leading up to 2009 had engendered a sense of security among Sydneysiders and this attack shook the city. In response, researchers at the New South Wales Department of Fisheries were instructed to drop whatever they were doing and to divert all their efforts into finding out just what was out there in this, the largest natural harbour in the world.

Lines were set all around the harbour bearing hefty stainless-steel hooks baited with chunks of eel. At dusk each evening, the fisheries, boats headed out to check the lines and, if necessary, undertake the daunting process of hauling a potential man-eater to the surface. Over several weeks, their efforts yielded twelve bull sharks, the largest of which was more than 3 metres in length. With a little careful manoeuvring and no little trepidation, each shark was rolled over onto its back at the surface, sending the animal into a kind of dreamlike state known as tonic immobility. Once becalmed in this way, the crew could take measurements of the fish and fit it with an acoustic tag before righting it and allowing it to resume its sharky meanderings.

The data that emerged from the tags over the following months provided an incredible insight into the behaviour of these mysterious and much feared creatures. The sharks remained in or around the harbour for the entirety of the Australian summer, spending each day cruising around its margins, often in close proximity to famous swimming beaches, and in one instance, passing right through a locale on the north shore that hosted confidence-building lessons for people with a fear of ocean swimming. Over the course of the study, the sharks would have had thousands of close encounters with bathers – encounters that those swimmers were entirely unaware of. Not a single incident occurred. Indeed, the attack on de Gelder was the first serious incident in the harbour since 1963.

I understood why my dad became anxious when he ruminated on the dangers that might lurk unseen beneath him as he swam. A fear of sharks is primal and widespread, and it's nurtured by the media circus that attends each attack. We know that sharks are out there and that there's always the chance that something might happen, but the likelihood that such a fate awaits us has to be put into context. According to the Florida Museum of Natural History's International Shark Attack File, which compiles data on the phenomenon, there were only around 800 incidents worldwide in the decade to 2020, of which 54 were fatal. Given the countless millions of hours spent by people in the sea, in the vicinity of sharks, the risks are off-the-scale low.

Those people I've known who have had a close encounter with a shark have emerged from it with a sense of wonder. As a case in point, John, a ranger in the Galapagos, described his experience of coming face to face with one of these apex predators. After seeing a huge shoal of bait fish in a bay in a remote part of the islands, he grabbed his face mask and went to investigate. As John swam slowly round the margins of the shoal, a pocket opened in the mass of close-packed fish, and there, right in front of him, was the unmistakable form of a tiger shark. 'One word came straight into my head,' he told me. 'Respect.'

Tigers have a daunting reputation; they're one of the three most deadly shark species, alongside whites and bulls. If John had panicked and splashed around, the outcome might have been different, but fortunately he held his nerve. The shark arched its back in the classic threat display, a clear signal that John should move away. There was no confrontation, no mindless savagery, just a warning that John was happy to heed.

My own brush with one of the 'big three' took place on the southern Great Barrier Reef some years ago. My colleagues and I were finishing a busy day by collecting our filming equipment from the camera rigs we'd set up in the lagoon. Dusk was beginning to fall and the rising tide was flooding over the reef crest, bringing with it a host of nocturnal predators. I saw a barracuda dead ahead of me as I snorkelled along and turned around to call to the others to come and take a look. When I looked back, the barracuda was gone, and in its place was a bull shark. It was a beautiful animal, stocky and deep-bodied, yet still with the characteristic sleekness of its kind. The bull cruised slowly towards me, seeming to stiffen slightly so that its measured tail strokes became almost jerky. Next, it began to circle me and my two friends, who by now had caught up. Perhaps I should have felt menaced by its attention, but instead of fear I felt a mixture of awe and intrigue. While unquestionably this shark could inflict terrible injury if it chose, I didn't gain any sense that it might. After less than a minute, the shark moved off into the gathering gloom of the twilit waters, leaving me feeling exhilarated by the encounter.

My subsequent encounters with sharks have followed a similar pattern. I've consistently gained the impression of sharks as curious animals, far from the cold, brutish killers of legend. And though there's no question that they can be dangerous, the fact is that they very seldom are. Nevertheless, politicians in global shark hotspots such as New South Wales are wary of the vote-losing potential of failing to act. In consequence, shark nets are installed off popular beaches. In 2022 alone, these nets killed several hundred non-target animals, including threatened species such as leatherback turtles. To me this seems like a heavy price to pay. Our claim on the ocean, in my view, shouldn't take precedence over that of marine animals. It would be far better if people could adopt John's watchword when thinking about sharks: respect.

* * *

Sharks were far from my thoughts as I revelled in my immersion in the cool embrace of Sydney Harbour, and I kicked down until I was gliding above the kelp forest's canopy. I came across a small clearing where I saw a trio of luderick, their stocky silvery bodies banded with vertical black stripes, rasping at the rock with fleshy mouths. With a flick of its fins, the larger of them edged from my sight until it was swallowed by the kelp, and a moment later all three were lost from view. I was looking at the place they'd just vacated when a new form emerged from the background of the fronds.

It's an animal that's so well camouflaged that, although not rare, I've never previously encountered in the wild despite hundreds of hours spent diving in these waters. It's slender and perhaps 30 centimetres long, and like the surrounding kelp, it's a shimmering green-gold in colour. Peculiar lobes sprouting from its body complete the disguise, concealing its fundamental form and giving the impression of some outlandish chimera: half animal, half plant. Take away these lobes, however, and what you're left with is something that looks akin to a mythical creature, and it's from one of these that it gets its name: the sea dragon.

This is no fire-breathing, mystical dragon, of course; like the luderick, it's a fish. More specifically, it's a kind of seahorse. Its slender, extended tube-like snout is a precision instrument, which it uses like a pipette to suck up tiny crustaceans like mysid shrimps. Effective predator and scourge of shrimps though it is, the weedy sea dragon is a poor swimmer. The evolutionary flourishes that have lent the sea dragon its exquisite camouflage also act to compromise its mobility, confining it to precision movements that are largely powered by a pair of restless pectoral fins that sit just behind the head. Not that this handicaps the sea dragon, whose typical modus operandi amounts to sitting among the algal fronds that it so beautifully mimics, swaying with the currents and hiding in plain view.

Its eyes trained intently on a shrimp, the sea dragon turns slightly and reveals a new and wonderful detail; a cluster of dusky pink hemispheres cradled against its belly. These are eggs, and now I know that the 'it' is a 'him' because, in common with all seahorses, parenting is the exclusive responsibility of the male. The next generation of dragons sit snugly within their eggs against his skin, their shape and colour giving the sense that dad's wearing an apron made of raspberry skins. When his partner first laid these eggs and transferred them to him in the touching face-to-face embrace characteristic of seahorses, the skin on his underside was soft and spongelike, allowing the eggs to embed themselves and to become, at least temporarily, part of his body. Responding to the deposition of the eggs, a membrane of skin closed protectively over the top of them and now he provides for his developing embryos, passing nutrients and oxygen to them over the course of a two-month vigil.

I returned to the place often in the days and weeks that followed and saw the expectant father again twice. I dearly wanted to see his progeny among the waving fronds, each a perfect miniature of their father, just a few millimetres long but already bestowed with the curious appearance that allows them to escape the notice of many would-be predators. Each time, I returned slightly disappointed. Then again, I consoled myself with the thought that with their extraordinary mimicry of their seaweed realm, the odds were always against such an encounter. Who knows: perhaps they watched me each time from the safety of the kelp forest? In spite of the fruitlessness of my searches, they were nonetheless accompanied by the profound feeling of contentment that comes from being part of the undersea world, sharing it, albeit temporarily, with animals such as this, and participating in the wonder of biology.

* * *

As the baby dragons began to make their way in life amid the kelp of Sydney Harbour, another drama was being observed 8,000 kilometres away in the winter seas off Honshu's Yamaguchi Prefecture. The observer in question was Tony Wu, in his own understated words a photo-naturalist. Despite the simplicity of his self-description, Tony has won prestigious awards, his exquisite portfolio comprising unique perspectives on celebrated animals as well as intimate portraits of the lesser known. Most are marine, but the one thing that unites them all is Tony's philosophy. As he puts it: 'To *take* a photo is aggressive. Instead, you share an experience.' It takes patience, typically lots of it, to build a rapport with an animal, and this is Tony's motif when he's underwater. Rather than rushing round 'collecting Pokémon' like some divers do, Tony's approach is to hang back, to wait. This isn't done passively, but consciously and respectfully. By following this credo, often as not there comes a moment when the animal drops its guard and starts to behave, and that's when the magic happens. This was never more evident than in his chronicle of the life and times of an extraordinary fish, one that Tony named Tako-kun.

Nobody ever called the Tanaka's snailfish beautiful. It has the head of a badly stitched sock puppet, with small, piggy eyes and a body that is the corporeal manifestation of dejection. Tako-kun, however, rose majestically above these shortcomings. His name – *tako* is the Japanese word for octopus – came from two perfectly circular scars on his pelvic fin, perhaps the result of wounds inflicted by the questing suckers of a predatory giant Pacific octopus. Tako was thus already a survivor, and Tony appended his name with the suffix *-kun*, an honorific used in Japan among friends; it's a humorous flourish but one that acknowledges the unlikely bond that formed between them.

Tako-kun was a hefty fish, roughly the length of a human arm, but unusually among fish species, whose lifespans generally correlate with their size, Tanaka's snailfish lives for just one year. Most of this curtailed existence takes place in the seclusion of deep and gloomy offshore waters, but in the winter both of the year and of their lives, they ascend to the shallows to breed. Males compete to establish and hold territories, then vigorously advertise their qualities to females in the hope of encouraging them to spawn. From this happy moment, the deteriorating males typically have ten days, possibly two weeks at a stretch, in which to usher their precious offspring to independence.

On 29 January 2024, Tony met Tako-kun for the first time, the latter having established his fiefdom by ousting the previous territory holder a day or so earlier. Though the fish look alike, it was the distinctive scars that identified Tako-kun as the brand-new owner of what can only be described as a strange choice for a piscine estate: a discarded fishing net. It's likely that the loss of seaweed beds to trawling drove the snailfish's decision, as these areas are usually their favoured breeding habitat. In any case, taking up residence in such a place wasn't the most unusual facet of Tako-kun's behaviour. That distinction was reserved for the way he interacted with Tony.

People have preconceptions about fish. Many dismiss them as inexpressive and unsophisticated creatures, capable of only the most rudimentary behaviour patterns. Tako-kun, however, defied these notions. After a characteristically unrushed introduction, Tako-kun first tolerated then accepted Tony, allowing him to approach his dais atop the net and, in time, sallying forth to greet him on his daily dives to solicit belly rubs. By degrees, Tako-kun took to sitting on Tony's arm or his shoulder during visits, like the fish equivalent of a pirate's parrot. We can never know what goes on in the mind of any wild animal, of course, but it was interesting that Tako-kun was very discerning with his favours and never treated any other divers in the same way.

Whatever Tako-kun was doing, it was working with the lady snailfish. His nest steadily accumulated ever more clutches until he was proud father to a truly stupendous mass of eggs. Tony's joy at seeing Tako-kun's successes, however, was tinged by the knowledge that their relationship would be brief. For a male establishing a territory at the end of January, surviving beyond Valentine's Day might be a reach, yet here again Tako-kun exceeded expectations. He was still going strong when Tony had to leave in mid-February and from then on reports of Tako-kun's wellbeing were relayed by other divers as his celebrity grew. On 15 March, fully a month beyond his predicted expiry date, Tako-kun was still holding court atop his net throne.

Like Caesar, however, Tako-kun had every reason to beware the ides of March. When divers paid a visit two days on from the last update, the nest with its few remaining, unhatched eggs was still there, but Tako-kun himself was gone. His demise remains a mystery, though it's possible that he was claimed by the same fishers who once discarded the net that became Tako's kingdom; if so, it would be a sad end for a remarkable fish. The silver lining is that, given Tako-kun's reproductive prowess, there's every chance that one of his progeny may follow in his finsteps next year and, if one does, Tony will be there to record it.

* * *

It's late afternoon in the Tasman Sea. The sun casts oblique rays through the forest, illuminating the olive-green fronds of giant kelp. There's a stillness as the currents slow in seeming reverence for the towering stands. Fish glide in and out of view, tracing a slow slalom among the ranks of tall, slender stipes, while far above me, the canopy is stirred by my rising bubbles and the cadence of gentle waves. In the subdued, mottled light and the quietude, there are echoes of pine groves, and I'm reminded of walking as a child through the muffled hush of upland plantations. This, however, is a richer experience, since it takes place in three dimensions. I feel as though I'm flying through the kelp; I'm not just in the forest, I'm part of it.

Each tree in this marine forest is a mimic, having taken the anatomical blueprint of plants for its own. The convergence can be seen in the leaf-like fronds, the stem-like stipes, and the root-like holdfasts that anchor each kelp to the seabed 10 metres below my floating form. But for all the similarity, kelps aren't plants but seaweeds. What differentiates these superalgae from plants is not so much size or appearance, but their relative simplicity. Their holdfasts draw no nutrients from the seafloor, their stipes are supple rather than supportive, and their fronds lack the network of veins that characterise leaves. Even so, the distinction that exists between plants and kelps might seem to be scientific hair-splitting, a botanic 'tomahto, tomayto' that ignores their commonalities, not least of which is their shared strategy of sequestering energy from the sun. If that seems to be the case, then consider that the last common ancestor of kelps and plants existed more than a billion years ago; they've spent an aeon going their own separate ways. Kelps themselves only appeared on Earth in their current form about 20 million years ago, and though they've spent much of the intervening time apparently evolving in homage to plants, they are entirely different organisms.

Not only are seaweeds not part of the plant club, but they're also not even particularly closely related to one another.[28] We lump them together clumsily as kinds of multicellular algae that grow in the sea, and just to put the boot in further, call them 'weeds'. The primary colours, if you're a seaweed, are brown, red and green. It's by these pigments that we group them, and each has its own distinct evolutionary history. Perhaps unsurprisingly, green seaweeds, such as sea lettuce, are most closely related to the land plants. The green seaweeds are also closely related to red seaweeds, which are most likely the oldest group, and with 7,000 species, are certainly the most diverse. The nori that's wrapped around your lunchtime sushi is red seaweed, despite appearances, as is dulse and the various types of delicate pink, coralline algae. Meanwhile, brown algae find themselves out on an evolutionary limb compared to the other two, and appeared long after both. Nevertheless, it's these brown algae, such as *Fucus vesiculosus* (known as bladderwrack) and *Sargassum*, that tend to be what we think of when the word seaweed

---

28 Formally, 'seaweeds' are a polyphyletic group, meaning that they don't share a single common ancestor.

is mentioned. They're large and have the most complex, plant-like structures; they are, in short, the most 'seaweedy'.

Though brown seaweeds are relatively new animals, and are only very distantly related to plants, it can hardly be said to have held them back. Kelps are grandiose, brown algae that exist on a scale that dwarfs many of the trees they're often compared to. In some parts of the world, giant kelps grow at more than half a metre a day to top out at a staggering 60 metres in height. And as I experienced in Tasmania, where they find conditions to their liking, they grow in dense stands of transcendent beauty. They're found in cool, temperate waters along coasts throughout the world, covering an area of inshore habitat that's collectively greater than Greenland in size. They're resilient, too, something that Darwin noted in his journal during his voyage on the HMS *Beagle*: 'I know few things more surprising than to see this plant growing and flourishing amidst those great breakers of the western ocean, which no mass of rock, let it be ever so hard, can long resist.'

The foundation of each individual kelp's success is its holdfast, a tangle of tough, fibrous tissues like a pile of spaghetti that grips the bedrock. The strength of this holdfast is prodigious, as it must be to anchor its own substantial biomass – a single kelp's weight approaches that of a person – against everything the sea can throw at it. The holdfast is also the starting point of a kelp's life, which begins with the settlement on the seabed of a microscopic sporophyte, a mere smidgeon of life that must very quickly secure itself. To do so, it produces fine threads that radiate outwards and adhere to the rugged surface with the aid of an ingenious glue known as alginate. Having done this, it waits. The trigger for it to develop further is light, which under an intact canopy of adult kelps is in short supply. If one of these mature individuals is lost to a storm, or is nibbled into submission by herbivores, the patient sporophyte springs into action and begins to grow. The larger it gets, the further its holdfast expands, cleaving to an ever-greater area of rock. Though the holdfast is central to their strategy, kelp also cope with the waves' buffeting by simply going with the flow, flexible stipes allowing them to bend rather than break in the turbulence. Even so, in a particularly savage storm, kelp can be torn from its moorings to accumulate in the aftermath in banks on the strandline. The misfortune of these individuals, however, is an opportunity for new sporocysts to take their place in the natural cycle of renewal.

The pile of broken kelp is also an opportunity for shore animals. Sandhoppers and seaweed flies make hay, which also draws in birds and small mammals to feed. And as it decays further, the kelp, washed ever further up a beach, provides the nutrients for pioneering plants that stabilise and strengthen dunes. We, too, have long made use of this windfall from the sea, and because kelp is packed with minerals, it makes an excellent fertiliser for improving soil and as a feed for livestock. Increasingly, we've been cutting out the middleman and eating it ourselves, albeit we tend to forage living plants rather than the festering clumps at the sea's edge. Seaweed has a venerable history on our menus – people have been gathering it for as long as records extend. At least 1,500 years ago the Japanese held it in such high regard that varieties such as wakame and nori were acceptable means of paying taxes, and it is now growing in popularity both as an additive and in its own right. Products as diverse as beer, toothpaste and ice cream often contain seaweed extracts such as carrageenan, while the seaweed in sushi and laverbread delivers a delicious umami hit.

Aside from the troubling smell, the property of stranded seaweed that seems to bother people most is its slimy texture. The mucus that creates this, however, is deserving more of admiration than squeamishness. Slime is kelp's secret weapon against a hostile world. It prevents kelp from drying out, acts as sunscreen to prevent damage by ultraviolet rays, protects the kelp from abrasion as it gets lashed about in the waves, and even traps nutrients from the water to allow easy absorption. There's one more problem that kelp faces, however, and it stems from being both appetising and stationary: it's prone to being eaten. Even a colossus such as giant kelp can be reduced to tatters by the army of rasping marine herbivores that it lives among, such as urchins and snails. Kelp partly solves this problem by hosting an army of its own. The surface of kelp is a highly desirable residence for lots of encrusting marine creatures, which benefit from being lifted away from the dog-eat-dog world of the seabed. Some, such as stinging hydrozoans, return the favour by defending the kelp against grazers and are sometimes present in such numbers that they've gained the name sea fur because of the fuzz they seem to impart to the kelp. The kelp gets an excellent deal from the relationship, getting not only a measure of protection from its colonists, but also receiving its entire supply of precious nitrogen from its lodgers at some times of year.

Valuable as it is against smaller grazers, the kelp's miniature militia is relatively powerless against either the guillotining mouthparts of urchins or infestation by disease-causing bacteria. Accordingly, kelp beefs up its protection with more slime, incorporating an arsenal of chemicals. Two in particular, phlorotannins and fucoidans, discourage herbivores by ruining their meal and simultaneously keep microbes at bay. Beyond kelp, both of these substances show enormous promise in a slew of different human medicine applications, including in the battle against antibiotic-resistant pathogens, cancer, diabetes and even cognitive decline. And if that wasn't enough, the fucoidan-enriched mucus traps carbon dioxide.

Regardless of their value to us as a food, fertiliser, medicine, carbon capturer and even bespoke building material, the 112 species of kelp are, first and foremost, incredible organisms in their own right, and crucial players in the ecology of coastal waters.[29] They calm the churning waves with their bulk, protecting the shore and providing animals with sanctuary from the currents. The dense stands offer innumerable hiding places for prey animals. And by dint of their size, they create habitat. Within a kelp forest, there's four times as much surface area for animals to occupy compared to the seabed alone. Taken together, all of these factors make kelp forests anything up to ten times richer in biodiversity than the surrounding waters, a fact remarked on by Darwin as he marvelled at the submerged forests of the Straits of Magellan: 'The number of living creatures of all Orders, whose existence intimately depends on the kelp, is wonderful. A great volume might be written, describing the inhabitants of one of these beds of sea-weed.' Around the world, kelp supports countless different species of animals, and they host stupendous numbers of individual creatures. One single kelp specimen, examined by biologists in Norway, was home to more than 600,000 animals. With populations of that size inhabiting them, each kelp is almost like a marine city.

Choreographed by the currents, the Tasmanian kelp fronds dance, alternately revealing and concealing small fry clustering by the central stipe, which look out warily as a fur seal cruises by. A little further on, I see a large fish, a trumpeter, smartly attired in horizontal stripes as it explores the shadows of the seabed. A spiny lobster finds its attempt at secrecy frustrated by its own antennae that poke like aerials from its hiding place. More successful in this game of hide-and-seek are the kelp urchins who wrap the fronds tight around their prickly bodies, turning themselves into the fillings of leathery seaweed dolmades. Despite the many opportunities for

29 In parts of China, kelp is used as a roofing material.

concealment, there are animals wherever I look; the kelp jungle is exuberantly alive. The same is true of equivalent habitats the world over, and many of the animals who live among the swaying stipes exist here and only here. They've become specialised, and, in turn, dependent, on the kelp.

In Europe, blue-rayed limpets migrate along the stripes, grazing on the alga and creating little pits in the kelp's tough flesh with their spiky, rasping mouthparts. Cautiously, they move down the kelp as winter, with its storms, approaches. But even if they should become dislodged, they have a solution. They cast a parachute of mucus as they fall, which prevents them from sinking to the mean streets below and helps them to navigate to a new stipe to resume their arboreal feeding. On the Pacific Coast of North America, kelp crabs scamper like steeplejacks among the towering trees of algae, which for the crabs act as both food and shelter. To blend in with the kelp and thereby stymie their predators, the crabs assume the same colour, and augment this by decorating themselves with snippets of the alga, which they attach to themselves using Velcro-like hooks on their backs. Such is kelp's draw that in New Zealand, there's even a terrestrial spider that makes its home in recesses of the holdfast, scampering in at low tide and then spinning a silk domain to hold a supply of precious oxygen when the waters return. For this spider, and the thousands of other species who are drawn to these places, kelp forests are essential to survival. The richness that Darwin noted in the kelp forests off Tierra del Fuego was no aberration; these are biological hotspots wherever they're to be found. Like so many such places, however, they're in danger.

* * *

Out in the Alaskan bay, a flotilla of kayaks rocks on the gentle swell. Around them, kelp blades poke above the sea surface, reminding me of dark chocolate shavings on a cake. As I watch, one of the boats transfigures sharply into a ball and then dives below the water. What I'd taken for kayaks are revealed as sea otters, the first I've ever seen, and they're here in impressive numbers. I feel foolish for mistaking them, but then I don't have my glasses on. Besides, I'm more used to European otters and these are much chunkier animals, about a metre and a half in length. I move around the headland to get a closer look at the furry armada with my binoculars. Now in focus, I see that they're resting, lying on their backs, their cream-coloured heads that I'd taken as prows raised slightly above the deeper brown of their torsos. Many of them have their front paws over their eyes, perhaps as an aid to sleeping in the bright sunlight. Their somnolence makes them easy to count; there are 21 of them – 22, if you count the absent diver – and hereabouts they're present in numbers that represent an astonishing recovery from just a few decades earlier.

Resurrection Bay, the appropriately named haven where these sea otters bob sleepily on the waves, was once a focal point of the 'fur rush', a sustained campaign of killing that extended for 150 years from the middle of the eighteenth century. The desirability of sea otter pelts is founded on their lushness. Among all mammals, they have the densest fur, with an average of 130,000 hairs on every square centimetre of their bodies, representing quite an upgrade on the 200 hairs that we have on the same area of our scalp. It's this that lends their fur a luxuriant glossiness, as well as the unsurpassed insulation that it provides. Sea otter pelts became the ultimate signifier of social standing and style in the courts of the Qing dynasty; anyone who was anyone wanted to be seen in clothes trimmed with – or ideally made entirely from – fur. It rapidly became the most expensive fur in the world, supporting an international trading network encompassing not only the Pacific rim but Europe. A single pelt in 1779 would command around $4,000 in today's money on the markets of Canton, while from 1743 to the end of the century, Russian fur traders realised profits estimated at $160 million from otters collected around the Bering Strait, from Kamchatka to Alaska.[30] These stupendous sums gave rise to the description of sea otter fur as 'soft gold'.

The killing continued throughout the nineteenth century, the dwindling populations of otters limiting supply and inevitably placing an ever-greater premium on the pelts. As Christopher Preston notes in his wonderful book *Tenacious Beasts*, just four sea otter pelts could buy a house in British Columbia in the early twentieth century. By this time, however, there was scarcely an otter to be found in the Pacific Coast of North America. In just over a century, more than a million sea otters had been killed; only a tenth of 1% of their original numbers remained and those in a few, remote pockets.

The same mercenary interests that had almost eliminated sea otters were now doing the same for northern fur seals in the region. Indeed, the paucity of otters was one of the reasons the fur traders had begun to covet seal pelts in the first place. The fact that these species weren't entirely wiped out is due in large part to the efforts of one man, the naturalist Henry Wood Elliott, who witnessed first-hand the scale of the slaughter in his role as an agent of the US Treasury Department posted to the remote Pribilof Islands in 1872. Though the US was able to claim exclusive access to the seals on the islands themselves, the bounty on offer drew hunters from other nations to pursue the animals in the surrounding ocean. Female seals, foraging to support the pups that waited for them on the beaches, were shot by the tens of thousands at sea. Unforgivably, the bodies of only around a fifth of those shot were recovered.

---

30 Until 1867, when it was purchased by the US government, Alaska was Russian territory.

By the time that Elliott returned to the Pribilofs in 1890, the beaches that had resounded to the clamour of millions of seals were quiet. Shocked into action, Elliott spearheaded a campaign to preserve what remained. It took 20 years of unrelenting advocacy in the face of sometimes extreme resistance for his efforts to bear fruit, but bear fruit they did with the signing of the first international conservation agreement in history, the 1911 North Pacific Fur Seal Treaty, which also provided in its fifth article for the preservation of sea otters.

The fur seals' recovery in the wake of the treaty was heartening. By the 1940s, they had returned to their former numbers. For the sea otters, things were much less straightforward. Numbers climbed slowly for the next 50 years, without ever looking like emulating the fur seals. Part of the reason is that, unlike the seals, the otters have relatively small home ranges and so are slower to recolonise lost areas of habitat. Having recognised this, the US government began a programme of translocations in the 1950s, moving hundreds of sea otters to locations that had been otter-free for a century. Finally, everything clicked. Sea otter populations rebounded, and the animals began to regain lost ground. A win for sea otters turned, unexpectedly, into a win for countless other species. In what has become the classic exemplar of top-down (i.e. predator-mediated) control of ecosystems, the returning sea otters helped to restore kelp forests, and their associated diversity, as they returned to long stretches of coastline.

The link between otters and kelp is the formidable, prickly relative of sea stars, the sea urchin. Though inevitably cast as the bad guy in the story of kelp's decline, sea urchins are fascinating creatures in their own right. Their delicate bodies are encased by a series of tough, interlocking plates that fit together into a globe-shaped structure known as a test. Apparently, this alone is insufficient for the cautious urchin, which backs up its armour with an array of vicious, mobile spines, much like the hedgehogs for which they're named.[31] They amble along the seabed slowly but surely by means of hundreds of tube feet, each a kind of hydraulic vacuum, alternately locking securely onto the substrate and then releasing to allow movement in a way that always reminds me of the eponymous garment in the Wallace and Gromit film, *The Wrong Trousers*.

Urchins' relevance to kelp is as a spectacularly effective grazer. A cavity on the underside of the animal hosts its formidable mouthparts, known collectively as Aristotle's lantern. The lantern is formed by five calcified plates, each the shape of an arrowhead, that taper to a sharp point. By articulating these, the urchin can pull, slice and dismember the tough, sinewy stipes of kelp. When large numbers of urchins assemble, they can deforest entire kelp beds, annihilating the macroalgae like a division of specialist mowers. Troublesome as this is, there's another facet of urchin biology that makes it near impossible for the kelp to recover. Most animals, having obliterated their food source, would either be forced to move on or contend with starvation. Not urchins. Faced with the absence of food, they first digest their own gonads as a stop-gap measure (which takes quite a while because urchins are very gonad-y creatures), and then decrease their metabolism sharply, so that they simply tick along in a state of emaciated suspended animation for months, or even a year, until the good times return. Kelp sporophytes have little chance of developing in the presence of an occupying horde of ravening urchins, and so a phase shift occurs, and a kelp forest becomes a bare-rock urchin barren.

Urchins take a great deal of blame for the parlous state of the

31 Urchin, from the old French word *herichun*, was the Middle English word for the hedgehog.

world's kelp forests, which is akin to prosecuting a bullet for killing someone. We're the ones holding the gun and we're the reason that urchins became so dominant that they cleared the kelp. We took out the predators: we overfished the crabs, the lobsters and the fish that ate, and thus controlled, urchins. And across the north Pacific, we exterminated their local nemesis, the sea otter. As the populations of these mammals recovered in the years following their reintroduction to Alaska, they feasted on the abundant urchins, focusing especially on the largest, most succulent ones. Sea otters eat about a quarter of their own hefty body weight in urchins every day.[32] With the urchins driven into retreat, kelp forests flourished once again around the shores of Alaska.

Today in Seward, the main settlement of Resurrection Bay, the sea otters are numerous and, judging by their portly bodies, well fed. The kelp fronds that poke above the waves at low tide, alongside the thriving marine life that occupies the forest below, are testament to the otters' top-down role in this ecosystem. This tripartite interaction between otters, urchins and kelp became the epitome of a trophic cascade, one that has been used as an example in lecture halls across the world for decades; it was straightforward, intuitive and, best of all, it's a good news story for restoration ecology. It cautions us that we need to contemplate the broader impacts of our actions, to understand that we can't machete our way through the web of life without reaping the consequences, but it demonstrates that there's much to be gained from careful stewardship of the natural world.

Ecology, however, is more complex than we sometimes allow. Sea otters, on their own, can't solve the many problems that kelp forests face. Furthermore, since otters occupy kelp forests in preference to

32 A sea otter might weigh 40 kilograms – I told you they were big – so each can account for 10 kilograms of urchins a day.

barrens, and favour the kinds of plump, juicy urchins that are found living in such places to the gaunt, gonad-less urchins of the barrens, an otter reintroduction doesn't automatically mean that the urchins' time is up. In some parts of the otters' range, it's left to other predators to keep urchins under control. The sunflower sea star, a behemoth that can reach 5 kilograms in weight and measures a metre across, is an avid consumer of urchins. When one of these turns up, most of the urchins wisely depart; those that don't will very rapidly fall prey to the gigantic sea star. Sadly, populations of these sea stars have crashed recently in the wake of a contagious wasting disease to the point that they're now considered critically endangered. The inevitable consequence, an explosion of urchins, has stripped kelp cover by more than 90% in some parts of California. Overlaying this, urchin populations fluctuate naturally, booming and busting in accordance with factors other than predation, such as El Niño oscillations, temperature and upwellings. As satisfying as it is to think of the relationship between sea otters, their prey and the kelp forests as straightforward, the real world is messy, challenging to predict and dependent on diversity.

Once a kelp bed has transmogrified into an urchin barren, it's by no means certain it will retain the capability to flip back into its former, forested glory, even if the urchins are removed. With the sudden availability of seabed real estate and the increase in light that comes from the removal of the kelp canopy, the stage is set for fast-growing algal opportunists to move in. Coralline algae, in particular, flourish. Urchins are not averse to grazing this, too, but coralline alga isn't particularly nutritious and its ability to colonise exposed rock rapidly gives it a competitive edge. A covering of coralline algae in turn makes it harder for young kelps to attach and establish. With so many challenges to them, it's not straightforward to restore kelp. But by conserving the animals that eat urchins, or, as is done in some parts of the world, using divers as proxy predators and coralline algae clearers, we can hope to reclaim the barrens and allow new forests to thrive.

Sadly, overgrazing is only one of the challenges faced by kelp forests. Pumping nutrients into the sea in both agricultural run-off and sewage fosters the development of algal blooms that turn the water into a green soup and, in some cases, use the kelp as a convenient platform on which to grow. With their precious light dimmed, the kelp struggles to grow, while a heavy burden of algal passengers increases drag on the fronds and amplifies the risk that storm waves will tear them from their holdfasts.

The kelp that we don't poison, we poach. I returned recently to Tasmania to see the east coast kelp forest that had given me one of the most beautiful sensory experiences of my diving career just a dozen years ago. It was entirely gone. In its place was nothing but a void of open water. Marine heatwaves, plumes of warm water flowing down from the tropics, pushed the temperatures beyond the kelp's limits and with a violent suddenness, the forests were gone. Long before I travelled, I'd heard that the situation was dire, but I was completely unprepared for this. I sat for a time on the clifftop, looking out at this impoverished Tasman Sea with a sense of desolation. When we lose kelp forests, the macroalgae aren't all that departs. Thoughts crowded my mind of the seemingly infinite schools of butterfly perch, the fur seals, the crayfish and the pastel phantasmagoria of the sponge gardens, all pieces of the wonder that was the forest and all now gone, possibly forever.

The departure of this wealth of life, the extinguishing of its vibrant, seemingly irrepressible energy leaves me with an emptiness that I feel profoundly. Above water, it's almost as if nothing has changed. The waves still lap at the stalwart shores, and seabirds still fly above a majestic fringe of Eucalypts that presses close to the ocean. The stasis topside is an illusion, however. Beneath the waves, hidden from our view, is the stark manifestation of a shift in our planet's climate, a gradual ratcheting of the temperature which has carried away a jewel of our collective natural heritage. Throughout the world, kelps are disappearing at a rate far greater than that experienced by coral reefs, yet this is largely a neglected crisis. Perhaps it's the similitude of normality that blindfolds us to the problem; whatever the reason, we undervalue kelps at our peril.

* * *

Whatever the future holds for the world's kelps, our history with them lends a poignancy to their current troubles. Our distant ancestors on the coasts of southern Africa foraged among the kelp for shellfish and other marine creatures that yielded not only calories but also essential omega-3 oils that provided for healthy development. More recently, a theory has suggested that as *Homo sapiens* expanded its range from Asia onwards to Japan and the Americas, rather than tracing an overland route as was once thought, they instead followed a coastal path, one paved with seaweed. This kelp highway hypothesis describes how inshore mariners in rudimentary boats followed expanses of kelp beds. Navigation was facilitated by the kelp's ability to calm turbulent waters and their convenience as anchorages, while foraging was supported by the wealth of prey to be found among these marine meadows. Thus, these ancient pioneers, armed with spears and shells carved into fishhooks, followed the kelp around the Pacific rim to new lands.

Patchwork fields of undulating kelp provide the structure to temperate coastal waters in just the same way as corals do in the tropics. And though nothing underwater can quite compare to the exuberance, colour and frenetic activity of a coral reef, a lavish tableau of life exists among their leathery stipes and fronds. In the cool, clear late spring seas off Cornwall, I watch as a stocky, mottled brown fish emerges from a crevice and pauses, alert to danger, before descending to an expanse of sand below. The tompot blenny is a comical-looking creature, its rheumy eyes, topped with lavish eyebrow-like cirri, make it look like an elderly historian. The blenny moves in bounds, resting on the seabed for a few seconds between each hop as if to gain reassurance that it hasn't been detected. I hold my breath so as not to disturb it with a stream of bubbles from my regulator. Gradually, furtively, it crosses the clearing, making for the forest edge, where perhaps it hopes to find an unwary prawn. On the cusp of the kelp, however, something spooks the blenny; it streaks back to the crevice, leaving in its place a miniature tornado of sand grains.

A moment later, one of the ocean's most extraordinary animals comes hovering into view. It's a thing of science fiction, a cybernetic drone made flesh, one that pulses mesmerically with waves of colour. The wondrous creature is a cuttlefish, a common cuttlefish to be precise, but there's really nothing commonplace about it. The cuttlefish is a mollusc, part of a branch of the animal kingdom that includes cockles and snails, yet it's equipped with a pair of advanced eyes and an outsized brain. Its complex nervous system controls a network of millions of pigment cells that give the cuttlefish the most dynamically expressive skin on Earth, and that facilitate its niche as a hunter. Despite its soft, pliable body, the cuttlefish is a consummate consumer not only of fish like the tompot, but also of crunchy, reinforced prey, such as crabs.

The origins of the fantastical intelligence and colour-changing capabilities of this animal may be found in its deep, evolutionary past, when its ancestors rid themselves of the outer shells that are the molluscan mainstay.[33] To step away from the protective yet limiting security of a shell to embrace a new, and naked, future represents one of the most dramatic transformations of any animal lineage. That they did so during the Mesozoic Marine Revolution, between 160 and 100 million years ago, when new and terrifying predators such as plesiosaurs and ichthyosaurs, as well as 10-metre-long sharks, swarmed the seas make it appear especially puzzling. Why would evolution favour the abandonment of tried and tested fortifications just at the moment the undersea world became the reddest in tooth and claw that it had ever been?

One answer to this conundrum occurred to me in the aquarium of the New Caledonian capital, Noumea, as I watched a *Nautilus* bump around, its crown of arms and a pair of large eyes protruding from a spiral shell, tiger striped by puce on cream. This is an animal antique, representative of a lineage that stretches back for hundreds of millions of years. Like the cuttlefish, the *Nautilus* is a cephalopod, the only extant member of the group to have retained a shell, which it can flood with liquid, or drain, to control its buoyancy, much like a submarine. In the confines of the tank, however, the *Nautilus* is slow, almost clumsy, and for all that its natural habitat in deeper water offers greater space, it's obvious at a glance that it lacks the agility that's so characteristic of its kin cephalopods. By internalising, and reducing, their cumbersome shells, cuttlefish and their relatives gained the advantage of speed and manoeuvrability, becoming hydrodynamic and fleet like the fish that filled those ancient seas. The shell that had protected their ancestors offered less reassurance in a world of powerful predators, whose jaws could easily crunch

33 All that remains of the outer shell in squid is a rubbery, chitinous structure known as a pen, while in octopuses it has gone altogether, or is reduced to a small, thin, boomerang-shaped nubbin called a stylet. Cuttlefish retain the remains of their shell within their bodies as 'cuttlebone'.

a brittle casing, and what the cephalopods lost in armour, they compensated for in mobility.

My Cornish cuttlefish slides through the water, then hovers above the seabed, its body perfectly still but for an undulating skirt of fins. As it prospects for its prey, the cuttlefish's movements seem deliberate and considered, a signature, perhaps, of the cognitive complexity that characterises these animals. Cephalopods are extreme outliers in this respect, invertebrates with an intelligence that rivals some mammals. The most prominent theories of how intelligence evolved, particularly among warm-blooded animals, centre on either the cognitive demands of interacting with others in stable social groups, or as the means to cope with a fluctuating environment, caching food in times of plenty and remembering where it is when the lean months come around. Neither of these explanations works well for cephalopods, so what drove the evolution of brains in these animals? Here, too, the answer might lie in the cephalopods' abandonment of their shells.

The fleshy vulnerability of cephalopods is counteracted in part by their phenomenal grasp of camouflage, facilitated by aggregations of nerve-controlled pigment sacs in their skin. The cuttlefish before me has an astonishing 300 of these structures, known as chromatophores, in every square *millimetre* of its skin, a density greater than that of the pixels in the best laptop monitors. With a twitch of tiny muscles, a colour is expressed, or concealed, and by blending from a palette of three pigments – in this case, red, brown and yellow – the cuttlefish can adopt a new hue in a fraction of a second. Beneath the chromatophores, a deeper layer of iridophores and leucophores perform a slightly different role, and reflect the tones of ambient light, thereby expanding the cuttlefish's portfolio of possible colours.

Replicating the prevailing shades of its immediate environment is only one part of the cuttlefish's competence in camouflage, however. Its repertoire is broadened by its capacity to blend into its background by adopting mottled patterns of contrast across its body to mimic those of their surrounds, and by an array of papillae beneath the skin that can be raised or lowered to change the animal's texture dramatically, from smooth to spiky and anywhere in between. They even use posture to accentuate the disguise, so that a cuttlefish hiding among the blades of a seagrass meadow lifts its arms and waves them in time to the currents like a dancer at a silent disco.

Control of all of these interacting components is exerted by the cephalopod's nervous system, responding to a stream of visual input from the eyes. Effortless though it looks to a marvelling human, the energy expended by the animal in producing these displays is considerable, a patterned pulse burning roughly the same amount of energy as is used to run all its other resting metabolic processes combined. This gives a hint to the complexity of the cephalopods' pattern-generating capacity; all of this advanced biological hardware demands a powerful processor to operate it, in the form of a large brain, arguably explaining at least some of the cephalopods' cognitive sophistication.

The capacity to vanish within plain sight is likely one of the main drivers of the evolution of colour change in cephalopods, but it's by no means the only one. Although at some times the animal may wish to go incognito, at others it might use its interactive integument to communicate, as exemplified by an octopus that I encountered in Tonga. The animal was sliding across the reef when it detected me, snorkelling at the surface. The octopus's first response was to freeze and to adopt, almost instantaneously, the pattern and colour of its background. Then, apparently lacking confidence in its disguise, it darted beneath a ledge. What the octopus hadn't banked on, however, was that another of its kind was already in residence, and a moment later it bolted back out, having adopted a uniformly pale cream livery, in a gesture of apparent supplication. The other octopus, the indignant homeowner, emerged a little beyond the threshold of its home, coloured crimson in rage. The message conveyed, both again adapted their appearances to conforming to their surroundings. The entire sequence of changes took place within just a couple of seconds, demonstrating both the different uses to which the octopus puts its mutable skin, and the exquisitely precise and rapid command of the process that it maintains.

Cuttlefish, too, use their incredible capacity to adapt their appearance when threatened. Rather than taking a one-size-fits-all approach, however, they select the best option for the situation they're in. Camouflage is of little use against a crab, an animal whose world view is assembled from a mosaic of chemical cues, and which hunts by smell. Accordingly, the cuttlefish doesn't try to fox these crustaceans with intricate colour changes but simply uses its greater agility to outmanoeuvre them. It's against visual predators, like predatory fish, that camouflage comes into its own, but even here cuttlefish have a plan B. If an assailant draws too near, cuttlefish drop the camouflage and deploy shock tactics instead. One such involves a sudden and dramatic blanching of the skin, so that instead of being concealed, the cuttlefish is now dramatically revealed. Its entire body is now an even pale shade, with the exception of two eye-shaped blots of dark pigment on its back. From the fish's perspective, what had seemed an opportunity worth investigating might now appear to be a terrifying face. Even if the attacker doesn't flee, caution demands that it pauses. The ruse needn't last long to serve the cuttlefish's purpose; all that's needed is a brief window to jet to safety amid the confusion.

Defence aside, most of the colour signalling of cuttlefish is aimed at their own kind. In the breeding aggregations that take place in spring and summer, flirts and feuds alike are negotiated in the medium of skin colour. A dynamic medley of mottles, bands and solids convey meaning and intent, signifying aggression or sexual attraction, among much else. If shades and patterns are the lexicon of the cuttlefish's visual language, then postures are the grammar. Arms can be raised in defiance to danger, or extended, curled or splayed to modify the overall message. And if all of that wasn't enough, cuttlefish can also communicate by means of an additional, secret channel: polarised light. It's a medium that can be used by relatively few animals since most species lack the ability to detect it, but cuttlefish can, and they use their iridophores to signal one another in hidden code.

As we begin to piece together the meaning encoded in cuttlefish conversation, we're gaining an ever-stronger impression of the intelligence that resides in this most unlikely of animals. This is, after all, the slug's second cousin, and yet it can converse, confound and conspire with aplomb. As I watch dark bands of pigment rolling across the body of the cuttlefish before me, I'm mesmerised. And, it turns out, such a sensation may be exactly how the cuttlefish wants me to feel. This kinetic display, sometimes referred to poetically as 'passing clouds', seems to disorient their predators and prey alike. It's yet another facet of this fascinating and complex animal.

Passing clouds or not, the cuttlefish is having little luck in finding a meal. Worrying that perhaps I might be cramping its predatory style, I slowly move away from the clearing to fossick for other creatures among the kelp. Where the forest runs up against the shelf of a rock wall, I dive down to explore beneath an overhang. There, in a lair beneath the ledge, a huge lobster holds court. It's a steely blue, spangled with white flecks, and brandishes its gnarled claws at me like a boxer taking guard. These claws are the crustacean equivalent of bulging biceps; they're an honest signal of power and strength to rivals and threats alike. Each pincer is about the size of my foot, and packed with muscle, but just as with their crab relatives, one is more massive than the other and it's this asymmetry that reflects their different tasks. The hefty crusher can deliver a force equivalent to 30 kilograms to pulverise its prey. The comparatively slimline cutter, meanwhile, is a more of a precision instrument, nimble enough to be able to snatch prey as elusive as a moving fish.

The lobster made no move as I admired him, though he may have growled. That we didn't know lobsters did this until recently is simply because we haven't been listening, but they growl when they're threatened and make a rattling noise as they feed. Both sounds are made by muscles that vibrate their carapace, turning the animal into a living maraca. At around 70 or 80 decibels, about on a par with a vacuum cleaner, the sound is certainly loud enough to hear. In the case of fish predators, at least, the growl is enough to make them think twice before tackling a lobster. Fish exposed to the lobster's audible threat not only retreat but are sufficiently disquieted to avoid the snarling decapod subsequently. The growling days of the lobster in front of me now may well be behind him, since at more than half a metre long he's most likely too large to worry about any fish. Nevertheless, as he contemplates me, it's entirely possible that he's attempting another means of communication, that of scent.

Lobsters live in a sensory world that is alien to us. Whereas we rely on vision, their eyesight is comparatively basic, allowing them to distinguish movement but little else besides. Instead, the lobster's milieu is a chemical one, and they collect information dissolved in, and delivered by, the currents via receptors positioned across their body. These receptors are particularly numerous on their long, curved antennae, and on a pair of twitching antennules that lie between the antennae, but they're also to be found dotted across the whole of the lobster's frame. When the lobster ventures out on its nocturnal forays to find its food, it samples the seabed with taste sensors on its feet as it walks. But it's in the interactions that occur among lobsters that things get really interesting. The male that poses in front of me broadcasts a collection of odours that act almost like a lobster CV. The chemicals describe his sex and his status, even his individual identity, and it's this messaging that governs the social habits of lobsters.

Given his size, the titan in front of me is likely the dominant individual in this neighbourhood, and if so, when he leaves his den to prowl among the kelp each evening, he may visit the abodes of other lobsters in the vicinity and invite himself in. This is all done without the need for a fight, because lobsters have the capacity to recognise one another. Consequently, when the dominant animal goes on his rounds, the others will make way without fuss. They remember who he is, and if they've had a previous stand-off, they know that he's not to be tangled with. The settled pecking order, based on mutual recognition of odours, brings peace to lobster society.

When communication is based on chemicals, the conversations take a rather different form than those we're used to. When two lobsters meet, the first thing that they do is essentially to urinate in one another's faces. They perform this rite by means of a pair of forward-pointing nephropores, akin to water pistols, that squirt the liquid ahead of themselves and into the visage of the recipient. Though unsavoury to us, the urine is packed with chemical information and serves to announce the identity of each lobster. This is never more dramatic than when females come to call on a male in his lair, having followed a trail of his tinkle through the kelp. The female's appearance results in an explosive mutual welter of piddle and, if the smells are to one another's liking, the female moves in to take up residence next to her beau. She stays there until she moults, because it's only at this time, when she's cast off her gnarled and careworn carapace to leave herself briefly pliant and vulnerable, that she can mate. Soon afterwards, with tens of thousands of tiny black eggs clamped to her tail, she leaves to take on the responsibilities of parenthood alone.

It's hard to say how old the veteran campaigner in front of me is, since lobsters are notoriously difficult to age. His immense size suggests that he could well be a grandfather of three score years and ten, and his survival to this vintage represents the most improbable victory against the odds. He may have hatched, after months of languishing on his mother's undercarriage in an egg, at a time when Elvis Presley and Marilyn Monroe were in their pomp. In contrast to his current heft, he began life beyond the egg as a defenceless speck resembling an apostrophe in size and shape. His first few weeks were spent drifting weightlessly in the water column, among a soup of other microorganisms. When, finally, he abandoned those days of marine flight and settled on the seabed, smaller than a chess pawn, he was already one of just a handful of survivors from the thousands that started out.

From here, he somehow overcame the threat posed by a battery of predators, perhaps by a judicious growl, or possibly by the lobster's trick of jetting backwards with a flick of his young tail. He survived even as his peers succumed, expanding with each moult until he essentially outgrew most dangers. It is sometimes said that lobsters live forever, but despite being untrue, this colossal male does have an inbuilt means of defying the ageing process. Senescence is caused in part by the decline of cells. One reason for this is that telomeres, the protective molecular caps on our chromosomes, sometimes likened to the aglets that tip shoelaces, progressively wear down with each successive cell division. Eventually, when the telomere has eroded to a mere stub, the strands of DNA become susceptible to damage, the cell becomes incapable of dividing, and it ultimately dies. Evolution's response to this is an enzyme, telomerase, that continually restores the telomeres and, in effect, stops ageing in its tracks. While most animals are equipped with sparing amounts of this enzyme, lobsters are awash with it, in effect meaning that they never truly grow old in the way that we understand it.[34] As a result, the lobster poised in front of me is simultaneously a geriatric and a spring chicken. Either way, my chance encounter with this magnificent creature in British waters is why I love diving; it doesn't matter where you travel, there's always something wonderful to see.

* * *

The summer holidays of my childhood always featured a trip to the coast, and I spent hours every day in my rockpool playground. When the incoming tide forced me into retreat, I'd leave the shores of North Landing and scamper up the steep slope to the headland, dodging impatiently among families labouring homeward from the beach, encumbered by their picnic baskets and damp towels. At the top, an ice cream parlour and the beguiling lights of the amusements beckoned, but I was in a hurry. Turning my back to the brightly painted signs and the afternoon sun, I followed a thin strip of a clay path as it wound first through embankments of brambles, then up and out onto the open meadows of the clifftops. Across those I went until I came to the end of the land, where beneath my feet, sheer chalk escarpments fell away to the sea. The endless assault of the waves has wrought a coast of nooks and coves, delineated by limestone fingers that seem to reach for an impossible horizon. The still summer air is effervescent with the spindrift shapes and irresistible clamour of seabirds on the wing, soaring and circling,

34 What often accounts for them in the end is not the decrepitude of age, but the trauma of moulting, which gets ever more difficult as the animal increases in size.

gliding and darting. They settle like snow on the cliff ledges and in rafts on the surface of a gently undulating sea. They fill this place with their irresistible, exuberant energy, and I sit, a small boy on the clifftop, to rejoice in this feathered festival of biology.

Flamborough Head is a tongue of chalk that extends 8 kilometres from the coast of North East England, and it was here my younger self was first inspired by the magic of biology. Each year, squadrons of seabirds come here to breed, nesting on cliffs that climb 100 metres above the breakers of the North Sea. The features of this jagged coast, with their ancient, evocative names – Staple Newk, Selwicks Bay, Old Dorr and Breil Nook – echo each year to the call of more than a quarter of a million seabirds, representing seven main species. Among them are three kinds of auk, the guillemot, the razorbill and the puffin; two gulls, the kittiwake and the herring gull; plus a petrel, the northern fulmar; and, most majestic of all, the gannet. A few more species, the European shag and its relation the great cormorant, are found here in small numbers, as well as visitors such as the skua. Such is the pull of this place to seabirds that it attracted a single, lost traveller, a black-browed albatross, blown far off from its home in the Southern Ocean, and seen soaring incongruously among its smaller brethren for a season in 2017.

The seabirds that animate the cliffs of Flamborough Head each spring and summer are among just 350 such species worldwide, representing a surprisingly small proportion of the 11,000 or so types of bird alive today. Aside from their mastery of the marine environment, these species have much in common. Like almost all seabirds worldwide, those that flock to Flamborough are colonial breeders, which provides strength in numbers and, crucially, access to information. Finding ephemeral and unpredictable congregations of their clandestine fish prey beneath the silvered expanse of ocean surface can be a challenging task, one that's complicated by the urgency of chick provisioning. Nesting in colonies means that, as they radiate from their base out into the expanse of sea and begin individually to scout, the mass of birds acts like a supersensor. Seabirds are keenly aware of the behaviour of their fellows, and they weigh the successes of each to tailor their own actions. They might tune in to the excitement generated by birds feeding on a shoal close by, or gauge the fullness of a returning neighbour's crop as it returns from further afield and note the direction from which it came. In doing so, the birds can take some of the guesswork out of foraging.

The skills required to forage effectively at sea improve with age and experience. It's perhaps for this reason that most seabirds wait until they're at least five years old to reproduce, yet even with this unbirdlike stalling, the fact that seabirds are long-lived – 50 years or more is by no means uncommon – allows them plenty of time for a family. Even here though, seabirds take things steadily, usually laying just a single egg and that with a partner whom they've typically raised chicks with many times before. The decades-long devotion that seabirds show to their chosen partners is all the more impressive when you consider that they spend most of the year apart. When the adults leave these cliffs in July or August, each member of a pair goes in its own direction, and won't encounter its mate again until they return the following spring. Who knows, perhaps they are suffused with happy feelings at the sight of a familiar beak after months of solitude, or perhaps their fidelity is simply an appreciation of cold, hard numbers: established pairs, on average, successfully raise more chicks.

There's another trait that seems to unite them, one that's obvious to even the most casual observer: their plumage. Seabirds have converged on a predominantly greyscale palette, a tuxedo-like formal wear of white belly and grey or black back. This is most often explained by reference to aggressive camouflage, making it harder for fish to detect the foraging seabirds as they fly overhead, alabaster chests blending with the pale luminance of the sky above. The darker back, meanwhile, provides some measure of protection against their own aerial predators, since this merges against the background of the sea below. As with any such traits, however, the plumage of seabirds is subject to multiple colliding selective pressures. The blackened feathers on the backs of the animals gain the protective strengthening provided by the pigment, melanin, which is one reason why the primaries at the tip of kittiwakes' wings are jet black, even though this seems to compromise their stealth colouration. Darker feathers even trap the sun's heat to the extent that they improve the amount of lift that birds can gain, and hence their flight efficiency.

The first seabirds appear in the fossil record around 100 million years ago, but it was only in the aftermath of the mass extinction event, 65 millions years ago, that they began to diversify into recognisably modern forms. Since that time, evolution has equipped seabirds with the means to surmount the challenges of foraging at sea. For species like auks, this involves a trade-off between the competing demands of flying and diving, whereas for others, particularly those that feed at the water's surface, it entails a mastery of transoceanic flight.

A puffin, its beak a pageant of vivid orange and yellow, rides the swells that roll in from the north. Its time spent floating on the the surface, however, is just a brief respite between dives and, sure enough, seconds later it ducks out of sight. Beneath the waves, the puffin rows, its short, powerful wings becoming oars, and its tail, a rudder. It swims in pursuit of small fish, such as sand eels and sprats, to deliver to its single offspring, referred to charmingly as a puffling. Whether the parent finds its puffling quite so charming as it exhausts itself to satisfy the youngster's appetite is anyone's guess, but each adult performs hundreds of dives each day in pursuit of this goal. While each dive is brief and shallow – puffins seldom stray more than

15 metres down and usually spend less than a minute submerged – cumulatively, each adult spends more than four hours every day underwater in pursuit of the fish it needs to supply its demanding chick.

The puffin dives in bouts, gradually accumulating a collection of small fish. By the time the bout is done, the puffin may have as many as 50 of its victims held adroitly in its beak. Regaining the air from the sea surface is a challenge with sodden wings and a payload of food, and the puffin seems to run across the waves before finally it becomes airborne. The journey back to its nest, a burrow in the soil on the high cliff, is fraught. Many other birds have designs on the fish it carries, so the diminutive puffin flies low over the waves in the hope of avoiding detection. Like its auk relatives, the razorbills and guillemots, the specialisations that make puffins such expert divers place constraints on their flying abilities. Short wings and a dense body with heavy bones that, to counteract excess buoyancy, lack the honeycomb centre of those of most birds, mean that auks expend at least a third more energy on flight than birds of a similar size.

Guillemots, the dark cocoa of their backs offset by a brilliant white chest, are the most numerous of all the seabirds at Flamborough Head.[35] At a kilogram in weight, guillemots are more than twice the size of their puffin cousins and are about as large as a diver that uses its wings to propel itself underwater can grow before it has to sacrifice flight. When they do fly, their tiny, dive-purposed wings have a comparatively huge amount of weight to bear. Indeed, their wing loading – the relationship between weight and wing size in flying animals – is the greatest of any bird. Their enormous relative, the great auk, was necessarily flightless, a fact that led to them being

35 Guillemots, *Uria aalge,* are found throughout the northern hemisphere and are known as murres in North America.

hunted to extinction in the nineteenth century. Standing almost a metre in height and resplendent in their black-and-white plumage, the appearance of great auks converged on that of another, albeit unrelated, bird far to the south. The name given to great auks in Welsh, translating as 'white head', possibly refers to the large white patches by their eyes, and it was this name that early European explorers gave to a similar looking bird that they encountered in the southern hemisphere: *pen gwyn*.

Auks are expert submariners, pursuing their prey through the water column. By contrast, fulmars and the two gull species, the kittiwake and the herring gull, claim food from the ocean's surface, seldom venturing much further than a dipped head can reach. Though it looks superficially like a gull, the fulmar is a petrel, a distinction that's made obvious as you approach by its possession of the petrel characteristic of enhanced nostrils, perched on top of its short beak. That said, approaching a fulmar is unwise for all manner of reasons, not least because their defence mechanism is to regurgitate the fish oil that it uses as an energy store, and squirting the foul-smelling liquid with impressive accuracy at trespassers.

What sets fulmars apart, for me at least, isn't their projectile vomiting but the extraordinary accomplishment that they display as aeronauts and navigators, travelling tens or even hundreds of kilometres beyond the hubbub of the colony to hunt at sea. They achieve this through an expert appreciation of the winds, which act as the fulmar's passport out into the far reaches of its feeding grounds. It's not simply a matter of reckoning the wind direction, the strength of the breeze is also critical. Calm days are expensive for fulmars, necessitating energy-sapping flapping flight. So while kittiwakes venture on comparatively short forays out to sea, the fulmars remain behind at the cliff, patiently waiting for the winds to gather and provide deliverance. By contrast, when it's blowing a hoolie and the wind is racing through at 40 kilometres an hour or more, fulmars simply glide, scarcely ever having to beat their wings.

Watching these wonderful birds coursing the sea is to experience an animal that has achieved near perfect mastery of its environment. Fulmars read the winds, as albatross do in the southern hemisphere, adapting their flight patterns according to their assessment of what lies ahead. Faced with an array of different passages back to the nest from a far-flung foraging trip, fulmars show an almost unerring ability to choose the optimum route for energy saving, a fact that can be observed by relating satellite tracks of tagged birds to maps of capricious wind patterns. Exactly how they do this isn't known, since it seems to demand near omniscience not only of current conditions but also those to come. They seem to read the sea with a practised eye, foretelling developing weather patterns with an uncanny expertise akin to that used by sailors of old to travel across the oceans.

* * *

From early spring, the Flamborough cliffs begin to accumulate seemingly impossible numbers of birds, and the coves echo to the cries of a feathered flood tide. Kittiwakes and fulmars cluster at every scant recess and, as the spring takes hold, the arrival of ever more birds triggers intense competition as claims are disputed and quarrels break out. Later in the season, as their downy youngster claims what little space is available on their ledge, the parent birds gain little more than a toehold. Though the spats over housing die down as the season progresses, new challenges emerge in the form of larger gulls. Adult kittiwakes shriek their protests as black-backed and herring gulls patrol the cliffs in search of unguarded chicks. Adding to the cacophony, returning foragers call to their partners as they approach the nest with freshly garnered supplies.

Lower on the cliff, guillemots pack together, side by side on long, transverse ledges, striping the pale cliff with a horizontal band of their dark plumage. Each pair is squeezed into an impossibly narrow domain, but what may offer the birds some consolation among the congestion is that occupation of any given ledge is, at least partly, governed by relatedness in these auks. The birds return to these clans every year to mingle with blood relatives. When your neighbour is family, it's possible that the jostling that goes on may be felt less keenly. It may also be an explanation for why adjacent guillemots are less aggression to their neighbours and sometimes even deign to preen one another.

Standing apart from the chaos on the cliff faces, the magisterial gannets stand aloof, making their nests on wider ledges, and on the flat tops of stacks and islets a little way off the land. Gannets are handsome birds, their plumage a uniform ivory aside from the blackened tips of their primaries and the soft gold shading around their necks and head. They're also imposingly large creatures with the wingspan of a policeman and the mien of a murderer. Before I'm accused of anthropomorphising and besmirching the reputation of what are undeniably magnificent birds, let me add quickly that my thoughts are guided by their behaviour at the colony. With real estate in demand, the nests are packed close, but not too close. The guiding principle for home builders is to construct their nests just beyond reach of the sabre-like bills of sitting neighbours.

The aggression isn't restricted to rivals in adjacent nests either. Even within established pairs, the males are shockingly aggressive to their mates, jabbing at them and biting them. The females' response to this is typically to turn their head away, breaking eye contact in an attempt to defuse the situation. The adults are also more than ready to kill stray chicks, even those comparatively defenceless youngsters still coated in down. The notebooks of the great ornithologist Bryan Nelson are full of these instances, as this typical example recounts: 'One such chick, already almost dead from previous attacks, crept up to a nest, containing a chick of similar age and lay there. The adult on duty attacked it violently and killed it.' The hostility towards younger birds continues even after they've fledged and left the gladiatorial environment of the colony. Again, Nelson describes juveniles in flight being attacked and driven from the air, and instances of youngsters being savaged by up to 20 adults at a time as they sat, inoffensively, on the sea. Yet despite this, gannets do sometimes show a more tender side. There have been multiple instances in which gannets have adopted and fed guillemot chicks that have fallen among their numbers from ledges above their colony at Flamborough.

Oddly enough for a bird that displays such sociopathic tendencies, as adults depart on a foraging mission, they seem to leave the battleground attitude behind. Gannets travel in skeins, up to six birds flying in a line as they commute to feed. Further on, out to sea, the transformation of the gannets' behaviour from belligerence to artistry is complete. It's exhilarating to watch as the birds soar and climb almost 50 metres above the waves and then, with a half-turn, begin a steepling dive towards the ocean. In preparation for impact, the gannet folds its wings tight behind it and hits the water travelling at a speed of up to 90 kilometres per hour. And 'hits' is the right word; graceful as the entry is, the *thock!* as they split the sea is audible. A person smacking into the ocean at a similar speed, even feet first and perfectly aligned would be lucky to get away with only a fractured spine. That gannets don't suffer in the same way is thanks to the exquisite geometry of their dart-like bodies, and a system of air sacs in the head and chest that are thought to cushion the impact like the safety airbags of a car. Nevertheless, the deceleration is massive, the birds experiencing

G-forces of up to 23 – more than double that which would render a military pilot unconscious – and they repeat this dozens of times on every foraging trip.

* * *

As summer peaks and the year begins the descent to autumn, tens of thousands of chicks are compelled to face the uncertainties of independence. For anywhere between five weeks and three months, their every need has been catered for, but now a seabird Rubicon must be crossed. For the fledglings, this is a perilous moment, yet it's one that they're apparently keen to embrace. In the days immediately before they finally depart, kittiwake youngsters jump excitedly on their ledges, with an accompanying fluttering of inexpert wings as they do so. Their youthful clumsiness means that once they depart the ledge, they'll be unlikely to regain it, so their departure is definitive. The heady enthusiasm with which they embrace this fate sometimes ends in disaster, and it's a sad but common sight to see the sea surface pockmarked with the corpses of young birds, wings spread and animated by gentle waves in a terrible pastiche of the flight that is now beyond them. In a good year, perhaps two thirds of these birds will be alive a month after leaving the nest; at other times it may be as little as a tenth.

Some of the auks, like the razorbill, have a very different approach to fledging. The chicks take a death-defying leap off the cliff, part-formed wings whirring ineffectually, at the tender age of just two or three weeks. Having crash-landed into the sea, it's the job of the male parent to chaperone and feed the youngster for the next few weeks at sea. The result is the charming spectacle in June and July of an adult razorbill scooting around on the sea surface, attended by his little son or daughter. It reflects the lifestyle of these auks that they learn to dive long before they master flight, but it's also a response to the threats posed by predatory birds to the chicks while they remain at the cliffs.

For almost six months of the year, the cliffs at Flamborough are coated with an irresistible, electrifying throng of life. Gradually, as summer progresses, the carnival leaves town as the birds disperse out to sea. The residents of this seabird city scatter as far as Greenland and Newfoundland to the west, or Norway and the Baltic to the east. Gannets might travel down to Spain and beyond to the coast of Africa. Many trace a lonely path through the storms of the mid-Atlantic, leaving the cliffs bare, and the air devoid of their thrilling cries. There's a melancholy aspect to a coast unadorned with birds; without the charge that they provide, this is just stark geology. Still, it's all part of the natural cadence of life, and I know that the auks, the gulls, the fulmars and the gannets will return to reclaim Flamborough before long. But there are many parts of the world where this certitude is crumbling.

The sandstone massifs of northern Norway, carved by storm and glacier alike, tower over the dark Barents Sea. Pulsing riches of marine life in these high latitudes once brought seabirds in their millions to nest here. Photographs taken 40 and 50 years ago show a teeming mix of kittiwakes, guillemots, razorbills, puffins and gannets. Today, most of them are gone, the silent cliffs locked in permanent seabird winter. In less than one human lifetime, the kittiwake population in Norway has plummeted by 90%, bringing this once hyperabundant bird to the edge of local extinction. It's an alarming, and recurring, pattern. Make no mistake, this isn't an isolated case. Globally, seabird populations have collapsed by two thirds since 1950, a signal that something is fundamentally, terrifyingly wrong with nature.

There is no single answer to the question of why our iconic seabirds are disappearing, rather it is, as so frequently, a dreadful fusion of factors. Pollution is one horseman of the seabird apocalypse, and particularly the magnification of contaminants that occurs at each ascending level of the marine food chain. Guano-enriched soil samples taken beneath seabird colonies from the high Arctic to Antarctic wildernesses yields contaminant readings that would be more appropriate to a site of heavy industry. The guano itself, once a prized commodity harvested for agricultural land, is now in some cases so laden with mercury, pesticides and forever chemicals that it is, in effect, toxic waste. Plastics, too, shadow the seabirds, finding their way into the animals via their prey, or even by direct ingestion. In the latter case, what draws the seabirds is the tragic coincidence that some marine plastics emit the very same compound – dimethylsulphide, or DMS – that is released by phytoplankton and which birds like shearwaters, petrels and albatross use as an olfactory beacon to locate their prey at sea. There are few, if any, individual birds today that don't carry a payload of microplastics in their gut, their tissues, even their feathers. Within the body, these plastics are neither inert nor benign. Even those seabirds that survive the choking accumulation of plastics are severely compromised because of the poisons that leach from the particles into their bodies. So, the contaminants and the plastics go hand in terrible hand, damaging the seabirds' immune systems and impacting their fertility.

Climate change, too, is strongly implicated in the seabirds' fall. Partly this is to do with the role of warming in altering patterns in the distribution of the birds' prey species, and converting previously reliable foraging grounds into new and unpredictable states. While chronic trends can escape our notice, acute events such as a marine heatwave known as the Blob that hit the north Pacific in 2013 bring matters into focus. Hundreds of thousands of guillemots washed up dead on the coast of North America, having starved to death in the torrid, fish-stripped ocean. As a terrible adjunct to an overheating climate, storms become both more frequent and more severe. And though strong winds are a boon to many seabirds, gales that whip the sea surface into a ferment make it impossible for many birds to feed. Worse still, if the storms strike at the most vulnerable period of the seabird calendar, in the nesting season, entire cliff faces can be scoured of their nests. Of course, seabirds have always coexisted with storms, violent weather decimating populations every few years. The critical part of that relationship was never so much the losses, but the interval for recovery. As a seafaring analogy, a ship can ride huge waves with little difficulty, so long as they're far enough apart. It's only when a tempest pushes the peaks into ever closer formation that the ship founders. Now as high intensity storms strike at seabirds with escalating frequency, each nascent recovery of their populations runs headlong into the next decimation.

The burden imposed on seabirds by pollution and climate change is augmented by fishing. The focus of this industry in the North Sea was once species like cod, haddock, herring and mackerel that have long been a familiar part of the European diet. If we didn't eat it, we didn't catch it. And though people have dabbled in the culinary potential of sand eels, they're small, bony and unrewarding to the human palate. Our historical disdain of sand eels is not something that's shared by the kittiwakes, fulmars and auks, who are reliant on these silvery ribbons. The supply of sand eels in the North Sea once seemed limitless, enough to feed the huge colonies of seabirds that wreath the coasts thereabouts, as well as supporting countless fish predators of the kind our fishing vessels have long majored in. Nowadays, the picture is far less rosy.

In the aquarium labs at Cambridge University, I'm being shown around by my friend and colleague James Herbert-Read. Arrayed around us are tanks full of sand eels; it's the first time I've seen these slender fish in the flesh, but none of them are hanging around to be admired. A sand eel's life is full of threat, and evolution has honed them to quicksilver elusiveness. James is protective of his charges and doesn't want to risk injuring one by catching it for closer inspection, but even if he did, it seems like it'd be a challenge to catch one. I'm filled with new-found respect for birds such as puffins because I simply can't imagine how they'd ever secure a steady supply of these lightning-fast creatures armed only with a beak. Nevertheless, the birds can – and do – snag them in their millions, and numberless generations of chicks have been raised on a diet of calorie-rich sand eels.

We are, perhaps more than anything else, a resourceful species, and if it seemed like opportunities were going to waste allowing a huge stock of sand eels to thrive unmolested in the North Sea, it was only going to be a matter of time before someone seized the moment. Beginning in the late 1950s, we started to target sand eels, positioning ourselves as competitors to the seabirds. As the years wore on, the catches climbed, and we began to take an ever-greater share of the seabirds' resources. We still weren't eating sand eels; instead, we were grinding them up into fishmeal to feed to livestock and, incredibly, unforgivably, burning their oily bodies in power stations to generate electricity.

It's a truism that few animals have yet entered into competition with mankind and emerged unscathed and seabirds are no exception. By 1996, we were taking more than five billion sand eels a year out of the north and central zones of the North Sea alone. In just 12 years, from 1990 to 2002, the populations of kittiwakes nesting in proximity to one sand eel fishery in Scotland fell by more than 50%. Meanwhile, the effects of our depredations have been seen at their starkest among guillemot colonies. Under normal circumstances during the nesting season, the parents divide the responsibilities of protecting the chick and foraging, so that as one goes out, the other stays back to care for junior. When the inevitable happened and the sand eel populations in the North Sea crashed, the adult birds were forced to change tactics. With food scarce, they were compelled to forage simultaneously in a valiant effort to supply their hungry chicks. There's a cost to this, however; unattended chicks are at risk from the weather, from gulls and, in the extreme conditions caused by the fish famine, their own kind. A seabird colony can be a brutal place, and this was never more the case than among the starving guillemots, where two thirds of chicks died, the majority killed by other guillemots. Emaciated adults pecked the lone chicks to death or pushed them off the cliff ledges. It's an extreme and uncharacteristic response, and one that's driven by our catastrophic mismanagement of the sea's resources.

* * *

Mankind has always fished. Hooks fashioned from bone have been found that date back to the Palaeolithic Age. Evidence also exists that fish traps and nets woven from plant fibres have been used for thousands of years. As we honed our skills in boat building and navigation, we gained the ability to pursue fish beyond the shoreline and to exploit their ocean fastnesses. Longer voyages and larger catches brought with them the problem of how to avoid spoilage, and this we initially solved by salting, drying and smoking fillets, and later by canning and freezing fish. The development of powered ships further extended the reach of fishing fleets. It also allowed larger vessels to put to sea and drove winches that could haul unprecedentedly massive nets. We refined the design of those nets, using synthetic materials to replace cordage, making them lighter, stronger and more durable. To streamline the industry and to improve its efficiency, we converted fishing boats into floating factories, complete with huge, refrigerated holds. And to make sure the fish had absolutely no hiding place, we incorporated sonar and satellite tracking, and deployed drones and fish aggregation devices. In a few short generations, we transitioned fishing from the artisanal to its modern-day form, a grim pinnacle of industrial-strength ecological warfare.

In some ways, the evolution of the fishing industry represents a triumph of human ingenuity. The problem is that our ability to exploit fisheries has grown much faster than our ability, even our desire, to manage them, and we've stripped swathes of the ocean bare of fish. Though things are worse in the age of industrial fishing, the concerns are not new. In 1376, a petition was delivered to King Edward III, calling on him to ban trawling due to the 'great damage of the commons and the destruction of the fisheries'. At the beginning of the nineteenth century, so effective were the new, sail-powered Brixham trawlers that they were having to range ever further afield to sustain their catches. From the second half of the twentieth century onwards, however, the pace accelerated. Fleets of factory ships embarked on a sustained and intensive onslaught on fish populations around the world. One by one, fish stocks faltered and collapsed. South African sardines, Peruvian anchovetas, North Sea herring and bluefin tuna in the Mediterranean, all of them once considered inexhaustible, were brought, one by one, to the cusp of disaster. But it was the spectacular crash of the Atlantic cod of Newfoundland in the early 1990s that really made the world sit up and take notice.

The Grand Banks, a vast area of shallow undersea plateaus on the continental shelf of Canada, east of the port of St John, was one of the richest fishing grounds in the world. Here, the cold Labrador Current collides with the warmer waters of the Gulf Stream, the churning waters yielding a bounty of nutrients that catalyse a near unparalleled biological abundance. This was a place of legendary catches amid the sea fogs, one that had been fished for 500 years or more until in 1992, the cod ran dry. The failure of the Grand Banks seemed to contradict the laws of nature; it was as though trees had failed to flush with leaves in spring. At a stroke, 37,000 people lost their livelihoods, leaving the once hectic fishing communities of Newfoundland subdued, and huddled along the shores of a ruined sea.

With the fish gone, the blaming began. Politicians had vacillated, their actions shaped by the tides of public opinion and the timings of the election cycle. Rather than act on scientific advice to slash the allowable takes earlier, they'd closed their eyes and wished the looming crisis away because the alternative was too awful to contemplate. The fisheries scientists who had advised those politicians weren't blameless either. Assessing stock levels of a fishery is immensely challenging; you can't count the fish directly, so you have to rely on assumptions to fill in the gaps. But what if those assumptions are wrong? In the case of the Grand Banks, they were. Recruitment, the number of young fish that enter the population each year, is a critical part of fisheries models. Since a single large cod can produce getting on for ten million eggs, the scientists presumed that so long as at least some adults remained, their fecundity meant that recruitment would hold. It didn't. And for all the anger that Canadian fishers directed towards politicians and scientific advisers, at seals and at the fleets of foreign boats that had come to raid the Banks, they too were culpable. From the mid-1950s, the boats began to catch two, three or even four times as many fish as they had each year for the previous century, culminating in a take approaching a million tonnes in 1968. The cod stocks on the Grand Banks belonged to all of them, yet none of them. If a skipper left fish behind, those would only go to feed the next skipper. Lofty ideals of stock stewardship fell by the wayside in the scramble until the cod that supported them were gone.

When the Canadian government declared the moratorium in 1992, the hope was that it would represent a short-term hiatus and that the cod would recover to reclaim the Banks. As years passed with no sign that the cod were returning, the awful prospect emerged that they never would. When an ecosystem collapses, there are no guarantees that it will rebalance back to its previous state. The Banks became a hotspot for crabs, shrimp and forage fish such as capelin, and though these are valuable prey for adult cod, they're also voracious consumers of cod eggs and larvae. Cod populations have been driven down to less than 1% of what they had once been, and with an army of predators waiting to welcome new recruits, there seemed no easy way back.

Yet in the early years of this century, signs began to appear that cod were beginning to return to the cold, murky waters of the Banks, and by 2015 large spawning groups were seen for the first time in a generation. Promising as the recovery appears, however, not everything is as it was. These twenty-first-century cod are smaller, and therefore less fecund than their predecessors, and the process of recovery is a delicate one. Nevertheless, lobbyists have long agitated for a reopening of the fishery and, ultimately, the pressure told. First came an easing of restrictions in the mid-2010s. This precipitated a stark fall in the cod population after just two years of modest hauls. Nevertheless, in 2024, the Canadian government withdrew the moratorium, reinstating the commercial cod fishery off the coast of Newfoundland and Labrador. While it's always possible that things are different now, the rush to reopen the grounds seems to be based on wishful thinking.

Our response to ecological warnings has seldom been coherent, and never timely. When prohibitions and moratoriums are enacted, they tend to happen after stocks have all but vanished. At the same time, a tangle of rules and quotas are applied to other exploited stocks that, however well intentioned, can have disastrous effects. One such effect in European waters has been to make it ever harder for smaller boats to cover their costs. As a result, fishing rights are transferred from local, small-scale fishers who have a cultural affinity with their fishing grounds and a vested interest in protecting stocks, to a cabal of multinationals whose responsibilities are often far less attuned to sustainability. Bad as this is, it gets worse. The regulations require fishers to dump catches of non-target species for which they have no quota. Tonnes of dead fish and bycatch are poured back into the ocean each year to satisfy the diktats of bureaucracy.

One solution to the challenge of overfishing is to set aside conservation zones and it's an approach that delivers. Across the globe, established marine protected areas (MPAs) have as much as five times the biomass within them than in equivalent unprotected areas. They support greater diversity, higher populations and the animals within them are larger. Given time, depleted barrens regenerate into healthy, complex ecosystems. Crucially, MPAs yield benefits beyond their borders because of the phenomenon of spillover, where the greater recruitment and growth of stocks within the reserve provides a measurable boost to the waters beyond. It's a rare win-win scenario in which the marine environment within the protected zone improves, while fishers outside the boundaries profit from increased takes.

Given the substantial benefits that MPAs provide, you could be forgiven for thinking that governments worldwide would be falling over themselves to establish them. They're not. The sea has long been the poor relation of the land when it comes to conservation. Proportionally, less than half as much of the ocean is designated as protected compared to the land. And though there has been welcome growth in this area since the early 2000s, it builds from a very low base. Even now, less than 10% of the sea falls within an MPA.

Another problem is the P in MPA. Anyone who has even a passing acquaintance with sanity would realise that for the aforementioned benefits to manifest, a protected area has to actually be protected. In fact, far less than half of global MPAs proscribe fishing within their boundaries. Astonishingly, virtually all of the UK's MPAs allow the most destructive fishing practices to take place within them. Bottom trawling in particular is devastating for the environment. A deadly conglomeration of tonnes of steel doors, heavy trawl gear and steel cables gouges its way along the seabed, catching or killing everything in its wake and turning the seabed into a graveyard. The effects are long-lasting, with recovery from a single trawl potentially extending into the decadal scale. We'd never allow Venice to be used for heavy artillery practice, yet we let the equivalent happen at sea because we don't see its effects. Besides annihilating life, and compromising its ability to recover, trawling has another effect: it liberates greenhouse gases from storage in seabed sediments. According to a paper released in 2019, trawling releases more carbon dioxide each year than the global aviation industry. The value of the fish and shellfish landed through trawling cannot even begin to approach the price that the environment has to pay for it. And when it's allowed to occur within protected areas, it makes a mockery of conservation.

Worldwide, something like two thirds of marine protected areas are open to fishing in one form or another. The remainder, sometimes referred to as no-take zones, are nominally closed to fishing. Often, however, these lack the levels of monitoring and enforcement needed to uphold such restrictions. What look like positive attempts at conserving our precious marine heritage are often 'paper parks', protected in name only. The tragic irony is that by failing to conserve the oceans, we're not only missing a golden opportunity to move fisheries onto a more sustainable footing, but we're also catching less.

Indeed, the world's fishing fleets are running to stand still. As populations of target species decline, ever more energy has to be deployed to root them out. The key statistic is known as catch per unit effort, which describes the amount of work that it takes to land a given quantity of fish. According to Professor Callum Roberts of York University, we now have to work 25 times harder to catch the same quantity of fish as in the days before industrial fishing. With each passing year, it gets harder still. The number of motorised fishing vessels has grown by a third since 1990, and they're bigger and more powerful than ever before. They deploy ever larger nets and use more sophisticated electronic equipment to target the fish. Despite this, the catches over the last 30 years have been static at around 90 million tonnes per year. It's hard to get a feel for what a number like that actually means, so allow me to put it into context. The amount of life that we strip from the oceans each year is greater than the weight of all the people in Europe and North America combined.

It's both a stupendous quantity and a massive underestimate. Bycatch, the sanitised name for the collateral deaths attributable to the fishing industry when they kill animals that they're not targeting, is huge. Added to this, catches are routinely underreported for all manner of shady reasons. All of this makes it hard to pinpoint exactly how much we pilfer from the oceans each year. The Food and Agriculture Organization of the United Nations (FAO) produces a summary report each year on the state of the world's fisheries. It's an impressively voluminous, well-presented and freely available report. It's also inexplicably blithe and pollyannaish. In the most recent edition, for instance, among its hundreds of pages, bycatch is mentioned just five times and always in passing, with no attempt at quantification. The World Wildlife Fund took a bolder approach, running some numbers and concluding that the total global catch was closer to 120 million tonnes per year.[36] We're taking far too great a share, and as we do so, we condemn other species to starvation.[37]

The Hawaiian petrel is an elegant seabird, attired in the sober monochrome plumage of its kind, with long, tapering wings that carry it far from its home out into the Pacific to forage for fish and squid. Like many seabirds, it has suffered through its interactions with us; its nesting sites have been disturbed, and its eggs and chicks have been targeted by the animals that travel in our entourage. This isn't all, however. The Hawaiian petrel, in common with virtually all of its kind, and indeed all marine predators, faces another threat, one that is far more insidious. Using bones and feathers collected from the petrel's ancestral breeding colonies and dating them through radiometry, Anne Wiley of Michigan State University and her co-workers were able to piece together the last 3,000 years of the birds' history. Moreover, by comparing these chemical timestamps to the stable isotope signatures present in those bones and feathers, Wiley built up a picture of what the birds had been eating throughout their past. For three millennia, the petrels maintained a consistent diet for generation after generation but then, in recent decades, their regimen changed dramatically. They began to eat smaller, less valuable prey, and they travelled ever further afield in an attempt to find food. Why? The researchers considered a wide range of

36 Even so, this is likely a serious underestimate in light of the growing problem of illegal, unreported and unregulated (IUU) fishing.

37 The total production of fish in the world's oceans each year has been put at 240 million tonnes. That estimate was made over half a century ago and may now be optimistic in light of how the oceans have changed since. If it is accurate, what it means is that we're taking half of everything that's available for ourselves.

possibilities, ruling them out sequentially until there was only one logical answer left: industrial fishing.

Petrels rely heavily for their living on surface-dwelling species, such as flying fish, and on the actions of large predators such as tuna, which force bait balls of forage fish up from the deep to pick them off. By specifically targeting large predators, and by taking as many of these as we could catch, we set in train effects that cascaded through the food web and changed oceanic ecosystems dramatically. There are few tuna now to drive the bait balls. The birds have adapted to an extent, but their prey now is less rewarding, making it ever more challenging for them to feed and to rear chicks successfully. Their population is now declining steadily, a fate that befalls any species confronted by the most dominant competitor in the history of life on this planet.

We're wiping out marine life at an unprecedented pace, throwing more and more effort into the process to catch the same amount, which ought to tell us something, and what do we do? Two things. We build ever larger ships to break – literally – new ground. And we chase what fish are left to the corners of the Earth. Supertrawlers that are a third longer again than a football field now pillage the oceans. One, the FV *Margiris*, described memorably on Twitter/X as the Death Star of the oceans, tows a net so large, according to the *Sydney Morning Herald*, that you could fly a Boeing 747 into its open mouth. Pictures released by the environmental group Sea Shepherd, who shadowed the *Margiris* as it was busily emptying the Bay of Biscay, show a slick of dead fish, estimated at 100,000 individuals, floating at the surface. That's all in a day's work. If anyone imagines this to be sustainable, they simply haven't understood.

The proponents of supertrawlers like this argue that they're operating within quota limits with gadgets that help to avoid bycatch. The quotas, however, are based on a degree of guesswork; you can't, after all, count the fish in the sea. The process of determining quotas relies on fisheries models that, according to Daniel Pauly, one of the most well-respected scientists in the field, have an inbuilt flaw that results in them serially and severely overestimating the remaining stocks of overfished species. The FAO's most recent report claims that a little over a third of the world's fish stocks are currently overfished. A re-evaluation of that same data using an approach informed by past trends, and which corrects the intrinsic bias of existing models, suggested that fully three quarters of the world's fish stocks are being plundered towards extinction.

When stocks crash in one area, the world's fishing fleet moves on, ranging ever further to secure their catches. Fishing fleets from the EU and China move down the coast of Africa, as they do so emptying the sea of fish and destroying the livelihoods of local fishers. They buy permits to fish within the economic zones of poorer countries, and they pay a pittance in relation to what they take. Further off the coast, beyond the 200-nautical mile (370-kilometre) limits of those national zones, anything goes. Here, in what's known as the high seas, fishing cartels operate with scant regard for the law, fishing intensively around the clock for months on end. This is the new reality of distant-water fisheries, a blight known as illegal, unreported and unregulated (IUU) fishing. According to estimates made by the European Union and by the World Bank, up to 28 million tonnes of fish are caught in this way each year.[38] It's a growing problem, and one that seems unstoppable unless decisive action is taken. To date, governmental responses to it have been completely ineffectual, leaving these pirate fishing fleets free to scour the oceans with impunity, systematically stripping an asset that belongs, in common, to all of us.

---

38 This is in excess of the 90 million tonnes figure that the FAO reports as the global catch.

An armada of mostly Chinese fishing vessels besieges the rich waters of the Galapagos marine reserve for several months of the year. Though they can't be seen from land, rafts of discarded plastic from the flotilla wash ashore on the islands every day as a kind of extra 'fuck you' to the environment. Out at sea, the vessels hover at the boundaries of the park, their locations relayed by satellite from tracking beacons, known as automatic identification systems. It is, however, possible to disable these systems, allowing the boats to 'go dark' for several hours at a time, and to go who knows where. Into the marine park itself seems the likeliest answer, but it's hard to prove – Ecuador lacks the resources that would be needed to patrol effectively a marine park the size of England.

The operation that's being run is relentlessly efficient. As each vessel fills its hold, it transfers the catch to a giant mothership, then heads straight back out to fish some more. On board the mothership, the illegal catch is mixed with the legal take, then processed and frozen. At this point, it would be all but impossible to trace the provenance of the fish, even in the rare event that inspectors are permitted to board. Occasionally a vessel is caught red-handed, such as when the *Fu Yuan Yu Leng 999* was intercepted in 2017 inside the reserve with 6,000 dead sharks, including endangered hammerheads and whale sharks, in its hold. This, and the ongoing controversy caused by the fleet's presence, caused the Chinese embassy to issue a statement that China was a 'responsible fishing nation' with a 'zero-tolerance' attitude towards illegal fishing.[39] Nevertheless, it's only because the motherships are able to collude with 'friendly' ports to unload several months' worth of catches that the practice can happen.

39 Collyns, D. (2020) '"They Just Pull Up Everything!" Chinese Fleet Raises Fears for Galápagos Sea Life', *Guardian*, 6 August, https://www.theguardian.com/environment/2020/aug/06/chinese-fleet-fishing-galapagos-islands-environment.

Human rights abuses go hand in hand with IUU fishing, something that the pioneering journalist, Ian Urbina, founder of the Outlaw Ocean Project, has done much to expose. Chinese fleets are often crewed with forced labour, many from the Uyghur minority, who suffer violence and intimidation, and are often prevented from contacting relatives during voyages that may last years. In light of Urbina's revelations, US seafood importers cut ties with Chinese companies suspected of involvement in human rights abuses, and US lawmakers pushed for legislation to block Chinese seafood from the US market. The EU has taken a different approach, engaging with the Chinese government and supporting its recent change in stance on the issue of illegal fishing. In any case, the reasons that China operates such a huge distant-water fleet – by one estimate, 20 times the combined size of the European and US fleets – are not purely to do with fishing. In running what's been described as the largest maritime operation in history, China's goal is to project its influence as a show of strength. Consequently, there is currently a gap between Chinese policy and practice, so it remains to be seen whether the pronouncements from Beijing result in significant changes to the actions of its fleet.

Meanwhile, other illegal operatives have used the focus on China as a distraction and an opportunity. Vessels carrying the flags of Russia and Mexico have been spotted engaging in similar kinds of illegal fishing around the Galapagos. Native Ecuadorian fishers, too, have taken the chance to exploit tuna and sharks in these waters. Even as we fish the oceans to death, right now there's money to be made, and unscrupulous people ready to make it. What's needed, urgently, is coordinated international action to bring illegal fishing to a stop, and for that to work, it has to involve China. Closer monitoring of ships at sea, involving greater scrutiny of deliberate tracking beacon disablement, along with a much stronger presence of enforcement vessels would have an impact, as would greater transparency in the seafood supply chain. At the moment, we pay lip service to all of the above; the ocean needs us to do better.

It would be reasonable to wonder, with wild fish stocks in such a parlous state, how it's possible for the fishing industry to make a profit. The answer is that, often, it doesn't. Instead, governments around the world subsidise the industry to the tune of $35 billion per year, in effect paying fishing fleets to destroy the oceans. About half of this global total is paid by China alone, and without this support, its fleet couldn't afford to act as they do. Across the world, the overwhelming majority of the subsidies go to support industrial fishing, rather than promoting the sustainable actions of local fishers. And most of this goes to subsidising fuel, meaning that the fleets can range over enormous distances at sea to damage far-flung stocks. The madness encapsulated within the subsidies is symptomatic of our attitude to the oceans, and keeping the pressure up on exhausted fish stocks in this way is perhaps the single most damaging thing we could do. For two decades, discussions at the World Trade Organization have been taking place to end this bizarre state of affairs. Finally, in 2022, a new resolution, the Agreement on Fisheries Subsidies, was adopted. The agreement prohibits financial support for the fishing of overexploited stocks, for IUU fishing and for targeting unmanaged stocks in the high seas. A little over three years after its adoption, the agreement was ratified, entering into force in September 2025. Challenges remain in respect of ensuring the compliance of sovereign nations, and further negotiations are under way to strengthen its terms, but the agreement offers hope that we might finally be recognising the error of our ways.

If we had set out to design the least efficient and most costly means of fishing, we could hardly have come up with anything more exceptional than that which has been our approach in recent decades. We have sabotaged the ocean, and fish stocks in particular, in spectacular fashion, and we've squandered billions doing it. In fishery after fishery, we've followed the same destructive playbook. First, we drive a stock to the brink and then throw huge amounts of money and effort at the task of harrying the few fish that remain. In our demented scramble for those remnants, we compromise the ability of that stock to bounce back by capturing the breeding adults while destroying their nurseries and feeding grounds. We, the public, pay for this three times over. First through our taxes to subsidise the vandalism, second through higher prices at the supermarket, and third by losing the ecosystem services that more sustainable practices would promote. Our present approach to fisheries is irrational; to continue this into the future would be suicidal. The good news, however, is that the future of the ocean's fish stocks is still in our hands. With a few straightforward changes, we could both safeguard fisheries for generations to come and save billions into the bargain.

The ocean's capacity to recover and regenerate is extraordinary. We see clear evidence of this from marine protected areas, and on a much larger scale from what's been called the Great Fishing Experiment. Twice during the twentieth century, fishing in the North Atlantic shuddered to a halt as we waged war on one another rather than on fish. The effect on fish populations was dramatic. Fishers returning to their grounds at the end of World War I found themselves landing bonanza catches, consisting of fish of enormous size. After World War II, the same happened; fishers were routinely landing twice or even three times as much fish on each voyage as they had in the late 1930s.

We know that fish populations can rebound, and we urgently need to give them the space and the time to do so. Quotas need to be cut, perhaps sharply in the short term. Subsidies must be redirected away from industrial conglomerates to support small-scale fishers who operate in their own waters. These people, not faceless multinational companies, are the future of fishing and it's vital that they have a stake in their local fishing grounds to support effective, long-term stewardship. Bottom trawling needs to end and factory ships need to be decommissioned. Inevitably, some parts of the industry would be hard hit by the changes, and they should be supported to transition towards alternative livelihoods, or at the very least, more sustainable activities.

In late 2022, 190 countries adopted the United Nations' Kunming-Montreal Global Biodiversity Framework. Target 3 of this, the so-called '30x30' goal, sets out the need to protect 30% of the world's lands, freshwaters and oceans by 2030. It's essential that we not only meet the 30% quota for the oceans, but also that the marine protected areas, once they are established, are rigorously enforced. The siting of those MPAs needs to be informed by science to conserve the most critical parts of the ocean, the breeding grounds and nurseries of fish stocks, and they must be established on the basis of connectivity, so that the network of protected areas provides the greatest benefit for conservation.

Finally, the global fishing industry cannot be allowed to continue to dictate terms. They have been allowed to get away with too much, for too long, and their track record is abysmal. Currently, the industry exerts a disproportionate influence on governments. I recall reading an interview conducted by the BBC's Richard Black with a Spanish fisheries official, who said with remarkable candour: 'For sure we are friends of fish. But still more, we are the friends of fishermen.' This is an attitude that seems to sum up everything that we've been doing wrong. The fishing industry puts profit ahead of conservation, and the politicians put votes ahead of reason. It's high time that this changed because, at present, both the industry and its lackey politicians are doing a grave disservice to the rest of us, and to future generations.

This does not mean the end of the fishing industry. As fish populations bounce back from overfishing to reclaim the seas, yields will increase far above what's currently possible. In the medium to long term, as the Great Fishing Experiment taught us, it means obtaining larger catches at a lower cost, and in a way that's entirely sustainable. We know what to do, and how to do it, now it's time to show that we're capable of safeguarding the oceans.

5.
The Cold Heart of the Sea

The grandeur of marine life in polar regions

I'M ABOARD A VESSEL IN THE NORTHERN ARM OF Isfjorden, Svalbard, hoping to meet one of the Arctic's most enigmatic species, the beluga whale. After an hour of fruitless searching, at long last we hear them on the ship's hydrophones, a rich, other-worldly tapestry of squeals, trills and whistles. The growing volume announces that they're approaching until eventually the pod surfaces dead ahead of us. High and plaintive calls carry to me across the gentle ripples of Isfjorden. I hold my breath so as not to miss this song of the Arctic as it rises to claim the chill air of a Svalbard summer. There are around 50 of them, each around 4 or 5 metres in length, splashing and cavorting. For the next ten minutes, the belugas linger, their brilliant white bodies lighting up the sullen waters of the fjord. The swirling chorus grows ever louder until at some unseen signal they dive beneath the surface, leaving only the faint stirring of the wind in our ears.

Many whales visit the high northern latitudes, but these belugas, like their close relative the narwhal and a baleen whale, the bowhead, are the only ones who live in Earth's freezer all year round. Belugas are among the most vocal of cetaceans, their perpetual squeaking chatter giving them the alternative name of 'sea canary'. They're also excellent mimics, as evidenced by numerous recordings of captive animals reproducing human speech. Sadly, this facility, alongside their curious and friendly nature, and the semblance of a smile on their faces, only encourages the ongoing, dismal trade in live Russian belugas for transport to public aquaria in East Asia.[40]

The loquaciousness of belugas likely emerges from their sociability. The large gathering of these beautiful animals that I witnessed in Isfjorden, about 20 kilometres north of the town of Longyearbyen, is more the rule than the exception. And while most other toothed whales are similarly gregarious, belugas are unusual in having a social system that throws the conventions applied by other cetaceans to the wind. In beluga society, your sex seems not to matter, nor are groups confined to close relatives; instead, belugas interact much more freely. Rather than relying entirely on the kind of nepotism that's at work in, for instance, family groups of killer whales, belugas seem to form relationships that rely as much on cooperation as blood. That isn't to say that kinship isn't important to them. The bonds that develop between mothers and calves are strong, just as is the case for nearly all mammals, and these are central to the formation of family groups of belugas that span multiple generations. Yet belugas don't rely exclusively, or even mostly, on familial support networks. Their groups comprise relatives and unrelated partners, representing both sexes, and individuals seem to flit between pods, sometimes travelling long distances to pay a call, perhaps, on old friends. So, each beluga is part of a broader community that has distinct echoes

40 Dozens of individuals are caught each year to serve what amounts to a life sentence in captivity.

of our own societies, where family is just one part of an intricate and diverse mix of social interactions.

Belugas, like many Arctic animals, have historically been challenging to research because of the extremes of their habitat. Nowadays we have advanced tracking technologies that, in harness with genetic sampling and stable isotope analysis, have opened their world to us and belugas, along with many other elusive animals, are revealing their secrets.[41] One of the first dividends from this is an insight into their society, which is more complex than we'd imagined, helping to explain their extensive communication and their intelligence. As a general rule, group-living animals tend to have greater cognitive abilities than similar, more solitary animals, likely because of the demands of successfully navigating a complex social milieu. Belugas are clearly highly intelligent and curious creatures, possessed of impressive brains about twice the size of ours, with the convoluted, whorled neocortical surface that's characteristic of the smartest mammals.

Belugas most often encounter people during summer, when the ice retreats from the coasts and thousands of them converge on the warmth of lagoons and estuaries in the highlight of the whale's social calendar. Warmth is a relative concept in the Arctic, admittedly, but the shallow waters of inlets and the mouths of rivers might be as much as 10°C above their wintering grounds in the open sea. Here the belugas socialise in enormous numbers. Youngsters take the opportunity to play, fashioning games from driftwood or excitably blowing bubbles underwater; females give birth, relying on the balmy conditions to provide respite for the calves from the paralysing cold of higher latitudes.

---

41 Genetic analysis allows us to infer familiar relationships, while stable isotope analysis enables us to construct a back story for an animal based on the chemical profile of certain tissues, so that we can learn about its diet and the environment where it grew up.

Their sojourn in the shallows is also an opportunity to do some sprucing. Over the winter, belugas, like many cetaceans, experience a build-up of microbes on their skin including the diatom *Bennettella ceticola*, a whale-hide specialist that bestows an increasingly chartreuse tint to the animals over the course of the year.[42] This film of tiny passengers doesn't harm the beluga directly, though like a fouled ship it likely carries a cost to their smooth progress. Consequently, as the warmer waters stimulate skin cells to grow, the belugas get the chance to exfoliate, rubbing themselves on the gravel beds of shallow rivers, scraping off the grime of the past year and restoring the snowy dazzle of their bodies.

Large as they are, belugas are prey to the Arctic's two dominant predators, the killer whale and the polar bear. The danger presented by killer whales tends to be greatest when the belugas cross open sea. By entering the shallows, the belugas go beyond the reach of killer whales, which risk grounding should they follow their smaller quarry. This estuarine refuge, however, brings the belugas into contact with polar bears. It's still no easy matter for a bear to lay its claws on them, but video footage of bears at the belugas' summering grounds shows the lengths that they go to in order to capture a white whale. In a tactic which calls to mind that used by brown bears fishing for salmon, the bear takes up a position on a vantage point, such as a boulder in midstream and then it simply waits. A rising tide encourages the belugas further into the river. If one strays too close, the bear leaps from its platform to seize it. A large beluga, while representing a meaty jackpot, would be challenging to subdue underwater, since the whale's weight is three or four times that of the bear. A calf, however, is eminently

42 The diatom's name 'ceticola' means whale inhabiting. It's not obvious on the darker skin of many cetaceans, though the tint on the bellies of blue whales with such a coating caused whalers to call them 'sulphur bottoms'.

manageable and it's the youngsters that the bears tend to focus their attentions on.

Later in the year, when the pack ice returns, polar bears face a different challenge to capture a beluga, but if they find a polynya, they have every chance.[43] The belugas' reliance on breathing atmospheric air means that they are forced to surface at intervals. There's little subtlety to the bear's approach, which relies on delivering a massive blow to the beluga's head and then using its prodigious strength to haul the stunned victim out onto the ice. Reports from the Canadian Arctic and from Novaya Zemlya describe bears taking as many as 13 belugas in sequence, many of which suffered fractured skulls. We can only imagine the hellish existence of the whales under those circumstances, being compelled to swim into the embrace of death every time that they need to breathe.

Today, in midsummer, the belugas of Isfjorden are far beyond the reach of polar bears. Nevertheless, on the far shore of the fjord, a mother polar bear is walking swiftly and purposefully, her two cubs having to trot to keep up. All are thin, reflecting that summer is a season of scarcity for these hunters, a perspective that is heightened by the stark, bare scree and rock of the mountainous backdrop. As I watch, the bears quicken their pace, and the reason for this is soon revealed. Two kayakers have come ashore by the glacier at the head of the fjord – the only people I've seen outside Longyearbyen – and the bears are on a trajectory to intercept them. The bears seem aware of the people, but not vice versa, and the distance is closing. I know that the kayakers must be armed, since it's illegal to leave the limits of Longyearbyen without a gun, so now I'm a spectator to an unfolding crisis. The gap closes to just a few hundred metres; the bears will soon be upon the unwitting

43 A polynya is a hole in the ice, which allows access to oxygen for marine air breathers.

kayakers – we have to intervene. We get as close to shore as the ship's draught will allow and my guide calls out to the kayakers over the tannoy. At first, the woman on the shore doesn't hear, cupping her hands to her ears. She gets it the second time and panics, shouting for her friend. By now the bears have disappeared behind a huge boulder, and they're perilously close to the people. We hear two gunshots and my heart sinks; have I just witnessed the death of one of these beautiful, endangered creatures? The kayakers have gained the comparative safety of an isolated shack, but of the bears there's no sign. Five or six minutes pass before, with a wave of relief, I see the bears mounting the glacier and moving away from trouble. It's a reminder that this is wilderness, a place where the comfortable norms of our lives are void.

* * *

Back in Longyearbyen, I return to my lodgings and past a notice on the outside of the inn, which reads 'Please do not park your dog here'. There isn't much sign in summer of the sledge dogs whose owners this is aimed at, but the reindeer that provide the meat to feed them are grazing the hillside among the delicate, downy white seedheads of the rampant Arctic cottongrass. An Arctic fox, the size of a domestic cat, skulks past, searching perhaps for an unguarded nest belonging to one of the profusion of seabirds that glide through the blue skies. Though we're used to thinking of the polar regions as places of shortage and privation, for these seabirds, as well as the marine animals they feed on, nothing could be further from the truth.

Strong currents, in conjunction with storms that rage in these wild places stir the water layers, bringing a gamut of nutrients from the depths. This, in addition to the perpetual daylight of the polar summer, provides the perfect conditions for phytoplankton to flourish. These provide the foundations for one of the richest food chains on Earth, and the waters swarm with copepods, shrimps, krill and fish. Millions of birds flock to these Arctic waters, drawn by the abundance of food. The seabirds seem to divide the banquet equably, and while there's overlap between them, each species has its eye on a particular delicacy.

Out at sea, little auks skim the waves, driven both by their relentless quest for prey and the threat of attack from piratical gulls swirling in the air above. Soberly attired in the seabird uniform of black uppers and white belly, the auks are stout but diminutive, half the size of their relatives and fellow diners at this Arctic feast, the puffins. Periodically an auk splashes down, transforming in moments from aeronaut to aquanaut as it dives in search of zooplankton. Copepods are their favoured prey, but even for a bird as small as an auk, catching enough of these miniature crustaceans to supply its own needs as well as those of its chicks is a gruelling task. Each bird needs to capture 10,000 or more every day at a rate better than one every second that they spend submerged. Uniquely among seabirds, they feed almost like avian hoovers, sucking copepods up one at a time with their dainty bills.

Fish such as sculpins and Arctic cod join the auks in their pursuit of copepods, and in turn form the basis of the diets of guillemots and kittiwakes, while fulmars forage for squid in the fjords. The birds themselves are on the menus of Arctic foxes as well as that of a polar giant, the glaucous gull. I watch from beneath the dark, towering cliffs of Fuglefjella, to the west of Longyearbyen, as the menacing approach of this bird, with its 6-foot wingspan, drives the massed ranks of smaller seabirds into screaming cries of panic, the air suddenly full of a chaotic blizzard of anguished parents. The gull, three times their weight, quarters their nests with impunity and, having gained a foothold on a crammed ledge, takes a chick and swallows it with two heaves of its enormous head. Further out into Isfjorden, a pure white apparition circles high above the boat. Ivory gulls are the seabird embodiment of the far north; they seem to be formed from snow itself, their keening, screeching call as cold and as brittle as the ice itself. These gulls have adopted an ingenious approach to foraging, trailing polar bears as they hunt and alighting to seize stray chunks of flesh and fat from slaughtered seals.

The Arctic, however, is changing. The flank of Operafjellet, one of the commanding mountains overlooking Longyearbyen, is embellished with a great natural mural of a champagne glass tens of metres in height, cast entirely in ice. The glass reveals itself as the seasons change and the snow melts from the exposed areas of the mountain to leave this distinctive shape, the bowl of the glass formed by the depression of a glacial cirque and the stem below by a ravine on the mountainside. They keep an eye on the champagne glass in Longyearbyen, watching it dwindle as spring progresses. The day the stem breaks – *stetten går*, as they say – summer has arrived. Historically this has happened in late July or early August. The year I visited, 2022, the stem broke in June for the first time ever. It's a signifier of change, of a new era for Earth's climate. The heat of the Anthropocene exerts itself to its greatest extent at the high latitudes where the warming is occurring at a rate more than twice that of other parts of the planet.

This is the front line of climate change, and the ecosystem of the Arctic is being rent from its icy moorings. The Barents Sea, which covers more than a million square kilometres of sea from the coast of Norway north to Jan Mayen and the Svalbard archipelago, is warming fast. The extent of the sea ice has decreased by half in my lifetime and, according to predictions, may be gone altogether before the end of this century. In such a scenario, the prognosis is bleak for the Arctic's most totemic animal, the polar bear. A female bear emerging from her den in early spring has lost half her weight in a fast that endures for several months. She's accompanied by her cubs, born in the den in midwinter and now blinking in their first exposure to the light. Polar bear neonates are tiny, just a fifth of the size of a human baby, but they've grown rapidly on their mother's rich milk. To sustain this supply, the mother now needs to find seals and the frozen seas are her only reliable hunting ground. Without sea ice, the cubs' situation is hopeless; they will starve as their mother's milk dries. The female herself may be able to scrape by for a while, but if the seals remain beyond her reach she'll be unable to lay down a supply of fat to sustain her through the next winter and she too will weaken and die.

The polar bear's reliance upon sea ice is far from unique among Arctic mammals. Far to the west of Longyearbyen, I'm crouched on a beach of grey, volcanic sand watching a huddle of walruses as they slumber and snort. There are five of them, and a mosaic of lumps, known as bosses, across their necks and shoulders identifies them as males. There's no hint of antagonism among them; instead, they're fully committed to resting. Occasionally one heaves itself ponderously, almost unwillingly, to the sea and disappears for a few minutes. Clumsy as they seem on land, beneath the waves they transform into sleek, relentless foraging machines who uncover and devour half a dozen or more clams every minute. They don't, as is sometimes assumed, use their impressive tusks for digging out their prey, instead they expertly pluck exposed bivalves straight from the seabed, or uncover the more recalcitrant ones from their muddy hiding places by squirting a jet of water from their mouths.

With each of these bulky creatures consuming the weight of a person in clam meat every day, however, they soon exhaust their supplies and are forced to move on. In Svalbard, where the walrus was almost driven to extinction by hunting in the 1950s, the small walrus population has sufficient shallow water clam beds for all. In other parts of the world, walruses are instead reliant on sea ice to grant access to feeding sites further from shore and as a resting platform between clam dives. As the ice disappears, however, the walruses are forced to adapt. Pacific populations are turning from their diet of clams to feed on other mammals, in particular ringed seals. Fewer females are becoming pregnant, and those that do show a concerning, stress-related pattern of abandoning their calves, some of whom have been spotted far out to sea, struggling and alone.

Now firmly in the crosshairs of the walrus, the ringed seal also has the unfortunate distinction of being the primary prey of polar bears. It might be imagined that the bear's travails would represent good news for these small, attractive seals, yet they too are threatened by warming. Ringed seals, like most of their tribe in the Arctic, rely on sea ice as a haul out while foraging, but it's in the cold and darkness of an Arctic winter when the ice plays a central role in the most delicate stage of the ringed seal's life cycle. Beneath the covering of snow that drifts across the sea ice, a ringed seal excavates a chamber. It's here, in a cosy cave of snow, that she'll give birth to her pup, supported by the ice from below and protected from the elements by the thick insulating layer above. This strategy has been the mainstay of ringed seals for millennia, but now, as the ice forms ever later, less time is left for the snow to accumulate on top of that ice, and the seals are increasingly left without this crucial refuge. Consequently, the newborns become obvious to become predators, including Arctic foxes and opportunistic glaucous gulls. A pup without a den also loses its protection against the harsh weather; it's a terrible irony that the warming Arctic places seal pups at risk of hypothermia.

The ice is as important for the smallest organisms in the Arctic ecosystem as it is for these large mammals. The spring explosion of phytoplankton each year is closely associated with the margins of the sea ice as it retreats. Meltwater creates the perfect growing conditions for microalgae and two thirds of the annual surge in productivity that feeds polar regions occurs at the ice edge. Each spring, the narrow strip of sea bounding the ice draws countless zooplankton to feed on its lush bounty of green, attracting a glut of fish and mammals. The wealth of food here is an obvious boon to these animals, but more than this it's the predictability of it that supports the copious marine life in this harsh environment. Ice isn't just a feature of the Arctic biome; it *is* the Arctic biome. If, as is suggested, we arrive at a point where the ice disappears for a significant portion of the year, the Arctic as we know it will no longer exist.

In late 2013, the huge mass of warm water nicknamed the Blob formed in the northeast Pacific, stretching from California to Alaska. The Blob pulsed and flared for the next three years and brought devastation to marine communities. It disrupted nutrient flows, initially causing phytoplankton populations to plummet, sending reverberations through the food web. Zooplankton abundance crashed as a result of both the dearth of food and the warmth. At the same time, fish, whose metabolism was supercharged by the additional heat, fed ravenously on what little food there was, pushing the local ecosystem quickly towards starvation. Compounding the situation, a record-breaking bloom of toxic algae flourished in the destabilised ecosystem. The flow-on effects of this were dramatic, and very visible. Forced to range further, species became dislocated from their usual range; warm-water sharks, skipjack tuna and sunfish moved into unfamiliar Alaskan waters. Thousands of emaciated fur seals, especially females and their pups, washed up in record numbers on the Western Seaboard. They were joined at the strandline by the corpses of seabirds, the death toll likely in the millions. Seven thousand humpbacks, one in five of the local population, are thought to have starved.

A concrete answer to the question of what caused the Blob is elusive, and can be found amid a mess of interacting factors. Anomalous wind patterns, unusually slow dissipation of heat from the sea surface, natural fluctuations associated with the North Pacific Oscillation and El Niño, and the Chinese government's efforts from 2010 onwards to tighten up on aerosol emissions are all implicated.[44] At the core of all these is one key factor: temperature. For decades, the oceans have been absorbing the overwhelming majority of the excess heat that we produce, and now, it seems, they're starting to give it back.

The Blob was one of the first, and the most dramatic, examples of a marine heatwave, a phenomenon that is now becoming uncomfortably familiar. Australia has been hard hit, with a string of heatwaves that devastated the kelp forests of Tasmania and killed corals on both the east and west coasts. And heatwaves in the Bering Sea in 2018 and 2019 have been identified as the driver in the deaths of ten billion snow crabs through starvation and the 2022 closure of the crab fishery for the first time in history. The likelihood is that these events are harbingers of what is to come; the more heat that we pour into the oceans, the more frequent extreme events will become. And while it's true that ecosystems are dynamic, and oscillate naturally, they tend to do so within limits that allow subsequent recovery. The danger now is that by charging the oceans with excess

44 Although cutting down on the production of aerosols (including particulate emissions from vehicles, industry and agriculture) is essential in the long term for reducing air pollution, there is one upside to clogging the atmosphere with toxic particles. Specifically, these particles contribute to cloud formation, ultimately leading to more solar heat being reflected back into space, thereby keeping the planet cooler. This, however, does not balance out the deadly effects of aerosols on health.

energy, the fluctuations tend towards chaos where rebalancing to a previous norm becomes impossible, and we reach a tipping point where the system flips into a different state. Normal, in other words, is in danger of becoming a thing of the past.

The dramatic nature of such events, alongside pictures of mass marine casualties, capture the attention of the media. Less grabby, yet every bit as problematic, is the creeping, incremental change that goes under the radar. In recent years, warmer, saltier water has encroached from the Atlantic into the Arctic Ocean, driving before it cold-adapted creatures and pushing them ever northwards. The Arctic cod is one of many marine fish whose range is contracting, the limits of its distribution pushing north by 150 kilometres each decade. There is, of course, only so far they can go. In Svalbard, native Arctic species are being supplanted by those from the North Atlantic, who themselves have been forced to shift polewards. Cod, mackerel and haddock are now commonplace in the seas here, and temperate blue mussels are establishing themselves on the rocky shores. The question is: does this matter? At first glance, many polar animals seem to be adapting. Seabirds, for instance, are transitioning to a new diet, eating the temperate fish instead of the Arctic species that sustained their forebears. This rosy picture, however, is complicated by some worrying realities.

The phytoplankton blooms that enrich polar regions are rich yet transient, bursting into life and then dying back within just two or three months. Copepods, like many Arctic species that feed on the phytoplankton, have adapted to the predictable sequence of feast and famine by storing away energy as fat reserves during the glut. This strategy allows these prudent savers to while away the leaner times over winter when the ice returns, and then to breed rapidly in early spring in anticipation of the riches to come. But as the Arctic changes, a reduction in sea ice extends the growing season for phytoplankton. This, along with rising temperatures, brings polar copepods into competition with invaders from the south. Though there remain plenty of copepods in the seas around Svalbard, increasingly these are the interlopers from the North Atlantic.

At face value, the usurpation of one kind of copepod by another might seem to be of little more than academic interest, but the potential repercussions are enormous. The Arctic copepod species are the pivot point of the food web in the frozen north and the reason for this lies in their life history. The ability of polar copepods to store fat frugally in times of plenty as a means to survive a long, dark winter is what gives them their value as prey. The invaders from the south, by contrast, shun such prudence, living fast and lean. The upshot of this is that these Atlantic copepods are not only smaller, but they also contain 20 times less fat. The calorific opulence of Arctic copepods made them irreplaceable in the diet of animals from fish to bowhead whales, and particularly in that of the little auks; without them, starvation beckons.

Something seemingly as insignificant as a shuffling of copepod species is illustrative of how even small tweaks to a natural system can have disproportionate effects. We often use machines as metaphors for ecosystems, relating the interacting species to components in an engine. It's an intuitive comparison, but not always apt because the science of mechanics is much cleaner and more predictable than that of ecology. Sometimes one species in an ecosystem can be supplanted by another with little or no effect to the functioning of the whole. And sometimes it can't. Ecology is complex, messy and beautiful. It is also fragile.

* * *

Though diminished in recent times, the Arctic ice still expands and contracts over the course of a year in a manner akin to the rhythmic beating of a great white heart. The Arctic's cyclical pattern is in antiphase with that of the Antarctic so that as the ice recedes in one, it expands in the other. This seesaw of freeze and thaw at the extremes of our planet powers the circulation of the oceans. As seawater crystallises into ice, it sheds its burden of salt, creating a dense concentrate of briny, cold water that sinks into the deep then cascades out beyond the poles, flowing like a river across the deep ocean basins. This is the beginning of the global conveyor belt, an arterial system of currents that winds its way through the depths, like a marine Möbius strip. The energy of these two pulsing, polar hearts is augmented in the tropics by the sun, and together they propel each drop of water in the ocean on a curlicuing journey of 100,000 kilometres that can take a millennium to complete. Slow it may be, but the conveyor's relentless action redistributes heat and nutrients throughout the oceans, and in doing so, makes it possible for life as we know it to exist.

The oceans are further stirred by the winds, under the influence of the Coriolis effect, to generate surface currents, including the five great oceanic gyres and the most powerful of all, the Antarctic Circumpolar Current.[45] This behemoth, a marine race comprising 100 times as much water as all the world's rivers combined, travels in a perpetual, clockwise loop around the southern continent, driven by powerful westerlies. Where the current funnels between South America and the Antarctic Peninsula, it intensifies, its power evidenced by the eastward warp of the tips of the two landmasses.

The Drake Passage, the 1,000-kilometre stretch of ocean that separates Tierra del Fuego from the northernmost point of Antarctica, is infamous among mariners as the most treacherous crossing on Earth. A monument at Cape Horn commemorates

45 The Coriolis effect is caused by the Earth's rotation.

more than 10,000 sailors who are thought to have died here. Of all the seaborne routes to the Antarctic, the Drake Passage is by far the shortest, but it's a journey of unparalleled and capricious hostility, featuring mountainous waves the size of five-storey buildings. It's a test for anyone, as Santiago Diaz, a postgrad student in my research group, testifies. 'There's no such thing as a person who doesn't get seasick,' says Santi. 'There are just people who haven't experienced seas that are rough enough.'

Santi's reward for the crossing is to experience the majesty of Antarctica, and with a harrowing journey behind him, is able to collect his thoughts from his seat at the bow of a research vessel. The boat anchors in Neko Harbour in the lee of the islands to the northwest of the peninsula, allowing Santi the chance to take in the dramatic scenery. Towering peaks dominate the landward scene, black escarpments rising above the continent's shroud of ice. From the boat, the foothills are obscured by the blue-white ice cliffs at the margins of this south land. The fury of the recent days in the passage has given way to stillness, the peace disturbed only by snorts of steam from lolling humpbacks and by the sploshes of chinstrap penguins as they race to shore, porpoising across the surface of the calm ocean. The reason for the penguins' hurry becomes clear when Santi notices a leopard seal lurking beneath the boat's hull. He drops a video camera into the water and the seal comes to inspect, its dark eyes gazing into the lens, outsized jaws giving its head an unusually deep profile. This is an impressive animal, a skilled hunter whose skull matches that of a lion in size. The leopard moves on, rewarding the camera with a view of its battleship grey flanks and the field of black spots that inspire its name.

Santi and his colleagues are here to dive, but they cautiously choose to wait until the leopard seal has gone before entering the water. Leopard seals have a mixed history with humans.

There has been a scattering of attacks over the years, the worst of which culminated in the death of a researcher 20 years ago as she snorkelled. Contrastingly, another leopard seal interacted with the Canadian photographer Paul Nicklen in a benevolent, almost supportive manner, presenting the diver with a freshly killed penguin. Even so, at twice the length of a person and six times our weight, it's wise to be careful around this apex predator.

Flurries of unsummerish snow melting reluctantly on a dark, forbidding ocean don't, perhaps, make the idea of getting underwater an attractive proposition, especially when coupled with a sea temperature of almost two degrees below zero. Yet with cumbersome drysuits overlaying thick insulative bodywarmers beneath, the divers are feeling uncomfortably warm as they busy themselves with pre-dive preparation, and they're keen to get in. Taking the plunge, the shock of icy water on the exposed skin of their faces makes them catch their breath; it's essential that they get this under control before descending otherwise they'll burn through their air in no time. At the thumbs down signal, the divers subside into the embrace of the Southern Ocean.

The cold and the starkness of the scenery above the water encourage the preconception that the seascape below the water will be equally austere. The reality is very different. The rocks below a depth of about 5 metres are a kaleidoscopic patchwork of brilliantly hued algae, from the pink of the coralline to the rich emeralds, caramels and oxbloods of feathery species. Delicate white nudibranchs slide over the canvas and cross paths with showier relations with tasselled peach bodies. Pycnogonids pick their way over stands of sea squirts on stilt-like legs and sea stars, tinted a deep plum, flow through the scene.[46] Fish watch from the safety of

46 Pycnogonids are more commonly known as sea spiders, despite not being spiders.

crevices, tempted by the movements of shrimps as they breeze by. This is a vibrant assemblage, packed with colour and activity, and few might guess from a picture that it's situated on the shores of Antarctica. Perhaps the only clue that this isn't a temperate shore, at least for those without a keen eye for taxonomy, is the bare rock immediately below the waterline. This is where icebergs gouge and scour the shore of life, leaving it entirely barren.

Over the next few days, they dive at sites around Neko Harbour before swinging round to the east of the peninsula to explore the marine fauna of Gourdin and Signy islands. At each, the underwater scene is rich and exuberant, a welter of life thriving in Earth's freezer. All good things must come to an end, however, and it's essential that Santi and his team leave before summer gives way to autumn and the weather worsens. They make the northward crossing of the Passage, this time in kinder seas, and return to South America. For all the grandeur of the Antarctic scenery, Santi tells me that he's gained a new appreciation for the greenness, for the grasses and the trees that he hasn't seen for weeks.

Back on the Antarctica continent, the days are shortening perceptibly and, before long, the continent will be consigned to the deadening permadark of winter. The ice that retreated south over the course of the Austral summer is regathering and preparing to push north once more. Ice crystals form and float in a slurry, known as frazil, on the sea surface. As the water cools, each crystal shyly extends delicate, glacial tendrils to its neighbours. For now, wave action continually sunders these brittle projections, but the broken fragments remain, joining and bolstering the slush. One more cold night might be enough to bind the fragments and form a solid canopy of ice, isolating Antarctica's mammals and birds from their foraging grounds, or their air supply. Like the expedition team, many of them are forced to move north. Some, like the humpbacks, decamp entirely, trading the fertility of these icy waters for the benevolent yet food-deprived seas of the tropics. Others follow the expanding ice edge or rely on fissures, leads or polynyas to sustain their residence in far south waters.

The frozen plains of fast ice are pockmarked by what look like boreholes, and within one of these a chubby grey face protrudes above the surrounding slush of glacial water. The face belongs to a Weddell seal, busily rasping and gouging at the walls of its icy well with powerful canines to prevent itself from being entombed. It's this energetic maintenance of breathing holes that supports the lifestyle of Weddell seals, most of whom live on permanent ice just a few kilometres beyond the shores of the Antarctic continent. The ice supports resting, moulting and birthing alike, and the Weddells' cultivation of breathing holes saves them from having to migrate northward in the winter alongside the other seal species. Moreover, it provides them with exclusive access to the fruits of the sea, the fish and krill that sustain them. Weddell seals pay a heavy price for their energetic gnawing, however. Over the course of about two decades, their teeth are worn to stubs and, no longer able to curate a breathing hole or to hunt effectively, they perish. Indeed, Weddell seals tend to die at a younger age than other seals, but by the same token their pups avoid the fate of those of similar species, which are decimated by leopard seals and killer whales. These predators sensibly stay away from fast ice, and consequently Weddell youngsters have a charmed puphood.

Along the coast, a few other hardy species remain behind to face the savage winter storms of the south land. Snow petrels nest in colonies, each pair taking ownership of a crack or fissure in the rocks, which affords them some protection from the biting winds. Inhospitable as this might seem, some of the colony sites have served the petrels for countless generations. One such, in eastern Antarctica's

Queen Maud Land, is thought to have been in continuous use for an astonishing 37,000 years. Snow petrels are captivating birds, whose bright white livery is offset by coal-black eyes and bill. A single egg is laid in early December and nurtured in a frigid scrape, perhaps augmented with a few pebbles to make the nursery more homely. It'll be seven long weeks before junior makes an appearance – life is often slower at high latitudes – so the parents take it in turns to scout for food and, occasionally, to spruce their feathers by bathing in the surrounding snow. Even once the chick has hatched, it'll be a further seven weeks before it's ready to fledge, and this extended developmental period compels snow petrels to remain behind on the continent when the onset of winter drives the exodus of so many other species.

In contrast to fair-weather Antarctic species, emperor penguins appear to actively welcome the harsh weather; indeed, they're the only penguin that breeds in winter. This, alongside a host of other peculiarities about their behaviour, led the British naturalist Edward Wilson to conclude in 1907 that their habitats were 'eccentric to a degree rarely met with even in ornithology'. He had a point; the hardships that the emperors put themselves through in order to raise their chicks are extreme. Nevertheless, the persistence of these behaviour patterns speaks of their effectiveness and, harsh though their habitat might seem to us, the penguins are perfectly equipped to cope.

The emperor's courtship, an exposition of exuberant clamouring calls, coupled with dramatic head shaking displays and bows, is timed to coincide with the onset of winter, and the single egg is laid around May, just as the weather begins to turn particularly vicious. In a reversal of the typical pattern seen in other birds and mammals, it is the female who departs, leaving the male holding the baby. Brooding is exacting work, requiring the male to support the egg on the top of his feet, buffering it against the ice, and to envelop it in a unique, pouch-like fold of skin on his belly. Naturally enough, this somewhat curtails his movement and with the females gone, the colony is a subdued place, ranks of expectant fathers statuesque and near silent. Gale-force winds assault the birds as temperatures plummet beyond 30 below. Unlike the smaller, feistier penguin species, emperors are supremely tolerant of one another. They have to be. Their main weapon against the storms that sweep in is the huddle, where birds crowd together to brace themselves against the cold.

The male's vigil with his egg can last for more than ten weeks, and by the time that the cossetted youngster emerges, in July or August, the father's last meal is no more than a distant memory. A fast extending to three or four months, compounded by his endurance of the worst conditions that the Antarctic can inflict, results in him losing half his body weight. In this weakened state, the male now has the job of feeding a chick on a belly full of nothing, a feat he accomplishes by producing a version of the mammal speciality, milk. Crop milk, however, isn't produced by some avian mammary gland but from the lining of his gullet. It's high in both protein and fat, which is crucial because emperor penguin chicks are astonishingly poorly equipped to face the rigours of an Antarctic winter. For one thing, they're tiny. At less than 1% of the adult's size, they're among the smallest of all bird chicks, in relative terms. For another, they have barely any down and are unable to thermoregulate. If they're to survive, they have to grow, and fast.

The male can't long maintain the supply of milk, but relief comes in the form of the female, returning from her marathon expedition to the ice edge with a crop full of food. She toboggans over the snow on her belly, arriving at the colony in an excited state and hailing her mate. Penguins make many different calls, sounding to my ears like a strange cross between a revving motorbike and a trumpeting elephant, but that which the parents give as they reunite is unmistakably enthusiastic. To add to the sense of a joyful reunion, their clamouring cries are accompanied by vigorous duetting displays of beak pointing, head bobbing and bowing. It's now the female's turn to assume the carer's responsibilities, yet so intent are the males on their duties that many seem reluctant at first to surrender the chick. Eventually, however, hunger tells, and one by one, they relent. The transfer made, each male sets off on his own trek to the ice edge.

By September, the chicks have developed a thick coat of down that provides a measure of independence from the protective insulation of their parents. But while the independence is welcome, a bigger chick needs more food, and now the parents are compelled to forage simultaneously to satisfy its burgeoning appetite. After a few more weeks of this intensive provisioning, the chick now stands as tall as its parents, around chest height to a typical person, its small head sat atop a body made barrel-like by a thick layer of fluff. The pampering, however, is about to come to an end. The parents must now be selfish, fattening themselves up in preparation for their annual summer moult, when, unable to dive, they face a month of enforced starvation waiting for their new feathers to grow.

In the face of what they might consider to be shocking parental neglect, the chicks now have to fend for themselves. They shed their down, replacing it with the sleek black-and-white plumage of the adults. In December or January, when the sea ice near their rookery has broken up, the chicks take to the water for the first time. They're clumsy swimmers, as might be expected since they have neither experience nor an adult to provide instruction, but most quickly get the hang of it, and begin their foraging amid the bounty of the summery Southern Ocean. It's this seasonal banquet that greets the young emperors that explains partly why the parents go through the rigours of raising their young in the harshest months of the year. Another benefit to their strategy is that the chicks can develop free from the notice of predators like giant petrels and skuas, which overwinter far to the north. Indeed, the privations of winter brooding that seemed so eccentric to Edward Wilson a century ago are, in fact, the hallmarks of perfect penguin preparation.

* * *

Even in summer, the freezing conditions of polar coasts make for an intense and energy-sapping experience. Despite the fetching orange immersion suit that I'm wearing on the open boat, the hostility of the frigid air still makes itself felt. The sea, which for me has always been a haven and a refuge against the cares of life, seems dark and unwelcoming here. I'm no stranger to cold, having learned to dive in an ice-covered quarry in the UK, but the polar seas are not for the faint-hearted. Such is the speed with which icy water drains heat from the body that the life expectancy of a person in these seas, without protective gear, would be mere minutes. The cold even sucked the life out of the submersible cameras I was using; an hour's worth of charge in normal conditions becomes enough for only a short video clip. For once, I had no desire to get in the water as the boat bumped and shuddered its way through a field of floating ice, each jarring thud and scrape of the ice against the hull seeming to voice the frosty antagonism of this place.

Yet alongside the starkness, there is beauty. The ice is tinged with an illimitable palette of blues, from the gentle pastel tint of a clear winter's sky through to an intense radiant sapphire that made the ice seem as though it was illuminated from within. The boat's skipper, a likeably eccentric Dutch woman, takes a diamond ring and reaches out to place it on an iceberg. It'll make a good photograph for her sideline jewellery business, she explains.

To me, the ice crystals overwhelmed and outsparkled the gem, which seemed lost amid the grandeur.

Blue was not the only colour expressed by the ice. Veins and bands of darker pigments ran through the frozen masses. Call me a philistine, but as a biologist, this was more entrancing than the glowing blues of the bergs because these veins are the signature of life. Within the matrix of the ice, tiny algal cells were not only surviving but thriving. Despite its appearance, much of the ice isn't solid, but instead is riven with microscopic channels and pockets of briny liquid, a residue of unfrozen seawater. This is where microalgae, such as diatoms, make their home and use what light filters through the ice to photosynthesise. The ice diatoms live under the threat of an especially cold spell freezing even this brine. Their ingenious response to this is to engineer their microhabitat by producing complex molecules that prevent ice formation and so maintain their liquid environment. When the melt comes, and these diatoms are released into the freedom of a comparatively balmy ocean, they seed the springtime explosion of phytoplankton that characterises polar regions.

After a long winter in the glacial deep waters of the Southern Ocean, Antarctic krill greet the emergence of the algae eagerly, rhythmically clapping their forelimbs together like tiny cheerleaders as they spiral through the clouds of phytoplankton. While seeming to signify delight, the clapping motion is in reality the krills' means of gathering food. The modified legs are coated in fine hairs, which when brought together create a basket that strains microalgae from the water.

At the Australian Antarctic Division's research facilities in Hobart, a series of holding tanks provide a home from home for these Southern Ocean superstars, as well as a crucial opportunity to research them. There are almost a hundred species of krill throughout the world, but the Antarctic krill is the dominant one. It exists in stupendous numbers, and though each is tiny – an adult is only a third the weight of a teaspoon of sugar – collectively their mass is equivalent to that of all the world's people. Antarctic krill are, like Robert Paine's *Pisaster* sea stars, the exemplar of a keystone species, but even among the select band of keystones worldwide, few bear as much as Antarctic krill carry on their shrimpy shoulders. The simple fact of the matter is that these finger-length creatures are single-handedly responsible for the continued health of the entire Southern Ocean.

Everything in the Antarctic depends on krill in one way or another. They're on the menu of the Southern Ocean's fish, and its birds, including penguins and albatross. All four species of seal to be found in Antarctica eat krill. Even leopard seals, the wonderfully fearsome half-tonne nemeses of penguins and other seals, eat more krill than anything else. Meanwhile, crabeater seals, despite their name, subsist almost entirely on krill. The crabeater's teeth are nothing short of bizarre, each one divided into multiple cusps that always remind me of a Swiss cheese plant's leaf. The teeth of the upper and lower jaw fit together snugly to form a sieve, which they use to separate krill from seawater. Given the hyperabundance of their prey, it's perhaps no surprise that crabeaters are the most common large wild mammal on Earth; 20 million or more of them thrive in Antarctica.

But it's whales, who cross half the world to gorge on the riches of the far south, that are most closely associated with the krill. Indeed, the fates of whales and krill are not just intertwined but conjoined. When one prospers, so does the other; and when one declines, the other suffers. When whales were driven near to extinction by industrial whaling, krill populations slumped. Even as whales eat krill, they lay the ground for the remaining krill to prosper, so that both organisms inadvertently engineer their ecosystem for the other's benefit. It's easy to see how krill support whales. A blue whale can get through more than 3 tonnes of krill in a single day, the unparalleled abundance of the crustaceans in the Southern Ocean summer underwriting the cetacean's great migrations. The surviving krill gain a boost from the whales less directly, but just as importantly. As the feeding whales continue to cram their bellies with krill, the inevitable result is whale crap, and lots of it. Unappealing as this is to us, it's manna to phytoplankton. Whale turds float and they're full of iron, a scarce element in the Southern Ocean, and top of these marine plants' list of desirables. This welcome boost to the phytoplankton supports the krill, which thrive on the abundant food and show their appreciation through faster growth and greater fecundity. It's a win-win for both of these iconic creatures of the south.

It's not just as food that krill are important. Though each is minuscule, the pulsing legs of numberless krill generate currents which stir and mix the Southern Ocean, bringing nutrients to the surface and supporting phytoplankton. And when the krill eat these microalgae, they trap the carbon contained within and send it plummeting into the deep in their faeces. In this way, krill collectively sequester 20 million tonnes of carbon every year. Seen in this way, krill not only bolster and engineer their ecosystem, but their swarms also represent powerful biological carbon absorption machines that yield enormous benefit to the planet as a whole. Having gained this appreciation, what do we do? Do we protect the Southern Ocean, or do we seek to exploit it? No one with any knowledge of our track record should be surprised as to the answer.

The *Antarctic Endurance* is an impressive ship by any standards. The custom-built, 130-metre krill vessel is equipped with a swathe of high-tech features that make it supremely efficient. It's engineered to be energy efficient. And its cutting-edge design is such that it can very efficiently attend to its main purpose: to catch vast amounts of krill. The ship's cavernous hold can accommodate more than 3,000 tonnes of krill and when it's full, it can offload them, efficiently, to a support ship so that it can then, also efficiently, get right back to catching another 3,000 tonnes of krill. This is just one among an increasing fleet of insatiable factory ships with their sights fixed on the riches of the Southern Ocean.

Operating in Antarctic waters has always been perilous; the roughest seas in the world offer a stern test for any vessel. On top of this, krill have a biochemical quirk that causes them to essentially self-digest and turn to sludge shortly after capture. Accordingly, conservation of krill stocks has long been at least as much a matter of the logistical difficulties of doing otherwise as a high-minded yen to safeguard nature. But the operational complexities once faced by the krill fishery are rapidly being addressed by technology, as embodied in the impressive form of the *Antarctic Endurance*, and with these out of the way, actually catching krill is uncomplicated. The crustaceans form obligingly into huge, dense, easy-to-find swarms, which can be encircled with gargantuan nets and then simply hoovered out of the ocean. The krill are then flash-frozen and baled into half-tonne lots in the hold. There's no need to pause operations because the vessels are capable of running round the clock. The annual take – I can't bring myself to use the industry's go-to word 'harvest' because the implications of this are absurd for a wild fishery – has quadrupled in the last 20 years.

The krill products that are now becoming ubiquitous on the shelves of supermarkets and health food stores are emblazoned with words such as 'eco-friendly' and 'sustainable'. The claim isn't unreasonable on the face of it; indeed, the Antarctic krill fishery is, in some ways, a model. Bycatch is relatively low, and it's less destructive to the environment than many other comparable fisheries. Managed carefully, Antarctic krill could make a major contribution to global food security, providing much-needed protein to feed the growing population as well as to sustain aquaculture. Additionally, the current levels of catch are far smaller than the precautionary limits that have been set, and there's scope for increasing the current take without necessarily having a detrimental impact on krill populations – so what's the fuss?

A good deal is made about the amount of krill out there in the Southern Ocean versus the amount that we take, but there's a rather dramatic caveat to this: we don't actually know how much krill is out there. This is true, to some extent, of all wild fisheries, but even so, krill are an outlier; counting them in the hostile expanse of the Southern Ocean is viciously difficult. Consequently, our estimates of krill abundance emerge from extrapolations and informed guesswork, and there are massive margins of error. Imprecise as the data are, one trend is clear: krill populations have shrunk dramatically in recent decades, perhaps by as much as 70% since the 1970s. Even as the downward trend continues, their populations fluctuate significantly year on year, so that the graph looks much like a wave slide. A high catch in a year when the krill are plentiful might pose little threat, but the same in a year when the krill are scarce is a different matter.

While our incursions into the krill populations averaged across the whole 20 million square kilometres of the Southern Ocean are small, the epicentre of fishing activity is a place known as Area 48, encompassing the Antarctic Peninsula and the South Shetland Islands, South Georgia and the South Sandwich Islands. There are good reasons for this. For one thing, Area 48 is close to the ports of South America. For another, krill flourish here at high densities and warm currents keep Area 48 ice-free for longer than in other parts.

The krill in this region, however, are as much of a draw for other animals as they are for us. Area 48 supports thriving communities of marine mammals, birds and fish, and the concentrated efforts of the fishery puts humans into competition with the local fauna. When fishers deploy their nets to engulf a krill swarm, they inevitably catch not only the crustaceans themselves but any predators that are feeding among them. Entanglements in the fishing gear often prove fatal, and in recent times birds, fur seals and even humpbacks have been among the casualties. More concerning still, at least in terms of the numbers affected, are the impacts of the krill fishery on predator populations. Penguins in particular are suffering from the depletion of their foraging grounds, with Adélies, gentoos and chinstraps in decline in most parts of Area 48.

The krill industry would argue, with some justification, that climate change is more damaging to the penguins' troubles than anything they're doing, but that doesn't negate the fact that there's a terrible synergy between the two. A warming Southern Ocean is especially bad news for krill because of their dependence on sea ice. The underside of the ice supports a thriving community of algae, a verdant plain populated by vast numbers of grazing krill, like miniature, inverted antelopes on a frozen Serengeti. This icy affiliation lies at the heart of the krill life history strategy, sustaining them while simultaneously providing protection from their predators. Sadly, in some of the most important parts of the krill's habitat, there are some three months' less sea ice cover over the course of each year than was the case in the 1980s, and still the retreat continues. The range over which krill populations once roamed is contracting, as the crustaceans flee the rising temperatures and cluster ever closer to the cold continental heart of the Southern Ocean. Fewer krill are recruiting, which means less food for the animals that rely on them. When the present-day extraction of almost half a million tonnes of krill each year is added to the equation, the result is that the inhabitants of the Southern Ocean are caught between the warming Scylla and the fishery Charybdis.

Setting quotas for a declining species against the backdrop of data deficiency and uncertainty is an unenviable task for the agency whose responsibility this is, the Commission for the Conservation of Antarctic Marine Living Resources (CCAMLR), and the result is a series of arbitrary limits on catches. These limits are on the low side, but commercial interests are pressing for more. In recent times, proposals to renew existing controls on the krill catch and to establish new marine protected areas in Antarctic waters were blocked by China, backed by Russia. This followed the decision of the Chinese government to expand its krill fishing operations, commissioning five new gigantic factory ships to plunder the Southern Ocean. Historically, the discussions among member states at CCAMLR have been characterised by harmony. This new development charts a worrying course away from cooperation as powerful national interests assert themselves. With the krill fishery already gearing up to expand their operations, it remains to be seen for how long the consensus for conservation will hold.

Our history is littered with examples of the disastrous mismanagement of wild fisheries and the insistence that 'this time, it's different' has to be viewed with skepticism. The fact is that whenever powerful financial interests come up against conservation objectives, the result is virtually always the same: a zero-sum game in which business wins and everyone else loses. Governments around the world set the scene for this – more than ten times as much money is spent on fisheries subsidies than is spent on fisheries research. Bad as this is for any target species, krill represent a special case. Given their unique ecological role in the Southern Ocean as the keystone species, and as drivers of the biological pump that neutralises so much carbon, they are irreplaceable. And if the krill fishery makes little sense from an ecological perspective, it makes even less sense from an economic standpoint. The entire operation is estimated to be worth $250 million annually, a fraction of what Premier League clubs spend on footballers each year. We should be doing everything we can to safeguard the long-term viability of krill, rather than placing the entire Antarctic ecosystem in jeopardy by persecuting them.

* * *

Peering into the crystal water of the tanks in Hobart, I follow individual krill as they shuttle ceaselessly in search of who knows what. They're hypnotic to watch as they glide and spiral, swimmerets kicking like the legs of galloping horses, but in slow motion, reflecting the gelid environment through which they swim. It's the temperature of their habitat that determines that of their body and, consequently, their metabolic rate and pace of life. In adapting to the conditions of the Southern Ocean, and with a pace of life every bit as glacial as the waters they live in, the creatures of the far south show some extraordinarily unusual life history patterns.

One is gigantism, the tendency of animal species to become supersized.[47] Take sea spiders, for example. In most parts of the world, a sea spider is, well, spider-sized; in the Antarctic, they grow large enough to span a car tyre. Then there are cup-shaped glass sponges on the seabed here that may be 2 metres tall and 1.5 metres wide. Even these pale in comparison to the standard-bearer for

47 The geography of gigantism is striking; modern marine giants tend to occur at polar regions and in the deep sea. This has led to speculation that these two great ocean biomes are strongly interconnected in regard to the flow of species between them. Certainly, the ability to embrace low temperatures and long periods of hunger preadapts animals to cope in either.

gigantism: the colossal squid. In 2007, a New Zealand fishing boat operating in the Southern Ocean hauled aboard a specimen that was almost half a tonne in weight and longer than a bus, with eyes larger than your face. Incredibly, this was likely just a youngster, and there are larger colossal squid in their home waters around the Antarctic, a fact that we can infer from beaks recovered from the stomachs of sperm whales hunting in these regions.

Krill don't have any truck with gigantism. Indeed, in regard to growth, they've torn up the rule book. The long, dark, starving winters in the Southern Ocean represent a challenge for all animals, one that krill cope with by shrinking. They downsize during their hibernal moult, using their own bodies as a larder to sustain themselves until the algae return each spring. As a result, mature krill seem to live a kind of Dorian Gray-ish existence, cycling between their adult guise and that of their younger, smaller selves. This is a superficial change, however; the animals don't actually grow younger. Even so, Antarctic krill are one of many cold-water animals that exhibit another unusual biological pattern: longevity.

Alan the krill was a celebrity at the Antarctic Division. As the last remaining individual of the stock one year, he was transferred to a container on a shelf for safekeeping while the aquariums were cleaned. When a new batch of krill were brought up from the Southern Ocean aboard the RSV *Aurora Australis* and introduced to their new home in Hobart, the plan was to add Alan to their number. Caution is vital, however, when it comes to the unique challenges of krill husbandry, and it was decided that Alan posed a contamination risk to the newbies. Consequently, Alan remained in glorious isolation as a kind of lab mascot, fed, fostered and feted in his little aquarium for almost a decade. He might still be going strong were it not for the intervention of tragedy. Alan's tank was accidentally knocked over by a person or persons unknown, resulting in the krilly pensioner falling into the sink below and then to an ignominious end in the drains. Though Alan's life was unusual, the extent of it wasn't; other individual krill in Hobart have reached their teens – 20 times as long as tropical krill species manage.

For a krill, Alan might have been ancient, but his tenure on Earth might seem like a flash in the pan in comparison to other cold-water animals. Greenland sharks and red sea urchins can each persist for 200 years, while the record-holder is the clam, Arctica islandica, which can live for more than five centuries. The comically stuffy scientific term 'negligible senescence' describes how these animals barely seem to age. This has a lot to do with their slow metabolic rate alongside long periods of enforced fasting, but many of these long-lived animals also seem to have incredibly impressive cellular defence and repair capabilities that stymie the ageing process. Interestingly, it's not just cold-blooded animals that enjoy an extended existence in the freezer: Arctic whales do, too. In 2007, a bowhead caught by an Inupiat hunting party on Alaska's north coast yielded evidence of a previous, unsuccessful attempt on its life. Lodged in the whale's body were antique harpoons and, based on historical records of the weapons used by the Inupiat, these dated to the nineteenth century. Based on this, it's estimated that the unfortunate bowhead may have been more than 200 years old when it finally succumbed, making it the oldest known mammal. Intrigued by this revelation, researchers dug into the bowhead genome and identified patterns in the genetic profile of bowheads that equip them to repair damage to DNA and so fight off age-related diseases such as cancer, if not harpoons.[48]

48 Strangely, whales along with other long-lived animals have very low rates of cancer. In species like elephants, this is because DNA damage triggers a genetic safeguard known as programmed cell death, so that the body eliminates cancer at an early stage. Bowhead whales, by contrast, have very slow cell division rates, which make it difficult for cancer to get a hold.

A subsequent examination of bowhead's genes in 2019 using a biochemical test known as a genetic clock shows that these Arctic giants could, in fact, reach the grand old age of 268 – more than three times the human average and based on very similar genetic architecture to our own. It's hoped that investigations of these durable cold-water animals will show us the way to a deep, scientific understanding of aging, to a point that therapeutic interventions become a reality in human medicine.

* * *

While low temperatures can cause life to become lengthier and larger, they imposed an almost insurmountable challenge to the first animals attempting to colonise the polar seas. The fundamental processes of existence become ever more sluggish as the mercury drops. Anyone who's ventured outdoors on a cold winter's day without gloves will be familiar with the numbness and clumsiness of chilled fingers caused by the dulling effect of cold on nerve activity. Moreover, the reason that makes a drop in body heat so dangerous is that the biochemical activity that sustains us works most effectively at around 37°C. With every degree you go below that benchmark, the less efficient life's chemical reactions become. A fall in our core body temperature of just two degrees induces hypothermia, which causes uncontrollable shivering and a lack of coordination. If the body chills by a couple more degrees, critical metabolic functions begin to falter and, unless we're very lucky, we die.

For the warm-blooded animals that thrive in polar regions, insulation is the key to survival. Penguins use layering to maintain their body heat. An outer layer of stiff contour feathers provides a protective barrier against wind and water alike, in the latter case augmented by the bird's judicious application of oil from its preen gland. Beneath this, up to 100,000 short, intensely fluffy feathers mesh together to make a kind of dense down undergarment for the bird. In combination, the layers conserve something like 90% of the penguin's heat, leaving just enough leeway for it to cool if it overexerts itself. Some mammals, notably sea otters and polar bears, rely extensively on dense fur to retain warmth. In the case of polar bears, the hair has an additional property; it's hollow, making it both more lightweight and more resistant to heat loss. This evolutionary fine-tuning of anatomy has given rise to bioinspired design in the field of materials science, and both the polar bear's fur and the penguin's down layer have provided a template for the manufacture of raw materials to make clothing.

Fabulous as their furry and feathery raiments are, a layer of fat beneath the skin is an essential for most marine mammals. Blubber, however, is distinct from the kind of fatty deposits that we build up after overindulging. It's a specialised tissue made up of fat cells, held in a framework of tough yet elastic fibres that covers the body of all whales and seals like a tunic, and can be as much as half a metre thick in some polar whales. Thickness is only one part of the insulation equation, however. Bowheads make extensive use of saturated fatty acids in their blubber because these thicker fats pack tightly together and are more effective at holding in the warmth, while the blubber of dolphins in the tropics is suffused with lighter, less insulative unsaturated fats. Crucially though, the animals can shuffle their fat pack according to the water temperature, so a whale migrating between the poles and the tropics can adjust for the different water temperatures that it will experience by changing the fat profile in their blubber.

Even with the assistance of blubber, keeping warm is an enormous challenge for marine mammals for the simple reason that water sucks heat out of a body 25 times faster than air. One way of offsetting this is to grow to a stupendous size, because larger bodies hold heat better than smaller ones, and indeed this is one factor that has played a part in driving the supersizing of many whales. Even so, as mammals that give birth to live young, newborn calves or pups must experience quite a shock on first leaving the constant warmth of the womb, especially since many of them are born with only the most meagre of blubber layers. To combat this, marine mammals have evolved a variety of strategies to cope. The shuttling of whale species between the rich feeding grounds to be found near the poles and the warm waters of the tropics is driven in part by the advantages of giving birth in benign temperatures.[49]

Seal pups, especially those born in colder climes, are born with a layer of fine hair, known as lanugo, in lieu of the adults' blubber. Though this works well as long as the animal stays topside, entering the water would be fatal for them. The pups must eventually swim, however, so their mothers prepare them for the rigours of cold immersion by providing them with the richest, fattiest milk imaginable. As a rule, the colder the sea temperature, the fattier the milk of marine mammals is. That of lactating seals and whales in polar regions can be as much as 60% fat, and has a consistency like cottage cheese, so it's more a case of eating than drinking it.[50] This intensely calorific diet quickly blows junior out, so that a harp seal pup goes from having almost no body fat at birth to half its weight comprising of blubber just days later. With its layer of insulation in place, the youngster can now embark on the next step in its development underwater.

---

49 The fact that not all whales do this suggests that there's more to this than simply seeking warmth. Killer whales, a major predator of whale calves, tend to be less common in the tropics, and the seas near the equator tend to be calmer.

50 For comparison, cow's milk and human milk both contain about 4% fat.

The challenges faced by warm-blooded animals in polar seas seem slight by comparison to the wholesale biochemical reconfigurations that their cold-blooded cousins have been compelled to adopt. Antarctic ectotherms in particular somehow manage to function in water approaching –2°C, a temperature that sends the molecules of life haywire and risks freezing animals solid. In the same way that butter that's soft out on the kitchen table goes hard in the fridge, the fatty cell membranes of animals start to solidify and lose function as the temperate drops. To counteract this and to allow them to operate normally, the cell membranes of Antarctic animals such as fish are packed with unsaturated fats that maintain plasticity in the cold. Similarly, the crucial proteins that provide structure and function have undergone massive changes in polar animals to prevent them from unravelling in the icy conditions. And mitochondria, the power plants of the cell, don't work as effectively in the cold; so, animals at high latitudes cram their cells with extra mitochondria to compensate. Even so, the dearth of thermal energy in these frigid waters means that the tempo of life is much reduced: polar animals dawdle by necessity.

We're used to the idea of blood being red, but in the ocean there's a rainbow of different blood colours, from the purple of lamp shells to the blue of octopuses and the green of some worms. And in Antarctica, there's one more 'colour': clear. There are some physiological advantages to living in cold water, not least that it's packed with oxygen, and it's this that is the basis for one of the weirdest adaptations of any vertebrate. Blood, just like the fats mentioned before, becomes thick and claggy in cold temperatures. To allow things to flow more easily, the dominant group of fish in the Southern Ocean, known as the notothenioids, have cut right back on their red blood cells and, uniquely among vertebrates, some have got rid of them altogether.[51] Icefish, as these are known, rely on the superabundance of oxygen in the chilly water to perfuse their tissues without the need for haemoglobin to grab it. As a result, their blood is translucent, almost clear with a slight milky tint. The adaptation of icefish to the extreme conditions of the Southern Ocean represents a triumph of evolution. Their reward is dominion over a huge swathe of the ocean, and a release from the intensity of competition and predation found in warmer climes. The discovery in 2022 of a vast breeding colony of one species, Jonah's icefish, off the coast of northwest Antarctica in the Weddell Sea gives some idea of just how successful these creatures have been in their frozen biome. An area of seabed, four times the size of Manhattan, was pocked with 60 million nests, each one containing over a thousand eggs and guarded by a single male.

The cold offers one further, rather dramatic challenge to life. Since animals are comprised of water, in the chill of the Southern Ocean there's a real danger they might turn into popsicles. The salt in seawater means that it freezes at –1.9°C, whereas the slightly less salty interior of fish means that they solidify at about –0.8°C. Here, too, evolution has conjured an answer – the bodies of many Southern Ocean species are flushed with antifreeze proteins that latch onto ice crystals as they begin to form, stemming their growth and, ultimately, lowering the freezing point of the animals below that of seawater.

Antifreeze represents one among a wealth of evolutionary adaptations that occur throughout the animal kingdom. From the deep sea to mountain peaks, and from desert springs to here in

51 The icefish was first discovered in the 1840s when one was discovered during an expedition led by James Clark Ross. The fish was thrown by chance aboard his ship, HMS *Terror*, in a storm. Before it could be preserved for science, the ship's cat spirited the creature away and scarfed it.

the Antarctic, the harshest places tend to inspire the most astonishing biological solutions. For me as a biologist, this is worth celebrating in its own right, a glorious exposition of the indomitability of life. Nevertheless, I recognise that others need a little more persuasion and it's certainly true that many of these adaptations can be harnessed for practical purposes. Fish antifreeze proteins, for instance, can be synthesised in the laboratory for use in fields as diverse as cryosurgery and preventing ice cream from getting crunchy. For the animals themselves, however, the extensive adaptations that allow them to thrive in the face of the Southern Ocean's trials come at the cost of flexibility. They're exquisitely equipped for this very particular, very cold niche, but in a warming world, the specialisation of these biological wonders may prove their undoing.

* * *

Two great spouts of steam rise into the chill air from the surface of Isfjorden, and as I train my binoculars on the spot, an enormous dark fluke emerges from the water, silhouetted against the white of a faraway glacier. It's a fin whale, cruising the fjord in search of schools of capelin or swarms of Arctic krill to feed its enormous appetite. Fin whales are second only in size to their close relative the blue whale, and both of these leviathans paradoxically rely on a diet of incredibly tiny animals. If we were to emulate fin whales' pursuit of krill, then (allowing for scale) we'd specialise on chia seeds. To come to terms with their minuscule prey, fin whales are among 16 or so species of baleen whale whose jaws are full not of teeth but of long sheets of keratin, which hang, close packed, from the upper jaw and are configured like the teeth of a comb.[52] This dental

52 Foetal baleen whales do have remnant teeth, a distant ontogenetic echo of a bygone, toothed ancestor.

arrangement is used by all baleen whales essentially as a sieve, though there are differences between the species in terms of how they feed. Grey whales roll onto their sides to suck up shrimp-packed payloads of silt from the ocean floor through the side of their mouths, while bowheads simply swim along with their mouths open, their curtain of baleen parted in the centre to allow water and food in. The fin whale here in Isfjorden, in common with humpbacks and blue whales, is a lunge feeder. In essence, the whale hunts by accelerating towards its prey and engulfing a huge mouthful of water complete with a cargo of luckless prey. It then uses its enormous muscular tongue to squeeze the water through its baleen, leaving only the krill or capelin behind. Lunge feeders have disproportionately large mouths and extensible, pleated throat grooves that increase the capacity of their mouth cavity dramatically. A large fin whale, for instance, can engulf 82 tonnes of water in a single lunge – more than its own body weight – from which they might extract tens of kilograms of delicious krill.

The baleen that serves whales so well in filtering miniature prey from a mouthful of water is tough, yet flexible and lightweight, and it was precisely these qualities that made us covet it. In the pre-plastic Victorian era, whalebone, as it was misleadingly called, was a wonder material. It was pressed into service most famously in corsetry and dressmaking, but also formed the basis for umbrellas, fishing rods and brushes. When soaked, it could even be made pliable enough to be woven into baskets. Useful as it was, however, the importance of baleen paled in comparison to the most valuable commodity of whales: their oil.

The first half of the nineteenth century was fuelled by whale oil. The clean-burning oil lit the parlours and streets of Europe and North America, and it lubricated the machines of the Industrial Revolution. Whale oil fuelled the boilers of steamships and factories, and drove the mass production of soaps, creams and cosmetics. Insatiable demand drove the price ever higher. At the market's peak in the mid-1850s, the oil from a single whale might realise almost a quarter of a million dollars in today's money.

Whaling itself was a punishing career, especially at this time. Months of exhausting, grimy toil were attended by the perennial risk of injury or death. It seems extraordinary today to think of pursuing such a colossal animal through inhospitable seas by sailing prowess alone. And it seems scarcely credible that, at the critical point, an open-topped whaleboat would be lowered from the ship, powered by oars, and armed simply with a spear to bring the animal to bay. If the harpoon stuck, the wounded creature would, often as not, rampage in fear and pain, dragging the flimsy whaleboat scudding madly behind it in what was known as a Nantucket sleighride. The superstition that was rife among whaling communities meant that few of the mariners learned to swim – to do so was to tempt fate – so if the boat was sundered, its crew would find themselves at the mercy of harsh seas. That whaling ships were seldom short of volunteer crew despite this speaks of the money that was available for those with sufficient courage.

The harpoons wedged in the whale's flesh were little more than anchors to prevent the animal from escaping. The killing itself was an unspeakably grim affair that could stretch out for the best part of a day, and relied on the whaleboat closing on the victim, so that the whalers could stab it to death with long lances. Whalers spoke of the 'flurry', the last maddened actions of a mortally wounded animal, swimming in confused circles and thrashing about in its torment, its huge tail threatening to smash whaleboats and their occupants alike.

Once the kill was made, hours of toil lay ahead for the whalers, flensing the corpse to relieve it of its blubber and rendering this down to oil in enormous cauldrons. The conditions endured by the crew are unimaginable to us now, forcing aching limbs to cut, cut, cut, while caked in a grime of sweat, blood, fat and filth. The oil was decanted into barrels, perhaps 50 or more, to be rolled into the hold for storage and the giant cadaver would be cast off, left to spiral into the depths.[53] This done, the lookouts would ascend the masts, signalling the start of the next hunt in a grisly sequence spanning a journey that could last for months or even years.

Sometimes a whaler might get lucky, and chance upon a place where the whales clustered in large numbers. Such a place was Ojo de Liebre, a sheltered lagoon on the Pacific Coast of Baja California. It was here that grey whales would gather to give birth, taking advantage of the calm waters to nurture their calves over the first days of their lives. In 1857, the whales' secret was discovered by Charles Scammon, captain of the whaling ship, *Boston*. Grey whales were sometimes called devil fish by whalers, in acknowledgement of their aggression once harpooned. Within the lagoon, Scammon and his crew suffered a series of reverses, losing boats and men in the melee. A change of tactics from harpoon to bomb lance turned matters in the whalers' favour. The bomb lance was a short gun that fired an explosive charge into the victim, turning their spouting breath crimson. It was a method not much used at sea because the stricken whale was likely to sink, but here in the shallow lagoon, it was horribly perfect. By the end of the voyage, the *Boston* returned to port with so much oil on board that it rode low in the water, attracting the attention of other keen-eyed captains. Over the next four years, Ojo de Liebre and the other lagoons on the Baja coast were besieged by whalers. Some used the bomb lance, others chased calves into the shallows, relying on the protective instincts

53 A barrel holds 42 US gallons of oil, which is around 159 litres.

of distraught mothers to follow them into the killing zone. In short order, the lagoons were cleaned out; 10,800 adult whales were slaughtered. Orphaned calves, too small to be worthwhile to the whalers, were said to circle the boats desperately, perhaps mistaking them for their mothers' bulk, until one by one they starved to death or were taken by killer whales.

The revulsion that many of us might feel for these practices comes so naturally that it's hard to put ourselves in the whalers' position. Yet our thoughts on whaling are largely a modern phenomenon; the nineteenth-century viewpoint was far less sentimental. As Richard Ellis, in his excellent book *Men and Whales* writes: 'It is only through the lens of hindsight that the whaleman's job becomes malicious or cruel ... Oil was needed for light and lubrication; baleen was needed for skirt hoops and corset stays. That whales had to die to provide these things is a fact of seventeenth-, eighteenth-, and nineteenth-century life.' Our perspective on whales has changed in recent times, partly because of our greater understanding of them, but it also emerges from an event that decoupled humanity not only from a reliance on whales, but which changed our world forever.

In 1859, prospectors struck oil at a well in Titusville, Pennsylvania. The thick, oozing black liquid that gurgled forth was to become the most significant fuel in human history. Refined products from what was then called rock oil rapidly supplanted whale oil for lighting and became a cheaper, more reliable source of lubricants. As rock oil became ever more dominant, its whale-derived equivalent was pushed to the margins, a relic of harder, dirtier times. The magazine *Vanity Fair* commemorated the shift in 1861 by publishing a cartoon showing cavorting whales in formal dress celebrating their deliverance from persecution at a gala dinner and toasting one another with champagne. Titled 'Grand Ball given by the Whales in honour of the discovery of the

Oil Wells in Pennsylvania', it seemed to herald a new future for the animals.

In the aftermath of the Pennsylvania oil strike and those that followed, the price of whale oil slumped. Whaling, once one of the most significant industries in the United States, went into a tailspin. But just as it seemed that it was to fade into memory, whaling was pulled back from the brink by two major developments. The first, just three years after the appearance of the *Vanity Fair* cartoon, was the development of a new whaling cannon by the Norwegian shipping magnate, Svend Foyn. No longer would whalers have to pursue their quarry with spears in a manner that recalled cavemen hunting mammoths. Now they could fire a lethal 50-kilogram harpoon at their target from the gunwales. Foyn soon refined his original design, adding a grenade to the harpoon's tip that would explode on contact with the victim. The second development was the transition from sails to propellers, driven by steam and, later, diesel. Mechanised ships could travel faster and further, they could chase whales upwind, and they could penetrate the whales' last refuges in the far south. Whaling wasn't dying, it was being reborn. Its new incarnation was hungrier, more intensive and far deadlier than ever before. The stage was set for one of the most dramatic extirpations of any mammals in history.

Just two centuries ago, it's thought that there may have been more than three million baleen whales in the world's oceans. In just the hundred years up to the 1960s, the global scramble for whales brought these peaceful giants to the very edge of extinction. Factory ships were built to process the animals at sea, hauling the carcasses on board and converting the majestic animals into component commodities. As petroleum superseded its role in lighting and lubrication, whale oil was repurposed as a cheap and abundant source of dietary fat. In the early decades of the twentieth century, half the population of Europe and North America were spreading margarine made from hydrogenated whale oil on their daily bread. Whale oil was also the primary source of glycerine, which when processed into cordite became the chief agent of munitions and destruction in the world wars. A use was even found for whale meat; instead of being discarded as it once had, it was processed and packaged into stock and soups, or ground along with bone as food for pets and livestock.

What whales had, the market wanted. Fortunes were made even as whale populations plummeted under the remorseless assault, the allure of the prize eclipsing concerns for the animals themselves. Not all the hands of whaling ships were able to rest easy in the face of this industrialised slaughter, however. In an interview given to the BBC, the Shetland whaler, Gibbie Fraser, recalled his conflicted emotions about the hunt: 'You couldn't help wincing when that harpoon went in because that was a living animal, feeling the same as we do, as far as pain is concerned. It was a brutal way of life.'[54] Fraser's colleague, Roddie Morrison, shared a similar perspective: 'They seemed so friendly, and they'd make a noise when you hit them – a cry, really. I felt that.' An account by the Scottish physician, Dr Harry Lillie, a medical officer on board a British whaling ship in the Antarctic in the 1940s, described his revulsion: 'If we can imagine a horse having two or three explosive spears stuck in its stomach and being made to pull a butcher's truck through the streets of London while it pours blood into the gutter, we shall have an idea of the method of killing. The gunners themselves admit that if whales could scream the industry would stop, for nobody would be able to stand it.'

Even at the time of these accounts, an awareness had already developed that the situation was deteriorating rapidly. In 1946, the 15 nations with the greatest interest in commercial whaling met in

54 *Britain's Whale Hunters: The Untold Story*, presented by Adam Nicolson and produced by KEO Films, was first aired on BBC Four in 2014.

New York to discuss what might be done. The result of the meeting was the establishment of the International Whaling Committee, the ostensible purpose of which was to regulate the industry and to set limits on the number of whales that could be taken. Much as this might have seemed to herald a new era in conservation, the whaling nations were at least partly motivated by the need to control the supply of whale oil onto the market. Restricting the number of whales being slaughtered kept prices, and profits, nice and high; protecting whales was merely a by-product of this. For the next 20 years, quotas were set for each season based on scientific advice, and each season those quotas were ignored amid international jockeying for the proceeds. Indeed, the net result of the control was to increase competition among whaling nations, generating a race to be the first to fulfil the quotas that became known as the Whaling Olympics.

Whenever scientists urged caution and tighter controls, the whaling nations expressed outrage and filed objections. Some, such as Norway and the Netherlands, resigned in a fit of pique. The inevitable result was that by the early 1960s, the great whales had been virtually wiped out. At its 1964 meeting, the IWC recommended that no more than 2,833 blue whales should be taken from Antarctic waters, already an improbably high figure relative to the likely number of whales remaining. After the meeting was adjourned without agreement, representatives of the whaling nations retired to more convivial surroundings and decided that the actual quota for the year should be 8,000. The following year, blue whales were all but eliminated from the southern hemisphere. With almost no blue whales left to hunt, the Commission voted to accord them complete protection, salving the financial pain by increasing quotas dramatically on other whale species. Even then, the Soviet Union, sitting outside the IWC, decided that it needn't take any notice, while Chile and Peru thumbed their noses at the moratorium and between them caught another 500 blue whales the following year.

A combination of greed and indifference had brought the whales to the brink. Across the entire world, the population of the largest one of all, the blue whale, was estimated as comprising no more than 2,000 individuals and perhaps as few as 600, distributed across millions of square kilometres of ocean. Their chances of survival were in the balance, and for the next 30 years, there was little sign of a recovery. But with the dawn of this century, better news started to filter through. Each year seemed to bring a small, but important, increase in their population until now it's thought that there may be something like 20,000 blue whales cruising the seas – ten times as many as at the end of whaling, yet more than ten times fewer than there were before we started to hunt them. It's too early to celebrate, however. These increases have to be seen against the backdrop of the slew of toxic chemicals that accumulate in the whales' bodies, not to mention ship strikes and global warming, but for now, blue whales are heading – just about – in the right direction.

Whaling persists in some parts of the world. The dogged defiance of Japan, Norway and Iceland continues in the face of international censure. The righteous fury expressed by many is something that I fully support, yet it has to be viewed in context. Although the stubborn remnants of the whaling industry slaughter around 1,000 animals a year presently, at least ten times as many are killed by ship strikes. Most of the time, these go unnoticed, but not always. A young humpback, named Sweet Girl by locals, suffered appalling injuries in 2024 after being struck by a ferry in Tahiti. Graphic and horrifying images show the whale labouring along in the aftermath, bleeding heavily, with her upper jaw torn off. In 2022, another humpback had her back broken in a collision in the waters off British Columbia. Despite her injuries, the whale, referred to as Moon, still completed her southward migration, crossing 5,000 kilometres of open ocean to reach Hawaii three months later. She'd made the journey by adopting a kind of breaststroke, using her pectoral flukes to propel herself along, but the supreme effort left her emaciated and plagued with whale lice. Unable to feed, her fate was to starve to death.

These two whales are the harrowing face of a growing, but largely hidden, problem. The vast majority of the world's commerce is carried by ships, and there are four times as many of them at sea as there were just 30 years ago. The skippers of these huge vessels likely wouldn't even notice a whale strike on the bows, which lie half a kilometre in advance of their position on the bridge. And even if they did see a whale surfacing ahead, their momentum is such that there's little they could do to avoid a collision. For their part, the whales may not even notice the juggernaut bearing down on them, since the ship's bulk masks the noise that comes from a churning, 6-metre propeller at the stern. There's no such thing as a glancing blow from a 200,000-tonne vessel travelling at 20 knots.

Attempts to address the problem by imposing speed limits along whale migration pathways offer a partial solution, but even at 10 knots (about half to three quarters the typical speed for ocean-going ships) the risk of a whale dying from a strike is still 80% from a container ship and 50% from a smaller vessel. Stopping at poorly enforced, and relatively minor, speed restrictions is an abrogation of our responsibility. We need to do more, and we can do more. The use of sonar to detect whales may be counterproductive, as I describe below, but a combination of passive acoustics, drones and satellites can be used to create a detective network to reduce the risks of ship strikes.[55] As yet, the deployment of these systems is rare, and highly localised. It's time that these measures were adopted universally.

55 Passive acoustics is essentially listening out for the sounds made by whales.

Shipping isn't the only threat to whales. I've already talked about chemical pollution, but there's another pollutant that compromises whales: sound. For animals that use complex phonations to communicate, and which have a highly evolved sensitivity to sound, our sonic invasion of the deep is a disaster. Low frequency sonar that overlaps with whale communication channels compromises the ability of the animals to locate one another and to interact. Worse still, military exercises involving detonations, and the airguns used in seismic surveys, produce sounds that can kill. To allow a means of comparison, I think anyone would agree that a jet engine, at 120 decibels, is loud. The decibel scale is logarithmic, so a sound of 130 decibels is 10 times louder than a jet, and 140 decibels is 100 times louder, and so on. The sound levels involved in surveying, and in military operations, can easily exceed 200 decibels and would be instantly fatal to animals in the vicinity. Since water conducts sound far more effectively than air, the damaging effects radiate outward for tens or hundreds of kilometres. As with whale strikes, we don't necessarily register the problem because the corpses of the victims are typically lost to the depths. However, in early 2023, thirteen Cuvier's beaked whales washed up dead on the shores of Cyprus, following a series of military exercises in the eastern Mediterranean. They were unmarked, but autopsies showed that the luckless animals had undigested food in their stomachs, a signal that they'd died suddenly. The cause of death in each case was gas embolisms, the marine mammal equivalent of what we call decompression sickness, or the bends. Under normal circumstances, this doesn't tend to happen, but the trauma of sonic shock destabilises their diving behaviour. Although gas bubbles were the cause of death, it was the military exercises that killed them.

Forty years on from the announcement of the moratorium on

whaling, the practice of harpooning some of the most incredible animals on the planet seems callous and indefensible. Yet the seas are busier, dirtier and noisier than ever before, and though the hunt has almost stopped, we've instead made the whales' habitat inhospitable to them. It may be that the demands of global commerce and defence are weighed as being more important to nations, but a wrong is a wrong. How we treat whales, how we treat any marine animal, is a test of our character and of our morality. It is not sentimental to love our fellow creatures, because if we dismiss concern for animals, we become less humane, and less human.

I've never yet seen a blue whale. So far, the largest of all animals has eluded me, but it's a comfort to know that they are still out there, despite us. Like their slightly larger cousins, fin whales, too, were almost lost. A million of them once travelled the world's oceans, a number that we reduced by 98% in our greed. Those few that survived might be counted lucky, and yet we have testimony to how they suffered. Whales, like us, produce earwax and this builds up year on year in layers that allow analysis of their stress hormones. The patterns are both clear and challenging. Those animals that lived through the whaling pogrom show levels of cortisol that closely match the intensity of hunting, and which fell sharply once we ceased persecuting them. One anomaly in the trend occurred during World War II, when whaling fell yet cortisol spiked, most likely because of the *Sturm und Drang* that we unleashed.

Far in the distance, I see the fin whale spout once more as it heads beyond Isfjorden, out to the open sea. We are so lucky to share the planet with these magnificent creatures; they are not so lucky to share it with us.

6.
The Jewel Sea

Swirling life among bright-painted reefs

IT'S 80 KILOMETRES AND A MILLION MILES FROM THE Queensland port of Gladstone, with its heavy industry, noise and trucks, to our destination, the coral fantasia of One Tree Island. Beneath the keel of the catamaran, a changing kaleidoscope of colour marks our passage. The murky, spoiled brown waters of the working mainland harbour gives way to the midnight blue of the Curtis Channel. I stand lookout at the bow, scanning the water in hope of seeing dolphins or seabirds, occasionally looking back to the white trace of foam in our wake, its fading trace pointing back to land, back to the clamour and dirt of civilisation. Ahead lies one of the greatest wonders of the natural world, the Great Barrier Reef.

There's a buzz among my research group: we're wishing this journey away; we want to be there, in the midst of a thrilling profusion of life. We know that there we'll each be billionaires in kind, surrounded by a limitless wealth of biology. Our eyes are fixed on the horizon, straining to see the first hints of distant reef islands, but it's the sea that is our harbinger. There's a gradual, almost imperceptible, transition from deep ocean blue to brilliant aquamarine as the seabed rises to meet us. This lightening of the sea and the intensification of its colours puts me in mind of a dawning, as though the sun were rising beneath the waves.

As the ship slows and enters the lagoon of Heron Island, I look down into the glassy waters and see the outlines of dozens of familiar species of coral, and the flotillas of fish that attend them. Eagle rays, their coal-black backs fizzing with white polka dots, fly through the open-water channel, then bank around a huge, bulbous brain coral as a school of electric blue damselfish huddle for safety among a forest of staghorns. It's a fantasia in aragonite from the house-sized bommies to the white sand on which they sit. Look closer and the scene is made up of countless separate colonies, each in turn a tessellation of myriad individual polyps. These anemone-like polyps are miniature sculptors, their collective efforts yielding colonies that embrace a multiplicity of forms, each as distinct as a human fingerprint.

Corals are such exceptional and unusual organisms that the question of exactly what they are was one that perplexed early naturalists as they tried to make sense of the world. To the touch, coral can feel like stone, and this was enough for Theophrastus, a student of Plato and Aristotle, to declare that these were not living organisms, but mineral formations. Later, he revised his opinion somewhat, covering his bases by remarking on their tendency to 'flower' when submerged, and deciding that they were in fact stony plants. Others firmly, if mysteriously, asserted that they were trees, while the Roman scholar, Pliny the Elder, suggested that being neither plant nor animal, they belonged to a third kind of organism, one that had characteristics of each. Indeed, it wasn't until the eighteenth century, and the advent of the first microscopes, that we learned that these organisms are incontrovertibly animals.

Nevertheless, the earlier speculations contained a degree of truth. Corals are multifaceted creatures, and though they are animals, each polyp fashions its home in stone and depends for its livelihood on plant-like organisms that live within its own tissues.

Over time, the efforts of these industrious polyps give rise to a reef that's crammed with idiosyncratic colonies. Amid the profusion of smaller corals sit some colossal and venerable giants, such as huge *Porites* colonies that can stand twice as tall as a double-decker bus and measure anything up to 70 metres around the base. Titans such as these likely established about 400 years ago, at around the time that the *Mayflower* departed for the New World, yet imposing though they are, each rose from the humblest of origins. Each colony is founded by a single oceanic rover, a coral larva known as a planula, casting around for a place to call home. A planula isn't much to look at. It's less than a millimetre in length, its oval form fringed with minuscule beating hairs that propel this tiny fragment of potential onward.

The planula's wanderings may last anywhere from a few hours to months. After all, this is a crucial judgment, and shouldn't be rushed To aid its choice, the microscopic traveller is armed with sensory apparatus that allows it to accumulate a dossier of information on its options. Much like a human homebuyer weighing the attributes of different homes, there are plenty of factors that the planula has to ponder. The perfect home for the planula is brightly lit, should feature the welcoming smell of certain preferred kinds of algae, and should be at the ideal depth: shallow enough to ensure uninterrupted access to light but deep enough to dampen the worst effects of storm waves. Oddly, mood music is important, too. Acoustics might seem an unlikely consideration for the planulae, but they home in on a distinctive cacophony of sounds that declare a robust and healthy reef. By contrast, sickly reefs lack this enticing siren song, and the planulae avoid them, just as they sidestep those plagued with the sorts of sonic disturbances made by boat traffic.

Having assessed its options, the planula makes its call and, forsaking the roaming of its youth, fastens itself to the substrate and adopts a sedentary lifestyle. It undergoes a transformation becoming a polyp, a central trunk bearing a crown of six tentacles that reach for prey in the water. But catching passing morsels provides only part of the polyp's diet: the bulk of its calories are donated by a particular kind of lodger, one that pays its rent in sugar.

The lodger in question is a dinoflagellate, a tiny single-celled alga that's fundamental to the coral. The more considerate among coral parents bequeath a starter pack of algae to their youngsters, diligently apportioning a few to each of their eggs. Most, however, don't cater so carefully to the next generation, so their offspring must attract their own. To do this, newly settled polyps put out a chemical advert for tenants. Dinoflagellates flock to the call and are enveloped by the welcoming polyps, establishing themselves in the surface tissues of the coral. Once there, they begin to do what algae do best, which is to photosynthesise, busily converting the sun's energy to glucose. The coral landlord provides the algae with the security of a home and feeds its partners with its own waste products; the algae reciprocates by delivering a sugary remittance.[56] It's one of the longest standing and most successful partnerships in the natural world, one that shallow-water corals can't do without. There might be as many as five million algal cells hard at work in every square centimetre of coral tissue, contributing anything up to 90% of the coral's nutrition.

The polyp, ensconced in a limestone fortress of its own making,

56 This, at least, is as the relationship is usually cast. However, the real world is a little more complex, and darker: corals are unsentimental algal farmers who get a fair chunk of their food by eating their excess crops.

can now turn its attentions to expanding its horizons, which it does by splitting itself in two and creating a clone. The new clone establishes itself next door to its progenitor and sets to work building its own fortifications. By this process of doubling, the polyp multiplies and becomes a colony. Depending on the coral, it might construct lobes or branches, in the style of trees, or expand outwards to form mats or plates. Many use their forebears as foundations, with new polyps growing over the top of their ageing colleagues, smothering them. In good conditions, the whole structure expands outwards at a rate of around a centimetre a year, a living layer of polyps sitting on top of a dead core, an ossuary of tightly bound skeletons.

Picturesque as Heron Island is, the crowds of holidaymakers mean that it's a throbbing metropolis compared to our final destination, the research station on One Tree Island, which lies beyond the horizon to the southeast. Aside from the station managers, Russ and Jen, we'll be the only people on the island. Not that we'll be lonely; we'll be sharing the space with a breathtaking diversity of life. Russ is waiting for us at the jetty, a smile on his bronzed face. He's lived a nomadic life, but it's here on a remote island on the Great Barrier Reef that he's found home and he's fiercely passionate about the reef. On one of our previous trips to the station, he'd spotted a vessel on the horizon, approaching the edge of the lagoon. This is one of the most stringently protected areas of the reef: no trespassing without a permit, and definitely no fishing under any circumstances. Russ suspected that the vessel was up to no good, most likely a spot of illicit fishing in the protected zone, a suspicion that seemed to be confirmed when he couldn't raise them on the radio. There are plenty of places where fishing is allowed on the reef, but the balance is maintained by strict no-take zones, and the idea of someone transgressing brought Russ to fever pitch. He scuttled, red in the face and muttering to the station's speedboat, and charged off into the distance. He returned not long afterwards, and though the trespassers had very wisely hightailed it at his approach, Russ had managed to scribble down their Hull Identification Number. The poachers could expect a warm welcome from the authorities when they returned to port, as well as a stiff fine.

Today, however, Russ is in a genial mood as he helps us board the boat, the same 4-metre aluminium charger that he'd ridden into battle against the poachers, and we set off, leaving the rest of the planet behind. Just a few kilometres of plain sailing now to the island, and I'm surrounded by luminous turquoise waters and beautiful reef. The sun and the gentle spray of the sea is on my face; I've never been happier in my life.

* * *

As the boat slows to a stop among the wavelets that ripple against the shore, we jump into the shallows and set to work carrying boxes of equipment and supplies up onto the coral rubble that forms this cay. It's poorly named, One Tree, because however many trees there are crammed into this tiny island, where you can't walk much more than 200 metres in any direction, it's a lot more than one. Salt-tolerant scrub like velvet soldierbush, beach cabbage and *Pisonia* trees thrive, their waxy leaves hanging on greedily to the rare and precious freshwater. The trees, in turn, provide lodgings for an astonishing profusion of seabirds, especially noddy terns who make their slapdash nests seemingly at every point where a branch forks. There are thousands of these elegant birds, whose charcoal-grey body is topped with a silvery cap. A rim of silver beneath each eye makes them even more fabulous, as though they were glammed up for cabaret. Their screeching, rattling cries, along with the boom of waves far away on the outer reef are the soundtrack to One Tree, while the challenging stench of the terns' fishy, guano-garnished, rotten leaf nests is the signature bouquet. Never having experienced persecution, the noddies are entirely unconcerned by our presence, and stare impassively at anyone who walks by.

This riot of noddies, along with bridled and roseate terns, egrets and even the dazzling sacred kingfisher are all here for one reason: fish. The plethora of birds gives an indication of the biological richness of the reef's waters, yet this richness is a paradox. The tropical seas that play host to coral reefs are starved of nutrients. Other parts of the ocean, particularly polar regions, are steeped in a wealth of minerals such as nitrogen, phosphorus, iron and silicon, which phytoplankton rely on. Consequently, primary productivity – essentially a measure of phytoplankton activity – is incredibly high towards the poles, and since this forms the foundations of marine food webs, these are seas of plenty. Much of the nutrients that fertilise phytoplankton arrive in the surface water via upwelling currents from the deep ocean. In cold climates, this works well, but the heat of the tropics results in a layer of warm water that floats on the dense, chilly water below, forming a barrier that prevents mixing and stops the rising nutrients in their tracks. So, although tropical oceans are blessed with an abundance of sunlight that would otherwise support phytoplankton growth, there's little to nourish these marine microalgae.

The crystalline waters of coral reefs and the astoundingly vivid and vibrant cobalt, emerald and ultramarine hues that draw us to them are that way precisely because there are so few phytoplankton. And for the diners at the next level of the reef food web that rely on phytoplankton as food, tropical waters are the thinnest consommé imaginable. And yet, somehow, this depauperate stage is the setting for the most dramatic explosion of life anywhere on our planet. A *quarter* of marine species call coral reefs home, despite the reefs occupying less than 0.1% of the sea's area. The equivalent on land would be if one in every four terrestrial species on Earth were found pretty much only in Cuba, or Kentucky. Also imagine that Cuba (or Kentucky) in this analogy is essentially a desert, and yet still teems with this unprecedented biological variety. It just makes no sense. The richness of coral reefs was the riddle wrapped in a mystery inside an enigma of marine biology; it has puzzled biologists for centuries, most notably Charles Darwin himself as he travelled on the HMS *Beagle*. Now, finally, we've begun to work it out.

Coral reef communities are astonishingly efficient. They're tenacious in grabbing what nutrients exist in these impoverished waters, and they keep hold of them. There are many different players in this thrifty operation; the the symbiotic dinoflagellates known as zooxanthellae within the coral's tissues are aided in their mineral grasping by other photosynthesisers. Coralline algae, bolstered and toughened by calcium, acts as the cement of the reef. It grows in rugged crusts, although it's capable of architectural flourishes in the form of lobes or even the kind of spikes you might associate with a bad hair day. Despite its sturdiness, it's handsome too, looking at times like someone's tipped a pot of pinky purple paint on the reef.

Cyanobacteria, too, are here. Like their algal counterparts, they manufacture their own food by photosynthesis, but they have an extra string to their bow; they make their own fertiliser. More specifically, they 'fix' atmospheric nitrogen, converting it into ammonium, which then forms the basis for the construction of proteins needed to make new cyanobacteria. This nifty trick also gives a boost to the rest of the reef ecosystem, yet the risk remains that these efforts could be lost to the starving waters beyond the reef, and this is where the reef's nutrient retainers and recyclers come in.

Typically, these crucial tasks are undertaken by unheralded creatures. As one of the simplest of all animals, sponges seldom attract much attention.[57] They might lack the charisma of, say, a dolphin, but sponges are extraordinary creatures, as any spongiologist will tell you. For most of their lives, which can span centuries, they sit in one place, inoffensively filtering water, drawing it through the bottom of their bodies and then expelling it from the top, minus its payload of tasty particles. So far, so underwhelming, but there's more to a sponge than being an animated flue. Take a sponge and turn it into a smoothie in a blender and the sponge will respond to the insult by painstakingly rebuilding itself, cell by cell. When sponges are threatened with something other than a kitchen implement, they produce some ingenious defences. Some strew their surroundings with spiky, star-shaped spicules similar to the caltrops used by the Romans. Ancient armies used these to impede the passage of horses; sponges use them to deter predators that glide over the seabed, like sea stars. Many of them augment this hardware with chemical defences that repel everything from bacteria to large predators, and which employ chemicals that have huge promise in human medicine. But it is as nutrient recyclers that sponges play a starring role in reef ecosystems.

As good as algae and cyanobacteria are at scavenging the precious nutrients they need to photosynthesise, both they and their coral partners inevitably lose some of these nutrients, as well as a share of the fruits of their solar-soaking labours, as waste products. This is where sponges come in, sucking up all of this turdy goodness and keeping it on the reef rather than allowing it to dissipate into the seas beyond. The sponge puts its unappealing diet to good use, converting the discards into sponge cells. To maintain the flow

57 One measure of animal 'simplicity' is the number of different cell types. Sponges have around 18 cell types compared to our 200. It's important to add, though, that although we are more complex organisms, we're not more 'highly evolved' than sponges. This is a nonsense term; sponges are an extraordinarily successful animal group.

of nutrient-rich water through their bodies, sponges employ cells known as choanocytes, each of which is equipped with a hair; it's by beating these hairs that the sponge generates a water current through its hollow body. That this is exhausting work is evidenced by the fact that choanocytes are replaced every few hours and expelled by the sponge. Millions of these used cells, underwritten by the sponge's peerless devotion to eating shit, go on to feed the hungry mouths of reef organisms.

Small, secretive fishes, such as gobies and blennies, are one of the beneficiaries of the sponges' largesse, and they're also central to the nutrient cycling. We seldom notice these fish, despite most reefs housing dozens of them in every square metre of the coralscape. They lie low in the fissures and interstices, living fast and dying young, and producing large numbers of offspring during their short existence. Tiny as they are, the larvae of these fecund fish account for almost 60% of all the fish eaten on the reef by weight, and since many of them don't disperse far, they're another demonstration of how the reef ecosystem hangs on tight to its precious resources.

No one organism is responsible for the smooth running of this incredibly effective biological machine, rather it's the result of the close collaboration of many different interacting partners that keeps it purring along. Collectively they boost reef ecosystems, making them an extraordinary 100 times more productive than the surrounding waters, resolving the paradox of how lush coral gardens exist in the nutrient-starved tropics.

* * *

The first thing I always want to do when I get to One Tree is to get in the water, so it's a scramble to dump the bags, grab the snorkel gear and rush to the shore. Splashing inelegantly through the shallows in my hurry to find water deep enough to swim while simultaneously trying to pull on reluctant fins isn't going to endear me to the local fauna. But I hope they'll forgive me for my haste because after all the travelling, I'm desperately keen to commune with them. Just moments later, my urgency gives way to a deep sense of calm as I glide along the surface above a bed of coral sand. I'm heading out to Shark Alley, a channel of deep water overlooked by a spectacular wall of coral and patrolled by the eponymous elasmobranchs, but that's a couple of hundred metres away. Besides, now that I've merged with this incredible marine scene, the fretful haste of my city-self has melted away.

The bed of the lagoon lacks the fishy richness of the coral outcrops, but there's one species making its home here that I'm always delighted to see. Glider gobies aren't the reef's showiest fish, but they have a charm that I can't resist. Roughly the size and shape of a carrot, the goby's dominant colour closely matches the sand that is its home and workshop. It then ruins its aspirations to camouflage by incorporating thin, brick-red stripes running the length of its body and pearlescent, sky-blue dots that fleck its cheeks. Its large head terminates in a pair of thick rubbery lips, like the unfortunate victim of an overzealous plastic surgeon. Aesthetics aside, it's their behaviour that I find captivating. Glider gobies live in pairs, maintaining a territory that's centred on a burrow. This is one of the reasons that the gobies need their reinforced smoochers; it's no easy task to dig down into the compacted sand. Excavation is largely the responsibility of the male, whereas the female takes things a little easier, saving her energy for egg development. He emerges from the hole every few seconds with a mouthful of grit, using this to build a mound by the burrow entrance. This isn't a haphazard construction; it's carefully positioned to divert the prevailing currents through the den, providing oxygenated water to the eggs that will be laid within.

The female watches her partner going about his work with an appraising eye. He has to show the requisite skill and hard work to convince her to commit her eggs to this union, especially since he'll also be the primary carer for the brood. And although these gobies are often described as monogamous, temptation, in the form of roving suitors, is never far away. Indeed, there's enough intrigue and infidelity on the lagoon floor to inspire a fiery soap opera. As I swim overhead, they watch me cautiously, lying cosily side by side on their bed of sand and looking for all the world like marital strife is far from their thoughts.

As I kick on, the depth gradually increases until I can no longer see the bottom beneath me. I've arrived in Shark Alley and if I'm lucky, the sharks will be along soon. Already here are trevally, dozens of them gathered in a loose shoal, each getting on for a metre long. These are majestic fish, open-water predators with deep, silvery bodies that are built for speed. They're also intensely curious and as I swim along, they crowd around me, some swimming right up to the glass of my mask to look me in the eye. I stop swimming and let them carousel around me for several minutes, and I'm struck by just how relaxed they are; like the noddy terns they show no fear. What a privilege it is to be accepted by animals in this way, on their own terms. I can't help wondering if all creatures were like this before we began our assault on them. Eventually, one after another, they peel out of formation and head off into the distance leaving me alone once more. Solitude isn't a common state of affairs in One Tree's lagoon, its 12-square-kilometre extent is crammed with measureless numbers of inhabitants. Sure enough, just a couple of minutes after the trevallies' departure, I look down and see perhaps the most unmistakable silhouette in the animal kingdom.

I can't think of a more beautiful animal than a shark. They are the perfect marriage of form and function; their sleek, torpedo shape trimmed with triangular, wing-like fins is a symphony of streamlining and power. The one below me is a blacktip and along with the diagnostic sable peaks of its fins, I'm struck by the sheen of its dove-grey back. I keep as still as possible to avoid spooking the blacktip because these are cautious creatures, despite being almost as long as I am. That said, remain motionless for long enough and their curiosity can overcome this caution; I once had a pair of blacktips approach to nibble my fins.

Deeper in the Alley's waters, whitetip sharks patrol. They're longer and thinner than their blacktip cousins, a form that lends itself to probing the narrow crevices of the reef where prey dwell. Over time, I've learned their routine, and the locations of their submarine cave dens, where half a dozen of them will doze the day away, jumbled together, cheek by jowl. There's a host of other sharks here, too. Lemon sharks, grey reef sharks and wobbegongs aren't uncommon, and occasionally bull sharks and tiger sharks put in an appearance, all drawn by the rich pickings on offer in this no-take haven. And as I finally reach the coral wall that marks the eastern boundary of Shark Alley, I'm rewarded by another aggregation, this time of groupers. It's unusual to see these hefty fish in a sociable mood – they're loners – but today around 20 of them are milling about in close formation. These are all substantial, each about as long as a suitcase and almost as wide, and looking conspicuously well fed. If this congregation is, as I assume, the prelude to a spawning event, then the future of groupers at One Tree is assured.

These species, the trevallies, sharks and groupers, are the top predators here, a collection of carnivores which, alongside other substantial fish such as barracuda, comprise a rich and abundant community of hunters in the lagoon. The presence of so many large fish makes One Tree seem unusual, an outlier among the other reef locations I've visited on the Great Barrier Reef, and more widely in the Pacific, the Caribbean and the Indian Ocean. The sad truth, however, is that One Tree is a rare example of how things should be; indeed, how things were at all these places before fishing sliced the top layers from the food web. It's scarcely surprising that fishers target predators; they're usually the largest fish and they command the best prices. The question is: what does it mean for coral reefs to lose their big predators?

It's a simple question with the kind of multifaceted answer that befits an ecosystem as complex as a coral reef. The dazzling diversity of reef communities sets them apart from most other ecosystems where the effects of predator removal have been studied. Unlike famous terrestrial exemplars of the so-called top-down impacts of predators on their prey, such as the return of wolves to Yellowstone National Park, apprehending clear signals amid the labyrinthine coral reef food web is challenging. One major reason for this is that one predator might eat dozens of very different species, and one prey species might have an assortment of predators. Nevertheless, one thing is clear: in marine reserves, such as One Tree, there are far more fish, and far more large predators than in nearby, unprotected areas where fishing is allowed and, effectively, anything goes.[58]

As predators become less common, their prey should, in theory, blossom. So why aren't there more of the miniature, jewel-like fish that we most associate with reefs in places where the fishers have wiped out the apex predators? One possibility is that, rather than being limited by the top-down effects exerted by predators, they're instead limited by bottom-up processes. On the reef, this would mean that the food supply isn't great enough to sustain yet more fish. There's some truth in this, but another reason is that removing

58 A common One Tree predator, the coral trout is five times more abundant here than beyond the reserve, where it's a fishers' favourite.

apex predators can result in what ecologists call mesopredator release. In other words, take out the largest hunters and, in the sudden absence of their chief persecutors, smaller predators, like snappers, inherit the reef. The decrease in threat emboldens them, so they become more active, more successful and, in time, more numerous. This represents a problem for smaller fish, like damselfish, who are too small to interest a large shark, but are a favoured target for snappers. It's proof that, in biological communities, your enemy's enemy is your friend. In this sense, a damsel's best friends are sharks, the bigger the better.

* * *

The clear, sun-dappled waters of the reef and the bustling activity of its vivid occupants seem to us to present a view of paradise. The peerless aesthetics of these cathedrals of natural beauty conjure a sense of tranquillity in our minds, yet the reality for the animals that live there is very different. This is the other paradox of the reef: what we imagine as Shangri-La is, to its inhabitants, a biological Hell's Kitchen. It's an arena full of aggression and competition, jeopardy and privation, that's a million miles away from the serenity that we infer.

The grief on the reef can be witnessed at every scale, but to see it played out in microcosm you often need to be patient. This virtue is often in short supply when people pull on a face mask and fins to explore a coral reef. Perhaps understandably, most zoom around in a thrill of discovery, revelling in the intoxicating scenery, ceaselessly searching for new wonders around the next outcrop. If they were to keep still, the reef would reveal itself to them in a richer, more rewarding way. This was brought home to me some years ago when I was doing some research at One Tree that required me to stay stock still in the shallows for hours at a time. After just a few minutes, a plethora of animals that had fled at my approach now begin to show themselves. A few minutes more, and they seemed to have classified me as harmless and were busying themselves with their daily lives right in front of my nose.

The caution that the reef animals display is entirely justified; predation is a fact of life here, and if you're small it stalks your every moment. Now, as I watch, a host of diminutive creatures emerge from hiding. One of the boldest is a glassy little fish, streaked with red stripes. It's a pygmy goby, and one that has the unfortunate distinction of having the shortest lifespan of any vertebrate. If it were to persist for a mere two months, it would be a Methuselah among its peers. A blenny squeezes out from its bolthole and begins to crop at the balding algae with fleshy lips, each little bite looking as though it were kissing the reef in gratitude. A pair of coral gobies catch my eye as they hop along a branch of a staghorn, keeping a wary eye on me, even as they peck delicately at morsels from the coral's surface. These secretive fish are among my favourites; fleshy little slivers of life, you could make four of them by cutting a grape lengthways into quarters. There are a few different species here, each with its own distinctive livery, from velvety black to sunshine yellow, striped in green and red, or blue blotched with crimson. As adults, they're homebodies. The single coral they've settled in is their world and they'll likely spend the remainder of their lives in the comparative safety of its branches. They swim in short bounds, spending most of their time stationary and watchful on a coral bough. Given how vulnerable they are, leaving the coral to find a mate would represent an enormous risk, but this is where evolution has equipped them for flexibility. If two males find themselves in a coral, then one will obligingly develop into a female, and, similarly, if a pair of females share the same space, one will do the decent thing and become a male. They're capable of flipping back and forth like this according to need, an adaptation that means that, so long as there are at least two of them *in situ*, they never have to run the gauntlet of the world beyond their coral home.[59]

Just as my stillness has given me a chance to see the reef in a new light, a closer inspection of these chunky little gobies rewards me with a view of an extraordinary society in miniature, especially in larger corals with more living space. The pair I'm looking at now are occupying the equivalent of a modest starter home, rubbing along together in a coral the size of a grapefruit. In contrast, those on the goby equivalent of millionaires' row in a colony rich in hiding spaces may have a dozen or more of these chunky little fish in residence. If so, there'll usually be one breeding pair and a long queue of wannabes and juveniles. The hangers-on will only get the opportunity to reproduce if one of the dominant pair leaves, or, far more likely, dies. The enforced celibacy of waiting has to be weighed against the risks of leaving to take up residence in a new coral, which in a world crammed with predators may be considerable. While a patient approach is pragmatic, there might still be friction between the dominant adults and the heirs apparent. Size is vitally important in determining social supremacy in these groups and if one of the underlings grows to the point where one of the alpha pair feels threatened, that friction will increase, potentially even leading to that underling being evicted from the group. Remarkably, the subordinates manage this by dodging meals to avoid growing to a size that might jeopardise their place in the group. So, the sex- and

59 The nonconformist approach to sex isn't exclusive to coral gobies, or even to fish. Even some of the corals on the reef do the same. Hammer corals, named for the shape of their tentacles by someone who can never have seen a hammer, like changing sex so much that they switch their orientations on an annual basis. Mushroom corals take a steadier approach to this kind of sexual freelancing, being male when conditions are less than ideal and female when life is more prosperous, on the basis that it's expensive to be female.

food-starved deputies play the long game, living a life of privation in the hope of one day inheriting the coral.

There is another factor that helps to calm the tensions within a goby group: blood ties. The goby groups within each coral are more closely related to one another, on average, than they are to groups across the rest of the reef. The fact that they share genes predisposes animals to be more considerate to one another. This is known formally as kin selection; we see it in our own societies and famously in insects like ants and bees, but for it to happen under these circumstances is little short of astonishing.

* * *

Reef animals have converged almost unanimously on a childcare strategy of Dickensian brutality centred on surrendering their offspring to fate. Some, like the corals themselves, simply launch their gametes into the water column and have done with it. The more attentive parents, including many fish, do at least tend their eggs to the point of hatching. But once this parental chore is done, they too leave their progeny to the whims of providence. The microscopic larvae are carried by currents off the reef and out into the open ocean where they must learn very quickly to fend for themselves. With so many predators in attendance, it's little surprise that the casualties are astronomical. At the tender age of one hour, the odds aren't great that a newly independent larval fish will even live long enough to see tomorrow. The exact numbers are hard to calculate, because the task of unobtrusively following the fortunes of a transparent fish the size of a cumin seed in the vastness of the sea is, at present, beyond us. Nevertheless, by reckoning how many start out, and how many return, we might estimate that anything up to 90% die in that first day.

The surviving larvae can boost their chances by getting as far from the reef as possible. The number of predators drops the further they go and this, of course, is why coral reef animals operate in this way. Strange as it might seem to us to simply jettison the kids, as reef animals do, the relentless meat grinder of coral reef predation means that to do otherwise would be to expose junior to even greater danger. Out in the comparative safety of open water beyond the reef, the larvae feed voraciously to grow and thus boost their chances of survival. Growth takes time, however, so the larvae attempt to stymie their hunters via their morphology, which is often dramatically at odds with that of their parents. Butterflyfish juniors are equipped with bony plates that serve as armour, while both minuscule groupers and surgeonfish sport a pair of enormous spiky fin extensions, the effect of which is to make them near impossible for smaller predators to swallow. The equivalent would be like trying to eat an olive along with the cocktail stick it's skewered on.

How long the larvae stay in blue water exile varies from species to species. Porcupinefish are in no hurry and might remain for months, whereas some damselfish are done with the open ocean in as little as a few days. Even so, the currents threaten to carry them far from their starting point, so as they develop, they begin the process of navigating back towards reef habitats. This they do by orienting towards an alluring blend of biological scent cues emitted by coral reefs, and by listening out for the characteristic sound of those reefs, which, when the reef is healthy, has been likened to the noise of bacon frying in a pan. It's a soundscape of biological exuberance, a chorus of snapping shrimps supported by backing vocals of fish with their buzzes, croaks and grunts.

Those larvae that have survived this long are truly extraordinary swimmers, capable of swimming at a speed that, size for size, is more than ten times as fast as an elite human freestyler, and these indefatigable, animated iotas can sustain this, without resting, or even feeding, for days on end. Finally, they approach their destination, perhaps gathering in an eddy of slack water in the wake of the reef in preparation for the next stage of their lives. Over the past few days, many of these larvae have undergone a transformation, discarding their youthful guise and becoming miniature versions of their parents, as though they wanted to dress properly for their formal introduction to the reef community.

That introduction comes at night, and typically during the darkest period of the lunar cycle, which gives these now fingernail-sized fish a fighting chance of surviving their introduction to the gladiatorial arena of the reef. For many years at One Tree, my research group and I worked on humbug damselfish, who live in colonies among the safety of a home coral. Often in the morning, we'd return to a group of these stunning black-and-white banded fish to find a new arrival in their midst, identical to its peers except for being just a centimetre long and less than a hundredth the weight of the colony's dominant fish. These are the lucky ones. Having survived the dramatic winnowing of their numbers in their open ocean phase, as many as two thirds of the new recruits are killed in their first two days back on the reef. They're undone by their lack of familiarity with the terrain and with what a reef predator looks like. Though it has to be said that the flow of young fish onto the reef plays a critical role in supporting the food web in this phenomenal ecosystem, it does seem cruel that they should overcome incredible odds, surviving the perils of blue water only to die at their journey's end, on the cusp of safety.

Before the advent of DNA testing, it had been assumed which reef a fish larva ended up was largely a matter of chance, dictated by luck and the vagaries of the currents. Nowadays, we know that rather than being aimless drifters, fish larvae, like coral planulae, are choosy and that some reefs are more desirable than others. About a third of them return to their parental reef, where they can at least be assured of finding conditions that served their parents well. This isn't quite the same homing behaviour seen in fish like salmon; it may instead be a result of so-called 'sticky water', where patchworks of reefs disrupt oceanic flow patterns, meaning that the larvae never disperse too far from their starting point. Whatever the reason, the fact that a sizeable proportion of them come home at least allows for the possibility of a family reunion.

That said, returning to your natal reef is not the same as returning to the parental abode. One Tree reef, for instance, is relatively small, yet still it spans several square kilometres and teems with tens of millions of fish. Seen in this light, the fact that minute coral gobies seem to find lodgings with relatives is nothing less than extraordinary. It's important to point out that, for the most part, it's more a case of finding distant cousins than their immediate family, but even so, it's remarkable that these tiny, vulnerable specks of life manage to find their way through the mean streets of the reef to set up home with their kin.

* * *

The risk for the newly settled fish recedes slightly as they gain experience of the reef and learn about its perils, but it never goes away. Coral reefs are quite possibly the most dangerous habitats on Earth. It's been estimated that, even as adults, some of the smaller species in these environments suffer a mortality rate from predation of more than 50% a *week*. The average length of fish on a coral reef is smaller than most people imagine, at about 4.5 centimetres, and with hordes of potential predators lurking, the threat level is never less than red. As a consequence of this, prey fish on the reef make use of a wide range of defences. Many invest in spines and spikes; others shoal to gain safety in numbers. Whiskery juvenile eel-tailed catfish form aggregations of up to a hundred individuals that, when threatened, close ranks and coalesce into a seamless ball of defiance. Presented with such an imposing spectacle, it's entirely possible that predators might miscategorise the group as a single large entity, one that's far too large to tackle.

Chemicals, too, are a crucial part of the defensive arsenal of many prey animals on the reef. Coral gobies, for instance, cram their skin with distasteful chemicals while pufferfish are laced with tetrodotoxin, a powerful nerve agent, which is fatal to most aggressors. Nevertheless, anecdotal evidence exists that some of their potential predators, including tiger sharks, are relatively immune to the effects of this. Indeed, young male dolphins seek them out, microdosing themselves with the poison to give themselves a high. Simply being toxic is a little too passive for another common reef animal, which is capable of fatally envenomating an attacker without having to suffer the indignity of being munched. It's a snail, and it's armed with a poison just as deadly as tetrodotoxin and, according to some, it's worryingly trigger-happy.

Some time ago, I was working at the fringe of One Tree's lagoon, looking at the creatures who make their homes among the dead coral rubble where the water flow at each tidal shift is most intense. I was pulling myself through the shallows using a makeshift blend of crawling and swimming to get into place to make my observations when I had the sense of being watched. Sure enough, when I looked around, there was a pair of googly eyes looking at me from the interior of a shell that I'd upended. My first thought was the eyes must belong to a hermit crab, but as the creature revealed itself, it became clear that this was a cone shell. Not just any cone shell either, but *Conus geographus*, one of the deadliest creatures on the reef, and I'd annoyed it! Looking around, I now saw dozens of them sliding along in water so shallow that I was having to drag my belly across the bottom. Apparently, they were looking for the same kinds of rubble inhabitants that I was, except that their idea was to eat them. If a single one of these was to fire one of its poison darts into my unprotected skin, then that'd likely be game over for me. It took what felt like an age to cautiously backtrack across the rubble into deeper water, being careful not to excite any of the menacing molluscs and wondering if this was what it felt like to walk through a minefield.

Notwithstanding its fearsome capabilities, the geographer cone is an attractive animal. Its shell is about a handspan in length and patterned in sepia shades resembling an old map, for which it gets its name. It hunts by firing a harpoon into its fish victim, injecting a spectacularly malevolent medley of chemicals. Neurotoxins block nervous signals to the muscles, insulin sends the fish into hypoglycaemic shock, and somatostatin retards recovery. Though it does seem a little like using a poison sledgehammer to crack a small fish nut, it does explain the success of these creatures, and it allows plenty of time for the burly snail to ingest its prey at leisure. When the geographer is threatened, however, it really ups its venomous game, using a very different toxin that is 350 times more potent than the one it employs to subdue its prey.[60] It's enough to stop any attacker in its tracks, even people, and indeed the geographer cone has claimed several human lives.

Very few species in the animal kingdom pack this kind of punch, but while they lack the chemical weaponry of the cone shell, other prey animals turn aggressor in an attempt to turn the tables on their attacker. Lizardfish are common predators on the reef, regularly

60 The chemical mastery of cone shells has considerable promise in human medicine, especially in more effective treatments for diabetes and hormonal disorders. Moreover, refinements to the neurotoxic components have yielded a painkiller that is vastly more powerful than morphine and without the addictive qualities of other, similar drugs.

seen resting on the reef with heads raised, watching out for the heedless. Their 30-centimetre-long body is a camouflaging mosaic of earth tones, fronted by a truly enormous mouth that, in turn, bristles with teeth. Like many reef predators, the success of lizardfish is founded on utter stillness and a reliance on going unnoticed. It's a strategy that works extremely well most of the time, but if they're detected by any of their regular menu items, a kind of mob mentality results. Damselfish and surgeonfish approach the lizardfish in small delegations and either chivvy it with bites (albeit being careful to avoid its colossal maw) or stay a little further away and beat their tails in its direction. Although no harm comes to the lizardfish from this treatment, it does at least signal that the game is up and usually the predator will retreat to find a new hunting ground. The award for the pluckiest among the reef's inhabitants, however, might well go to the diminutive blennies whose nests are threatened by sea snakes hoping to make an easy meal of the caviar within. Even with such a deadly foe, the redoubtable parents bite the snake, slap it with their tails, flick sand at it, and, if all else fails, stoically block the entrance to the nest by wedging it shut with their own bodies.

Few of the reef's prey animals are so resolute in defence, relying for their safety instead on the refuge offered by the maze of coral, and a highly tuned awareness of risk. This is never more evident than when twilight approaches because it's at dusk that the daily killing spree hits top gear. As the shadows lengthen, fish begin to retreat into the safety of the reef. The smallest and most vulnerable retreat first, but the rest of the reef's diurnal fish aren't far behind. So many fish seek sanctuary that fights sometimes break out, with the strongest trying to evict the weak; they're only too well aware of what's coming. As the sun begins to slip below the horizon, the legions of brilliant painted fish are all but gone, lending an eerie feeling to the reef. Large predators gather in the murk; moray eels, groupers and sharks have shaken their daytime torpor and are noticeably agitated. It's light, or the lack of it, that drives these changes. The reef's diurnal fish have excellent vision, but their eyesight is most effective in the brilliance of the daytime when they can detect approaching danger at long range. Every lost lumen handicaps them, and by degrees they yield their visual advantage to the predators, whose eyes are perfectly adapted to the gloom. Now these hunters course the reef, seeking those who have failed to find refuge, or sit low in the water column to use the fading light at the surface as a backlight to detect late commuters swimming overhead. The twilight period might account for only one twentieth of the time in the daily cycle, yet it's within this narrow window that most blood is spilled.

Although some of the reef's largest predators are twilight specialists, there's no let-up for their prey even when dusk progresses into night. Reef sharks, in particular, use the cover of darkness to roust sleeping fish from cover, making use of their exquisite electrosensitive capabilities to locate their quarry in a deadly game of hide and seek. Meanwhile, in the daytime, predators need stealth and cunning if they're to avoid alerting their cautious prey to their presence. Flatheads bury themselves in the sand of the reef floor, and a dextrous shimmy is all that's needed to disappear almost entirely from view, leaving only their ever-watchful eyes protruding from the seabed. On more than one occasion, I've been recording videos in One Tree's lagoon, completely unaware of the predators' presence until one launches itself from hiding to engulf a fish that was equally as oblivious as me. The whole sequence, from beginning the attack to swallowing its prey and disappearing back into the sand seldom takes more than a second.

Frogfish, close relatives of the deep-sea anglerfish, are among the most astounding of all fish to look at, though perhaps few would describe them as handsome. They're stocky, deep-bodied creatures whose appearance is devoted to camouflage, from a palette of mottled colours that permits a near perfect match to the coral on which they sit to a coating of small spines that breaks up their outline and makes them almost unrecognisable as fish, or even as living organisms, such is their skill in this regard. Perhaps most extraordinary of all is the fact that they barely ever swim, instead using modified, leg-like pelvic and pectoral fins to step stealthily across the reef. They're reluctant walkers, though, and I've only seen them do this on rare occasions. Last year, in Indonesia's Lombok Strait, I accidentally disturbed a frogfish when I approached too close to its fishing spot. I hadn't seen it, but that was little consolation to the frogfish, which stomped off grumpily. This one, in turn, almost tripped over another, and the two of them stalked off to find a little privacy elsewhere.

Under normal circumstances, frogfish move only enough to place themselves in range of their prey. Once in place, the frogfish waggles its lure, which is formed from modified fin rays and dangles just above its mouth. A small fish approaches, its interest piqued by what it might perceive as a free lunch, only to find itself confronted with a nightmarish, but mercifully brief, realisation. As the frogfish's mouth begins to open, the cavity within expands dramatically, creating a vacuum that draws the victim irresistibly to its demise. It's the speed of the attack that's particularly astonishing, however: filmed under lab conditions, using high-speed cinematography, the fastest confirmed capture took just 4 milliseconds, making it among the fastest reactions in the animal kingdom, and about 25 times faster than we can manage.

Stealth comes in many forms, however, and for my money, nothing can beat the extraordinary tactics of one of the most successful predators on the reef, the trumpetfish. These long, thin tubes of menace are expert piscivores, and like all stalking predators, they need to close to within striking distance before launching their attack. So the trumpetfish inches forwards, its slender physique presenting a minimal profile when seen from the head-on perspective of their quarry. Even so, experienced prey fish have every chance of seeing a large pair of lips as they loom out of the blue. For this reason, trumpetfish are often forced to improvise, using features of their environment to cloak their advance. This means the coral terrain in most cases, but in some parts of the world, trumpetfish have innovated by using other fish to hide their approach. The most cunning aspect of this behaviour is their use of large species such as parrotfish as their cover. The parrotfish pose no threat to the small fish, which in turn disregard them after cursory inspection. The genius of the trumpetfish's strategy borrows from the legend of the Trojan horse; it tucks itself alongside the flank of the parrotfish and shadows the larger herbivore as it moves innocently around the reef. If the parrotfish's peregrinations take the trumpetfish within a smaller fish's domain of danger, the predator springs from its mobile hiding place to seize its unsuspecting victim.

* * *

The intensity and the pace of natural selection in the reef ecosystem both demands and rewards innovations like these, but it's not solely in the sphere of predation that animals need to be inventive. Coral reefs are hotbeds of competition, and it's this very competition that has resulted in the stupendous diversity that we see today in shallow, tropical waters. As well as being a major driving force of speciation, competition also fosters creativity in behaviour and is responsible for one of the most important strategies on the reef: territoriality. With so many individuals trying to access the reef's finite resources, day-to-day life is fraught and aggression is the vital currency. To try and ensure a supply of food, the majority of reef fish species stake a claim to real estate. It's every fish for itself and once an individual secures a place of its own, it guards its space jealously against all comers.

Triggerfish are hefty animals, some approaching a metre long, with a deep, oval body and a face that makes them look as though they're whistling. Such carefree behaviour wouldn't be apt for a trigger, however, not least because they aggressively requisition huge spaces for themselves, often amounting to several hundred square metres. In securing exclusive access to such an expanse, the male guarantees access to the two things that he cares most about: food and sex. The trigger's diet consists mainly of crunchy invertebrates, with urchins and crabs high on the preference list. In both cases, the trigger brings its formidable intelligence to bear on the problem of how to disarm well-defended prey. In the case of urchins, the trigger flips them over by blowing out a jet of water, exposing the urchin's vulnerable underside to the mercy of the trigger's impressive teeth. This pressure-hose approach is also used for uncovering delicious titbits buried under sand, and as the trigger kicks up the dust of the reef floor, it can attract a flotilla of smaller fish. These are keen to explore the foraging opportunities created by the sand-blasting activity, and by focusing on the smaller of the organisms that are displaced, they don't compete directly with the trigger. It's this partitioning that is fundamental to territoriality; the titleholders of these plots typically confine their aggression to members of the same species, and to other direct dietary competitors.

There is one other prize within the territory that can cause the incumbent to become especially short-tempered, and in this case, it might focus its fury on all and sundry. That prize is a nest full of eggs and it's now that triggers in particular can pose a problem for divers. I've been attacked by lots of fish over the years, demonstrating that fish are excellent judges of character, but none with the sustained fury of a titan triggerfish that took exception to me off Lombok's Gili Islands. Titans are, as their name suggests, the largest of the triggers, growing as long as an adult human's leg, and with teeth designed to bite through armoured invertebrates, they're not to be trifled with. My crime was to swim within what it clearly considered to be its private domain, and it came charging up 5 metres from its seafloor residence to make its feelings known. The authors of a paper in the delightfully named *Journal of Wilderness Medicine*, devoted to the subject of triggerfish attacks, provide their advice in this eventuality: 'The recipient of such menacing behaviour is not apt to continue swimming towards the fish,' they write drily, before continuing: 'If, however, one does advance, the likelihood of the next rush of the triggerfish ending in an actual attack is increased.' The authors then document several case studies where triggerfish have inflicted flesh wounds on divers.

Typically when triggerfish attack us, it isn't a male who confronts a diver, but a female protecting her nest. The male's fiefdom is subdivided into a number of smaller territories, each of which is occupied by a female. While he devotes his efforts to parading around self-importantly and making sure he isn't cuckolded by any of the other local males, she does most of the hard work of preparing the nest, a depression in the sand almost a metre across, and, in time, looking after her brood of eggs. This is when things can get fraught because mama trigger doesn't simply consider the surrounding seabed to be out of bounds, but the column of water above it. In my case, the incandescent trigger appeared right in front on my mask, giving me an excellent view of her teeth, before withdrawing for a moment. I used the opportunity to change course, but without apparently satisfying the trigger, who charged again, this time wafting straight past my ear and then circling round to charge me from behind.

At this, I picked up the pace with the trigger in pursuit. When finally I looked back from the far side of a reef crest, this strange fish with its whistle-ready mouth was poised in the water column, watching my ignominious retreat with apparent satisfaction.

Triggerfish are one of the many fish that divide the reef into a patchwork of territories, and the result is an uneasy coexistence of neighbouring stakeholders punctuated by furious border disputes. Each fish wants the biggest territory that it can gets its fins on, but at the same time it doesn't want to have to spend all of its time furiously prosecuting trespassers. The result is a compromise; the plots tend to be big enough to supply the needs of their owners, but no larger. Just how large a territory depends to an extent on what the fish eats, and how many competitors it faces. Triggerfish need lots of space because the chunky invertebrates that they feed on are relatively few and far between. At the other end of the scale, planktivorous fish like humbug damselfish have their meals delivered to them by wafting currents. Accordingly, a gang of them can sit tight in a single, easily defendable coral head the size of a football.

The territorial picture changes constantly, however. A boom in population means more competition, in turn leading to a lot of quarrelling and potentially smaller territories for all. At the same time, if conditions are good, the food supply improves, and each fish can manage with less space. So, the territorial map of the reef is perpetually in flux and watching how the fish balance the competing demands of home ownership is one of the most fascinating aspects of reef animal behaviour.

One of the best studied, and most interesting, of all of these are fish like the dusky damselfish. Unlike their damsel cousins, the humbugs, these duskies are strict vegetarians and along with a few dozen closely related species across the world's reefs, they defend patches of lush, fluffy algae. The hard work doesn't stop at defending the perimeter of their algal field, however. They very carefully tend their crop, weeding out less favoured species to encourage the growth of delicate filamentous varieties that provide the best nutrition. Just how effective these interventions are is demonstrated by the yields that the damsels achieve. Their algal pastures can be three times as productive as untended patches of similar size. The obvious parallels with human agriculture are what gives this group of damsels their alternative name of farmerfish.

Part of the reason for the damsels' success comes down to choosing the most valuable algae to grow, but in some parts of the world, they bring in help. Caribbean longfin damsels appear to encourage swarms of rice-grain-sized shrimps to join them on their farms. The shrimps benefit from the protection of the damselfish and, in return, their presence seems to bolster algal growth, most likely by providing fertiliser in the form of their own waste products. In farms with resident shrimps, the algae are more nutritious and the damsels themselves are in better condition.

Patchworks of damsel farms, akin to the arable fields that pattern rural land, account for two thirds of the surface of some reefs. The largest and most dominant fish get the best plots, with plenty of direct sunlight for their crops and somewhere to hide if danger threatens. Exactly how much space varies enormously, but a reasonable amount for a high-ranking damsel that would fit in the palm of your hand would be about the area of a sofa cushion. Arranged around the margins of these highly desirable claims are slightly smaller properties, in the possession of correspondingly smaller farmerfish. With each successive step away from the hot properties at the centre of the territorial mosaic, the plots get smaller and less rewarding, and face greater risk from crop thieves. An army of prowling herbivores covet the damsels' allotments, and it's a Sisyphean task to keep them at bay. If they approach one at a time, the farmer can usually repel them. Surgeonfish, however, gather into raiding gangs for precisely this reason, rushing the defender en masse. All the farmerfish can do is to watch morosely as its hard work disappears down the gullets of the mob.

The high-ranking fish in the sunlit uplands of the central territories are largely insulated from these raiders by the repulsive effects of a cordon of aggressive farmers in outlying plots, yet they too have their own problems. As their neighbours grow in stature and station in the damsel milieu, they seek to expand their own holdings by encroaching on their neighbour's land. This results in the sorts of local feuds that are the bread and butter of local courts in human jurisdictions. Fish obviously have no such recourse; nevertheless, they show the kind of wisdom sometimes lacking in our societies and adopt something known as a 'dear enemy' strategy, which is their version of an unspoken truce. Though the mere sight of the fish next door might well be enough to make their fins bristle, they seem to respect the shaky status quo and instead focus most of their ire upon outsiders that stray into their realm.

The aggression displayed by reef fish is apparent to all, but it pales alongside that of another reef animal, one that cloaks its belligerance in secrecy. These are animals that we might not even think of as behaving at all, let alone being combative. They are the heart and soul of the reef itself: the corals. Disputes between adjacent colonies have similarities with those that occur between fish. What the corals want is space to grow, and what they don't want is to have their sunlight blocked as a result of being overgrown by a thrusting upstart neighbour. Being shaded out means death, and in defence of their place in the sun, adjacent colonies go to war.

The success of some species, such as plate corals, is founded on rapid growth. Admittedly this is relative – they grow at about the same rate as our hair – but the threat they represent for nearby corals is very real. Slower growing corals might seem to have settled for a steady pace of life, but their inoffensive, sedentary manner belies a savage determination. They gain intel on the doings of nearby rivals on the basis of chemical cues in the water, and if they sense that one of these has crossed the line, they let loose with an extraordinary, and highly effective, arsenal. Some lash out with specialised, whip-like sweeper tentacles that are studded with stinging cells, others extrude filaments from within each polyp that digest the living flesh of encroaching colonies.[61] Typically, these attacks take place under cover of darkness, as though the stony corals were reluctant to unveil the full extent of their savagery to the daytime reef community. Whichever Machiavellian approach they use, the result is usually the same. At first light, that part of the neighbouring coral that threatened to overgrow its neighbour is a deathly white. The living tissue of the coral has been eradicated, leaving only the bony, limestone structure behind. It's a harsh lesson for the trespasser, and it's a reminder that among this scene of beauty and, to us, serenity, coral reefs are among the most competitive environments in the world.

* * *

A familiar trope in films from *Slumdog Millionaire* to *Star Wars* is the story of the underdog who rises against the odds to triumph. It's inspirational because we can relate to the struggle, and it's interesting because it's rare. It's rare, too, on the reef. Paradoxically, however, pretty much every one of the animals that you see on a reef is the heroic protagonist of such a story. It's already survived where thousands, or even millions, of its brethren have succumbed. This ability of animals to not only cope, but ultimately to succeed in

61 Their close relatives, the anemones, even bundle packages of stinging cells into balls of mucus and fling them at their enemies like hand grenades.

this arena of threat and aggression is testament to a diversity of traits and strategies that shield them from the worst that the community can inflict upon them.

Take, for instance, a newly settled reef fish. It faces not only the considerable risk of becoming a predator's hors d'oeuvre, but also the relentless pressure of competition for food and for space. That competition doesn't just come from other species; it faces bullying from its own kind. In an environment where size determines status and access to resources, the youngsters are at a major disadvantage. The secret to minimising competition from larger peers is to occupy a different niche. Accordingly, juvenile reef fish often live away from the adults, on different parts of the reef, and eat from a different menu. Convict tangs, for example, are confirmed herbivores as adults, but as youngsters they get a high proportion of their energy by eating zooplankton. On its own, however, this might not be enough to deflect the aggression of larger fish, so some juveniles adopt disguises.

The beautiful colours of reef fishes go far beyond aesthetics; they're signifiers of identity. The problem for youngsters is that the moment they adopt their team colours, they open themselves to hostility. Confronted with the prospect of facing the relentless fury of tyrannical creatures such as the jewel damselfish, young chocolate surgeonfish go incognito. More specifically, it mimics another species, such as a pygmy angelfish. The reason is that while the damselfish seems to see red every time a surgeonfish hoves into view, it seems to have less of a problem with angelfish. Accordingly, by mirroring the angelfish's colouration and morphology, the mimic surgeonfish slips under the damsel's rage radar. The subterfuge has to end at some point since surgeonfish grow much larger than pygmy angelfish, but in the meantime, the juvenile gets some respite from its potential persecutor. And when it does outgrow its disguise, it's in much better shape to withstand the attacks of the angry damsel.

As impressive as the surgeonfish's ingenuity is, a more widespread strategy underwrites the success of a large proportion of reef animals. Over the course of evolutionary history, individuals faced with the many challenges that afflict them on life's path turn to others for assistance. Getting by with a little help from my friends is a motif for many in this crucible of conflict and competition; indeed mutualisms are not just abundant but integral to life on reefs. The partnership between corals and dinoflagellates is the most significant of these, but there are many more. For instance, the coral gobies and humbug damselfish that I described previously, living precariously within the protection of coral heads, in turn provide a valuable service to their hosts. Some of these are simply by-products of the fishes' presence, such as improving by means of their activity the flow of water through the dense coral branches, and fertilising the zooxanthellae with their fishy faeces, but coral gobies in particular make a more direct contribution to the corals by deterring their predators.

A high proportion of butterflyfish make their living by eating coral, a diet that's facilitated their extraordinary forceps-like mouthparts, enabling them to pick at the miniature polyps that coat the coral colony. Corals can cope with some degree of attrition, but the butterflies can be persistent, biting at each colony anything up to 3,000 times a day. Onslaughts like this are unlikely to be sustainable for the coral, and what's bad news for them is bad news for their goby tenants. Consequently, the gobies take it upon themselves to defend their Cnidarian sponsors with vigorous head-shaking displays of aggression and, if these fail, the bullish little residents dart out of the coral to remonstrate with the trespassers. The net result of this direct action is that corals with gobies *in situ* are, in some cases, almost completely bypassed by the corallivores;

even just the distinctive smell of gobies is sometimes enough to discourage them.

Anemonefish, famously, have adopted a similar life to the coral gobies in the sense that they live within the confines of a Cnidarian, but they've opted for a host that, if you're a fish, is utterly deadly. It's an arrangement that, through familiarity, we pay little heed to. We shouldn't; this is an exceptional state of affairs. Anemones are ruthlessly efficient predators, among the oldest venomous animals on Earth. The barbed arrows that they shoot are laced with compounds that destroy red blood cells, cause uncontrolled bleeding and compromise the nervous system; clearly, they're not to be trifled with. Anemonefish, however, seem to occupy their forests of lethal tentacles with impunity, and in doing so, gain protection. The anemone, too, is well served by the relationship, since their lodgers provide services that mirror closely those provided by coral gobies to their own hosts. Though it's true that there are few predators capable of taking on a sea anemone, certain kinds of butterflyfish, as well as turtles, are prone to tear into them. Anemonefish don't stand for this, racing out of their tentacular home to attack any such threats. In the Red Sea, the experimental removal of anemonefish triggered a butterflyfish free-for-all that left the anemones permanently withdrawn into a defensive globule or worse, in tatters. Clearly both partners benefit from the association, but this still leaves the question of how this phenomenal arrangement works: why don't the anemonefish get stung?

Surprisingly, they do, at least in the early stages of their relationship. Over time, however, the two develop a kind of molecular understanding. We still haven't cracked the exact secret of how anemonefish avoid being stung, but years of painstaking work have narrowed down the possibilities. The fish have to acclimatise to their chosen anemone, so rather than jumping straight into the midst of the tentacles, they take things slowly. A popular first port of call for an uninitiated fish is the middle of the anemone, the site of the creature's mouth, known as the oral disc. This area is a central clearing in the anemone's forest of tentacles, where the fish is somewhat safe from the stings. The introductions need to be made cautiously, at least on the part of the fish, because it's entirely possible that it could be stung to death. Indeed, at this early stage, the anemone responds to the presence of the fish in a rather unhospitable way, curling its tentacles threateningly towards the intruder, so it pays the fish to be watchful.

Gradually, the anemonefish starts to interact a little more with the tentacles, gently mouthing them and brushing them with its fins. What seems to be happening here is either that the anemonefish is coating itself in anemone mucus, which contains antigens – molecular identifiers – from the anemone, or it's priming its own cellular machinery to produce recognition proteins in their own mucus. In essence, the anemonefish is making use of the same mechanism that the anemone uses to avoid stinging itself when its tentacles contact one another. A simple creature like an anemone divides the world into 'self' and 'non-self', and by changing the character of its mucus coating, the anemonefish dons a molecular disguise, becoming 'self' as far as its host is concerned.

Even with this spectacular chemical camouflage, most species of anemonefish are specialists, capable of living within only one type of anemone. The introductions are made even before the fish hatch, when their exposure to the water-borne chemicals of their host anemone facilitate mutual recognition between the fish and the particular species of anemone. In any case, switching partners is unlikely to provide much benefit, since an anemonefish ensconced in its tentacled home has one of the most cossetted lives of any reef fish; some can live to more than 30 years of age, which is vastly longer than most. In turn, the anemone gains the resolute defence of its feisty lodger against marauding predators as well as a ready supply of fertiliser. The latter is especially important, because, like their coral relatives, anemones often engage in another partnership, this time with photosynthetic internal algae from which they gain a steady supply of food. Simply having fish nestled among their tentacles and periodically squirting out waste means that the anemones are able to culture twice as many zooxanthellae, making them healthier and allowing them to recover far faster from periodic stressors than fishless anemones.

Beneath the anemonefish, on the reef floor, a very different partnership is being played out. It's a strange pairing because the two animals involved seem at first glance to be temperamentally ill-matched. Lying on the coral sand, its tail flirting with the hole that is its burrow, lies a goby. Every few minutes, its partner, a perpetually industrious pistol shrimp, emerges from their shared home with an armful of grit. This is deposited beyond the burrow before the shrimp retreats to continue its spring cleaning. The goby, meanwhile, appears to gaze into space. It's this seeming asymmetry in workload that led my research group to dub the arrangement a '1950s household'. This is decidedly tongue in cheek, however, because the goby clearly earns its corn. As the shrimp stays busy keeping the burrow free of obstructions, the goby is on lookout duty. This is essential, because while the shrimp may not be blind, as is sometimes claimed, the goby has by far the better vision. Touchingly, the shrimp maintains contact with its fish partner as it goes about its work by means of resting a long antenna on the goby's flanks. If a potential threat appears, the goby quivers its tail fin, at which the shrimp scoots safely below ground. And if that threat comes in for a closer look, the goby, too, will disappear down the burrow. The relationship between the two is further enhanced by the odd fact that the goby feeds its partner, albeit not in the most appetising way since the pistol shrimp feasts on the fish's excrement. Regardless of its unsavoury diet, the shrimp has hit the jackpot with its goby partner and vice versa. They're an odd couple, certainly, but their relationship is a winning strategy for two small and vulnerable reef animals.

The lengthy roll call of cooperators on coral reefs is a powerful evolutionary response to the rivalries and threats that exist here. Along with the examples I've already provided, there are the extraordinary cleaning symbioses that exist between fish like the cleaner wrasse or the cleaner shrimps and their clients, who in many cases are amply equipped to make a meal of them, but who desist in recognition of their services in removing ectoparasites. The unscrupulous tendency of the wrasse to supplement its diet occasionally by biting a small chunk from its customers means that it's not always a smooth relationship, but the political adroitness of the cleaner wrasse as it subsequently fusses over and caresses a wounded party in an apparent attempt at restitution means that both parties are ultimately satisfied.

Even predators get in on the cooperative act, as exemplified by the teamwork shown by moray eels and groupers. In this latter case, the grouper recruits the eel to the hunt by approaching its bolthole and signalling by vibrating its body. If the eel is persuaded, the grouper leads the way to a trove of fish hidden away in the interstices of the reef where the grouper can't reach, but its colleague can. As the eel disappears into the labyrinth, the grouper waits outside, ready to snatch any fish who, disturbed by the eel, make a dash for safety. Teamwork is also much in evidence in pairs of harlequin shrimps, specialist sea star eaters who, in order to subdue a victim many times their size, first need to snip off its sucker-like tube feet with their pincers. The sea star has thousands of such feet and puts them to use in an attempt to frustrate its persecutors, meaning that the harlequins have to work quickly, and together, if they're to succeed. The strange sight of two shrimps bearing a huge sea star off to a secret larder is the final act in this collaborative play.

A final word on this topic belongs to the sea cucumber, one of the reef's most common, and most underappreciated animals. It's easy to see why they're underappreciated, at least on first appraisal; they barely move and have little aesthetic appeal. Even when you realise that sea cucumbers are echinoderms, a fascinating group of complex invertebrates that also includes urchins and sea stars, it only seems to make them poorer in comparison. Not showy creatures, sea cucumbers are effectively just tubes and aspire to being little more than a section of bowel. Flourishes, if you can use that term, come in the form of a mouth at one end and a bum at the other, while stubby little tube feet propel them at a glacial pace across the seabed and tentacles around the mouth sluggishly collect food.

Uninspiring they may be, but you can't deny that they're successful. The floor of One Tree's lagoon is liberally sprinkled with them, and censuses taken in the 1960s in the Philippines and at remote Pacific atolls put their numbers at a staggering 50 cucumbers per square metre. But with so many hungry predators around, how do these slow-moving packages of protein manage to survive? The answer is that they're extremely well defended. Sure, there are some predatory snails with a yen for them, and a small number of other foragers might turn to them *in extremis*, but most carnivores give them a wide berth. Cucumbers are impregnated with toxic chemicals, known as holothurins, and some even resort to the neat expedient of deliberately rupturing themselves and explosively spewing sticky guts over their attackers. The defensive mastery of sea cucumbers has been exploited variously by pearlfish, a slender excuse for a creature that shamelessly takes refuge in the safety of a cucumber's anus, and, historically, by fishermen in some cultures, who would squeeze a cucumber to release a small amount of its deadly payload into a tide pool and so stun the fish within.[62]

Where the sea cucumber really excels, however, is in its specialist field of bioturbation. If you haven't given much thought to bioturbation, indeed if you've never bioturbed, you're not alone. Yet this critically important function, most familiarly performed by earthworms on land, involves the processing and working of sediments. Sea cucumbers each chew their way through several kilograms of sand per year, the larger ones working through a suitcase worth of the stuff, extracting the food within and pooping out the clean sand. It's been calculated that the combined efforts of the sea cucumbers across the whole of the 19 square kilometres of Heron Reef, on the southern Great Barrier Reef, shift the equivalent weight of well over 100 Airbus A380s in sand annually.

This slow and steady turning over of the seabed might not be a glamorous occupation, yet it's one that is critical to the overall health of the reef. For one thing, the relentless cleaning and purification of the substrate by the cucumbers seems to control pathogen populations. This has come into particular focus recently with the devastating impacts of disease on coral around the world, and particularly in the Caribbean, where coral losses due to infection approach 50% in some places where cucumber densities are low. In addition, the chemical environment within the cucumber's body dissolves a proportion of the calcium carbonate in the sand that they ingest, which, when excreted, becomes available for calcifying organisms,

---

62 A cucumber's anus is its weak point, much like the exhaust port on the *Death Star* in *Star Wars*, and represents a portal for the reclusive pearlfish. Some species of these fish overreach themselves by eating the innards of the sea cucumber from within and, in response, some cucumbers have evolved anal teeth with which to defend themselves. So: Bum biter bites the biter.

such as corals, to use for home building. This, and the ammonia that this army of substrate slurpers produce, is at a higher pH than seawater, so buffers the water chemistry of the reef against the rising threat of ocean acidification, caused by rising carbon dioxide.

This is all wonderful news, and we ought to recognise the vital role that these animals play. The problem is that we don't. In fact, the capture of sea cucumbers, mainly as food but also as 'traditional' medicine, is at epidemic proportions. An estimated one billion of them are taken globally each year, the majority from shallow tropical waters. Some species command such high prices – more than $1,000 per kilogram in Hong Kong – that thriving black markets have sprung up. With these kinds of rewards on offer, it's perhaps unsurprising that across large parts of the world, sea cucumber species have been wiped out. The oddest thing about the whole business is that sea cucumbers are essentially flavourless; if you're keen to experiment with their culinary possibilities, but don't have any to hand, simply boil a washing-up sponge for six hours and serve. Why are they popular? Like shark fin, which is also wanting in taste, it's a cultural thing, a dish that connotes status. And, also like shark fin, the trade in sea cucumbers causes enormous ecological damage. One, though by no means the only, effect of the mass looting of sea cucumbers relates directly to the coral disease, discussed previously. In the absence of sea cucumbers, corals are 15 times more likely to die of disease.

I've described in this section some of the incredible cooperative interactions that happen on reefs. It could well be argued that sea cucumbers don't fit this picture since they're not engaged in any obvious mutualistic partnership. Nevertheless, the reef ecosystem as a whole, and every animal in it, gains enormously from the presence of the humble sea cucumber. Few reef organisms play a more important role in this habitat. The fact that we're – again – busily kicking the supports out from under our most important natural resources is a kind of madness that is hard to rationalise.

* * *

Exploitation of reef animals has a long history, but in recent decades it's been turbo-charged by greed. Sea cucumbers have been collected by artisanal fishers for hundreds of years with relatively little impact to their populations. Then, from the 1960s onwards, the picture changed in a way that is almost as if we declared war on them. Similarly, reef fish have long been exploited on a subsistence basis, but this turned decisively, industrially against the fish in the 1950s and 1960s. The first fish to be extirpated are nearly always the apex predators and with those gone, commercial fisheries have a problem. Their answer is to fish down the food web. They go after the mesopredators, such as snappers, and they also target herbivores. The impacts of the fishers have been exacerbated by disease. Sea urchins, vital cogs in the reef's algae-munching machine, were almost entirely wiped out by a parasite in the Caribbean in the 1980s and, following something of a recovery, another wave of disease has struck in the last couple of years.

To get an idea of what happens when herbivores are removed from a reef, Terry Hughes of James Cook University in Queensland ran an experiment using cages to exclude fish from sections of the Great Barrier Reef that had been damaged by bleaching at Orpheus Island. The results were astonishing; freed from the nibbling of grazers, the algae ran riot. Stands of *Sargassum* – almost completely absent from any other part of the reef – not only established but grew to a height of 3 metres within the cages, and encrusting red algae swamped the rocky substratum. At the end of the investigation, there was up to 20 times the amount of algae present in the cages as in the controls, and coral failed to re-establish in the face of this algal onslaught. The experiment was a beautiful demonstration of the crucial role of herbivorous animals on reefs, and the likely fate of those reefs if grazers aren't present in sufficient numbers to curb the explosive potential of algae.

Healthy corals have a neat trick to avoid being overgrown and starved of light by films of algae. Just when an algal spore, newly settled atop a coral colony, thinks it's found the perfect place to go about its photosynthetic business, the coral sloughs a layer of mucus and the disappointed alga slides off like snow from a pitched roof. Nevertheless, the battle for territory on the reef surface is relentless, and if algae get a foothold, they produce chemical cues that attract others to come and join the party. Despite their ability to withstand this botanic siege under normal circumstances, weakened corals can get swamped. Luckily for them, this is where the reef's battalion of herbivores comes in. Alongside a host of invertebrates, such as urchins, gastropods and crabs, a large number of fish crop the algae. Surgeonfish, named for the viciously sharp scalpel located towards the tail, bring their comb-like teeth to bear on algal turfs and rely on a unique supporting cast of gut bacteria to do the heavy lifting in the digestion of cellulose. Damselfish and rabbitfish, too, are among the ranks of avid algivores, but there is one group of fish that are true icons of the reef and make an outsized contribution to reef health: the parrotfishes.

Swimming lazily among the alleyways created by One Tree's huge coral outgrowths known as bommies, I'm delighted to be in the company of parrotfish. In what is admittedly the crowded field of the reef beauty pageant, these fish are for my money the most spectacular animals here. The flanks of the largest are a kaleidoscope of colour, a medley of crisp pinks, vivid yellows, oranges and turquoise, like a rainbow made flesh. While the name parrotfish references their impressive beaks, there's also something avian in the way that they swim with flaps of their wing-like pectoral fins.

Some of the parrotfish here approach a metre in length, and at this size, they're males with a coterie of smaller females in tow, although if these females live long enough, they, too, will become male. Sequential hermaphroditism, as this is known, is a trait that's common among reef fish. Anemonefish start out as males and graduate to become females as they grow. More commonly, however, the pattern is the other way around, with an initial female phase giving way to masculinity. This is the strategy adopted by parrotfish and it creates a population where there are a few large males and a greater number of smaller females. It's usually reckoned that the whole sex-swapping system is underwritten by territoriality. Only larger, stronger males have a chance of gaining a territory, thereby freezing smaller males out of the mating game. Under these conditions, the best strategy for a young adult hermaphrodite is to be female, switching to embrace its maleness later in life, when their greater size allows them to compete.

When parrotfish change sex, they don't merely undergo a subtle, covert rearrangement of their gonads, but rather they broadcast it from the reef tops, in technicolour. A newly transformed male has undergone a cascade of physiological changes that affects everything from his gonads to his behaviour, but the most obvious outward sign is his adoption of a completely new livery. The gaudy glory of this male stands in contrast to the subtler hues of the female. After all, he's now very keen to be noticed, and to draw the approval of females, in whose ranks he once stood. While the male has become a strutting peacock of a fish, it's left to the juveniles to be sober and sensible, especially in terms of their apparel. Well-turned-out whitespot parrotfish youngsters sport 1970s business wear, with cream stripes against a dark-brown background, whereas filamentfinned parrotfish minors favour a mottled appearance; others are blotched, or plain, or mimic other species of fish entirely. To add to the confusion, they change colour often, making it a tough job to tell the juveniles of different species apart. A researcher I once discussed this with, perhaps the world's foremost expert on this tribe, admitted that he'd only be about 60% confident of identifying the species of a young parrotfish from its picture.

Enigmatic as they are, I've spent many enjoyable hours watching small shoals of sausage-sized parrotfish youngsters making mischief in the shallows of One Tree's lagoon. They charge along like a gang of children heading out of school before all of them alight on a patch of grubby-looking algae and peck at it frantically as though starving. The reason for their haste is that they're trespassing, and before long a furious damselfish charges at them with murder in its eyes. The flock of parrots takes flight at this and dash off together in close formation, sometimes with the damselfish in hot pursuit, scooting along until they find another patch with an unwary guardian. This game of punctuated pilfering goes on all day, every day, and while I feel for the farmers, the ebullience of the little parrotfish is hard to resist.

Although parrotfish are usually classified as herbivores, this isn't always accurate. We see them biting at patches of algae and, not unreasonably, reach the conclusion that it's the algae that they're eating. They might enjoy a little salad, especially crunchy, calcium-reinforced algal crusts, but what really stimulates the appetite of many parrotfish species is fresh cyanobacteria. The more familiar name for these microbes, blue-green algae, gets the colour right, but the creature entirely wrong; it's a bit like calling an elephant a grey pine. Cyanobacteria do photosynthesise and have been merrily doing so for over three billion years; however, they are very much bacteria and very much not algae. The distinction may seem a fine one, but the difference between these bacteria and true algae is arguably greater than that which exists between plants and animals.[63]

There are thousands of different kinds of these remarkable bacteria and, as mentioned earlier, they play a critical role in reef ecosystems, where millions of them can be found on and even in every square centimetre of the reef's hard surfaces. A single cyanobacterium is a minute organism; a queue of 2,000 *Prochlorococcus* would stretch for only a millimetre. Although cyanobacteria thrive on the surface, where they combine into the complex microbial mats that form a fine skin over the reef, their diminutive size allows them to penetrate the porous rock of the reef itself, therein to find a degree of safety. It's a tactic that protects them from most predators, but parrotfish are equipped to access their refuge, crunching through the stony parts of the reef with their formidable mouthparts.

The beaks that give these sex-swapping superfish their name are formed by the fusion of up to 1,000 individual teeth. The beak is effectively self-sharpening since parrotfish rotate then replace their teeth, shedding the worn ones at the top and so rejuvenating the cutting edge to the beak. Even allowing for this constant renewal, to bite through rock the teeth have to be stupendously strong. The coating of enamel on the teeth that make up the business end of the parrotfish's beak is one of the most remarkable biomaterials ever discovered. It's the stiffest material yet found among living organisms, and also one of the hardest, outdone only by the teeth of certain sharks and the metal radulae of the beaded chiton.

63 We owe to these unheralded microorganisms the courtesy of getting their name right, not least because of their dominant role in generating the oxygen that we breathe, a process that began with the Great Oxidation Event 2.5 billion years ago, which transformed the Earth's atmosphere and made it possible for life as we know it to develop. If this fails to impress, and if you drink *Spirulina* smoothies, you also have them to thank because *Spirulina* is a form of cyanobacterium.

Together, the adamantine properties of the parrotfish's beak mean that it neither yields, nor fractures, when brought to bear on the rugged surface of the reef. Impressive as this is, biting off a chunk of stone is only stage one of the parrotfish's job; to access the goodness within it needs to crunch the hard mineral. To enable this, they have a second set of teeth in the backs of their mouths which mills the rock to a kind of stony, bacteria-infested flour before passing the whole mix to the gut. Here, the microorganisms in the mix are digested, and the remainder is excreted as fine white sand, which over time comes to form the majority of the world's picturesque tropical beaches.

Alongside the many herbivores that live on coral reefs, the unique role played by parrotfish is critical. The very same environmental challenges that have driven corals into retreat are manna for algae and cyanobacteria. Consequently, the battle that has raged among the competing claimants to reef real estate for millions of years is currently being lost by weakening corals. As they advance, prolific algal growths cast a shadow over the corals, while in the Pacific cyanobacteria have teamed up with encrusting sponges to form a collaboration capable of outcompeting and overgrowing living coral colonies.[64] Regardless of the aggressor, the result for the coral is the same; the sunlight that is so fundamental to coral health is blocked and colonies die. Moreover, both algae and cyanobacteria consolidate their positions by exuding toxic chemicals that weaken established corals and prevent new coral planulae from settling.[65]

In dire times such as this, the corals are ever more reliant on algae grazers and parrotfish alike to tilt the odds back in their favour.

---

64 The sponge in question is *Terpios hoshinota*, sometimes described as a cyanobacteria sponge because half of its mass can be made up of cyanobacteria cells.

65 The term for this is allelopathy, and one of the chemicals involved is neurymenolide A.

The grazers are incredibly effective at keeping algae in its place, as evidenced by the James Cook University cage study; the algal jungles that developed on the reef inside the cages were nibbled down to nothing within a month of the cages being removed. Parrotfish of various species play their part by either browsing on the algae themselves or crunching their way through the reef surface, and although they're often described as eating live coral, relatively few species actually do. This isn't some kind of self-denial enacted by the parrotfish, it's simply that a bite of dead reef rock is estimated to be three times as calorific as one from a live coral because of the richness of the bacteria that exists within that rock. And by taking chunks out of the reef surface, the parrotfish break the stranglehold of the microbes, creating virgin territory for opportunistic corals to settle.

Just how important parrotfish are can be seen in the aftermath of a bleaching event. As the coral teeter on the edge of death, algae and cyanobacteria capitalise on their rivals' distress by expanding to carpet the entire reef in a shroud of grimy green. Those animals that can leave do so. Others, excluded from the refuge of the coral, are picked off by opportunistic predators, and still more starve as the food web collapses. Left unchecked, this can, and sometimes does, lead to a permanent shift in the habitat. I've seen dead reefs in heavily fished parts of Kenya and even New Caledonia and they remain horribly memorable, the bleakest natural spectacles I've ever observed. There's a liminal quality to these places, where thriving communities of transcendent beauty become dismal mausoleums, populated by pitiful remnants of vagrants and refugees. By degrees, the reef loses its integrity and crumbles, the intricate contours that once fed, housed and sheltered generations of animals levelled by the remorseless actions of time, tide and storm.

It doesn't have to be this way. Reef ecosystems have a chance of fighting back and recovering their former glory and this is where parrotfish come into their own. For them, this wasteland of microalgae and cyanobacteria is a banquet, and it's one they set about with relish. Parrotfish populations boom and by degrees bring the green wildfire under control, opening up opportunities for old corals to recover and new ones to establish. The persistence of corals on tropical, shallow water reefs is indemnified by parrotfish. It's for this reason that intense fishing of parrotfish is such a terrible idea; without them, a pillar of the reef's life support system is lost.

* * *

The gentle swish of sparkling waves against a white beach and the stirring of palm fronds to a soft breeze provide the quintessential backdrop to coral reefs. The whole scene, land and sea, is as one in our minds, yet this holistic perspective is too often lacking in our planning. This applies to everything from development to research, where a focus on the wet part of marine biology prevents us from seeing the full picture. The land that fringes coral reefs can be as important to them as the surrounding seas and even as we strive to protect the reefs themselves, we hammer the adjacent land.

Tourism, with all of its supporting industries, has brought ever more people to the tropical shores around the world. Plantations, especially oil palm, have mushroomed, replacing rich and diverse native flora with monocultures. The immediate impacts on reefs are painfully obvious, ranging from pollution and littering, to trampling of delicate corals and even poisoning them with run-off from roads and even snorkellers' sunscreen. Alongside this, however, our actions have fractured the critical links between land and sea that have sustained reefs for as long as they have existed.

The value of pristine terrestrial habitat to reefs is exemplified by the glorious remote Indian Ocean islands of the Chagos Archipelago. While these islands have largely been spared the damage wrought by development, there's a pest in paradise: the rat. We were the rats' vector, and they've taken to the islands with relish. One of the most obvious impacts of the rats on such islands is that they tend to drive out seabirds. Only one or two lonely terns per hectare are typically to be found on rat islands versus a bustling 1,200 crammed into the same space where rats are absent. This translates into differences in the flow of vital nutrients back into the ocean because of bird guano; 250 times as much nitrogen feeds the reefs off bird islands. The corals grow twice as fast and are far more resilient when they're supported by this nutrient subsidy and those benefits flow on to the fish, which grow faster, and larger. The overall biomass of marine life around healthy, ratless islands is vastly greater. I can't imagine a more powerful demonstration of the crucial interrelationship between habitats. An estimated 90% of the world's island groups now have rats and at each place they're found, it's the native biota, seabirds more than most, that pay the price, with flow-on effects for corals. Bleak though this is, we're at least realising the towering importance of rat control, and we know what must be done.

I've lost count of the number of times I've heard the phrase 'the balance of nature' used in documentaries, with its implication that the natural world exists in a steady, harmonious state. This is seldom, if ever, true simply because unpredictable and sometimes drastic things happen with a regularity that issues a challenge to natural habitats at every scale from local to regional to global. Storm waves devastate a reef crest, El Niño scalds the eastern Pacific, or climate change sabotages global weather patterns. The result is that the natural world is in a constant state of flux. The resilience of ecosystems to these challenges is founded on a host of complex interdependencies and feedbacks, involving an almost endlessly diverse cast of organisms, like the guide ropes holding a tent in a squall. Biological communities can bounce back from most insults that fate can throw at them, but only if these supports are in place. When we unthinkingly cut away at the guide ropes, when we overfish, pollute, develop insensitively or transmit invasive species, we compromise the whole, and often in ways that we didn't anticipate or even imagine. In the past, ignorance has our defence. The progression of science means that we can no longer claim to be blind to our impacts.

Back on One Tree Island, I return from my exploration of the lagoon and make my way back to my dorm, water dripping from my hair and onto the towel around my neck, I stop to admire one of the noddies. Its perch is level with my face and we regard each other, its berry black eyes peering into mine across the distance of a handspan. I love the fearless composure of this bird; it seems exquisitely delicate, its entire body weighing about as much as an apple. Just as I'm wondering what it makes of me, it appears to answer with a squirt of guano, which drops wetly onto the hot coral rubble at my feet. This bird, along with the 15 others sharing the same salt-bonsaied tree, and thousands more of its ilk, is performing a vital role in the ecosystem of One Tree. The same can be said for virtually every other organism here, from the microbes to the heftiest sharks. They're bound together in a spectacular natural synergy, the evidence of which can be seen in the breathtaking biological opulence of this part of the Great Barrier Reef. Yet despite its robust health today, the future looks ominous for One Tree and the bustling occupants of this coral metropolis. In 2024, the Great Barrier Reef Marine Park Authority downgraded its outlook for the future of this, the world's largest coral reef, from 'poor' to 'very poor'. The relentless upward trend of global emissions, and the concomitant rise in global temperatures means that all the world's reefs face an imminent and existential threat.

It's heartbreaking to know that if we don't do anything about it, coral reefs may be no more than a memory within our lifetimes.

* * *

Coral bleaching is the dread phrase of this marine age. I've seen senior researchers at scientific conferences break down while relating their latest observations, such is their despair. There are some who paint these as histrionics. Such critics underestimate what it is to experience, first-hand, the death of the sea's most iconic environment. Furthermore, it isn't the slow, wasting decline that characterises so many of our natural habitats. In the case of bleaching, it can be sudden; shockingly so. A vibrant kaleidoscope of colour and life can be transformed almost overnight to a winterscape of monochrome death. It's scarcely surprising that emotions should run high among people who have devoted their lives to understanding coral reefs, only to find themselves chronicling their collapse.

Faced with this unprecedented crisis, initiatives are under way to try to arrest or even to reverse the damage being done by coral bleaching. The first step, as ever, is to recognise what we're up against. Bleaching represents a breakdown in the relationship between the coral and its dinoflagellate partner, leading ultimately to the expulsion of the alga by the coral. Since it's the dinoflagellates that typically lend corals their colour, their departure yields the ghostly appearance of the coral in the aftermath. Until now, the coral was getting the vast majority of its nutrition courtesy of its algal lodgers, but with them gone, it faces starvation and death if the situation can't quickly be resolved.

What might at face value seem a pyrrhic gesture on the part of the coral reflects the fact that the symbiosis between coral and alga represents the most delicate of balances. Under normal circumstances, the coral provides everything that its dinoflagellate guests need to photosynthesise, including CO_2 and a slew of micronutrients. In return, the alga rewards its host with a plentiful supply of delicious sugars. The complex physiologies of both organisms are closely, though not completely, intertwined. They're ultimately in this for themselves, and each maintains a little control of the other. The alga suppresses the coral's immune system to reduce its chances of being evicted, and the coral rations the nitrogen it gives to the alga, thereby preventing the lodger's population from exploding and consuming more resources than the coral can budget for.[66]

When conditions are just right, this partnership, with its checks and balances, works well. There are, however, plenty of triggers that cause the two parties to shift from cooperation to conflict. Waters low in oxygen or salt, or overburdened with nutrients, can unravel the collaboration, as can too much ultraviolet light. The one that we're painfully familiar with at present is, of course, too much heat. As the mercury nudges upwards, the carefully calibrated relationship starts to falter, and the longer the heat persists, the worse things get. The cellular machinery inside both partners starts to falter and critical proteins break down. High temperatures and bright sunlight initially cause the algae to photosynthesise like crazy, outstripping the ability of the host coral to supply them. Without the basic materials to work with, the algae are forced to shut down their most fundamental operations. Both partners are now under pressure, and stressed cells produce unstable, reactive molecules.[67] Unless these can be stifled by the organisms' inbuilt

---

66 This is one reason why coral reefs struggle when wastewater from human activities floods the environment with nutrients; released from the restrictive diet of nitrogen provided by the coral, the algae proliferate and turn parasitic, stripping the coral of its supplies.

67 Reactive oxygen species, essentially molecules produced either as a by-product of metabolism, or by environmental stress, can cause serious damage within cells.

safety mechanisms, the result is a kind of cellular poisoning, which cascades to trigger more damage. The relationship rapidly becomes, very literally, toxic.

In response either to the uptick in harmful reactive molecules, or to try and balance its gaseous budget and bring the demand for CO_2 under control, the coral begins to expel algae. The remaining algae retaliate by withholding their sugar rent, using it instead to prioritise their own growth and repair, and to rebuild their numbers. At this point, the alga is still making plenty of demands on the coral's other nutrient resources, and, deprived of these, the coral finds it progressively harder to supply its partner. The situation thus spirals out of control, all semblance of harmony in the partnership unwinds. In a desperate final act of self-preservation, the coral, drained by the algae, embarks on the mass eviction of its now troublesome guests and the result is bleaching.

The short-term respite that the coral gains from the separation comes at a heavy price. To save itself, it's abandoned the partnership and flung itself into a lifeboat, but it's a leaky lifeboat, since its food supply in the absence of the algae is now severely limited. The coral may yet survive, depending on whether it can forge some kind of reconciliation with its erstwhile partners in the coming days and weeks. But if the conditions that precipitated the break-up persist, the chances of a return to normal are slim. At the same time, the weakened coral finds itself at risk of falling victim to disease, overgrowth by rampant algae or attacks from predators.

Climate change is not the only challenge faced by coral reefs. Indeed, we've engineered a perfect storm for them. In addition to rising temperatures, we've polluted the marine environment with industrial chemicals and with fertiliser, both of which tip the odds against corals and, in some cases, towards their chief competitors. Fishing, too, has had powerful effects, some of which I've mentioned previously, but there's another creature that has benefitted from our mismanagement. It's a coral's worst nightmare: a voracious biological juggernaut known as the crown-of-thorns starfish (CoTS). In common with what I suspect is the majority of people, I had the idea of sea stars as relatively small creatures. My first encounter with a crown-of-thorns completely changed this perspective. I was among the corals in One Tree's lagoon when I noticed a broad streak of white across the reef surface that terminated in a huge, multi-legged creature, the size of a dustbin lid and bristling with wicked spines. It's a coral digester, the trailing slash behind it evidence of its acid destructiveness, dissolving the coral's flesh down to the underlying rock. A single one of these is shocking, yet oftentimes they descend like a plague to strip an entire reef, as happened at the Polynesian island of Moorea earlier this century.

It's a bit rich of us to vilify an animal whose impact on reefs is orders of magnitude less severe than our own, perhaps, but for those people charged with protecting what is already an incredibly vulnerable habitat, CoTS are a potentially serious problem. The giant adults, whose spines are loaded with a powerful venom, are immune to most predators. An enormous predatory snail, which glories in the name Triton's trumpet, is equal to the task but sadly the populations of these watermelon-sized molluscs have been decimated by fishing and demand for their shells as trinkets. It's not mere chance that the beginning of mass collection of these molluscs in the 1960s coincided with the first recorded mass outbreaks of CoTS. The 1960s also saw the ramping up of commercial fishing on reefs, which eradicated snappers, rock cod, wrasse and emperors, fish that were only too happy to make a meal of juvenile CoTS, which are less formidable than their adult counterparts. Our customary mismanagement lit the fuse on an ecological disaster that has since blown up in our faces. Corals have existed alongside CoTS for millions of years and healthy reefs are resilient to their effects. Now, however, the destabilising effects of our actions means that we have to intervene to cull CoTS. An intervention of a different kind is needed for the corals themselves.

In an ideal world, we would respond to the alarms that are blaring from all parts of the natural world by cutting our emissions drastically, and soon. The political class, however, along with large numbers of their electorates, seem deaf to the escalating warnings. There are any number of reasons why, from a fear of what limiting emissions would mean for short-term economics, to the actions of powerful lobby groups with a stake in the status quo, to a misguided acceptance of the pseudoscientific denialist drivel that infests the debate like intellectual herpes. Given the fiddling that's going on as the world burns, and the certainty that a truly global response will be far slower than it needs to be to constrain the process of warming in the near future, it's left to scientists and devoted laypeople to tackle the plight of corals.

Although the despondency that I've witnessed among fellow biologists is not uncommon, it's often followed closely by a determination to engineer solutions. The result over the last few years is an upsurge in creative ideas about how best to tackle the problem. Variability, it's often said, is the raw material of natural selection and in the wake of devastating bleaching events, there are often some corals that escape comparatively unscathed. In parts of the Pacific nation of Palau, for instance, certain corals appear to have developed a degree of resilience to the blistering temperatures that have cooked the reefs there with increasing frequency over the last few years. These survivors are the targets for research, so that we can answer the question of how they've survived when, around the world, so many corals have been devastated by temperatures lower than those that Palau's colonies have ridden out.

The answer seems to lie at least partly in the corals' genetic code. Palau was hit especially hard by an El Niño event in 1998, and again in 2010. At both times, the bleaching and subsequent coral mortality pushed the reefs to the limit, yet in 2017, when similar ferocious temperatures returned, the corals somehow managed to cope. The ones that did are genetically distinct from those that succumbed, which might suggest that some corals at least are undergoing a form of stress-based, accelerated evolution. In the Red Sea, where corals thrive in temperatures that are routinely higher than those that induce bleaching in other parts of the world, a more gradual evolution has taken place. We urgently need to understand exactly what underwrites the heat tolerance of the corals here and in Palau, since this might well provide critical insights that can be applied elsewhere.

Once we have a grasp of the genetic basis of heat tolerance, we can bring our expertise to bear. This might involve selective breeding, such as we've used throughout our history in species from crops to domestic animals. In the case of coral, we can select for strains that are resilient to heat and to disease, while also displaying fast growth and high fecundity. The progeny can then be shepherded through their early existence in nurseries before being outplanted on reefs. A less involved approach involves the simple expedient of locating colonies that have successfully rebounded from a bleaching event. By carefully snapping off small pieces of such colonies in a process analogous to taking cuttings of a plant, damaged sections of reef can be replanted with stoics. Like plants, the fragments have the potential to regenerate and grow into their mature glory.

Either way, non-profit NGOs around the world are hard at work replanting damaged reefs with coral fragments fixed to all manner of structures, from frames to 3D printed ceramic blocks, and even upscaled versions of the kinds of multi-hole planters popular with patio gardeners. Though good news in the age of bleaching is in short supply, it's great to be able to report that some of these cossetted corals are thriving. At Fitzroy Island on the Great Barrier Reef, outplanted *Acropora* corals recently spawned for the first time, only four years after being introduced to the reef.

The coral itself is only one part of the resilience equation, however, and other research has looked beyond the animal to consider the other partners in the symbiosis. Some of the dinoflagellates for instance, are especially heat tolerant. We now know that corals can shuffle their algal partners, evicting those that can't take the heat and replacing them with hot water specialists. Clade D dinoflagellate species, and, in particular, *Symbiodinium* D1a, are particularly well adapted to withstand the feverish conditions that are now becoming commonplace. In the aftermath of a bleaching event, the corals that survive are typically those that are richly infiltrated with these superalgae. The flip side is that heat-loving dinoflagellates aren't as good at producing sugar for the corals when temperatures normalise. Hence, the shuffling enacted by the corals, where colonies seem to perform a kind of algal pitstop, changing the make-up of their partner populations according to the weather.

Taking this as a starting point, we can foster the further development of heat-tolerant dinoflagellates by culturing them in laboratories at high temperatures. When these are added to live corals under experimental conditions, they have a strong and positive effect on the survival of their coral hosts in simulated bleaching conditions. It's a long way from this to the natural environment of the reef; nevertheless, ways of providing these to free-living corals as a kind of emergency probiotic are being investigated. And speaking of probiotics, there's another biological factor that has a strong influence on coral resilience. Corals, ultimately, are only the outward expression of what is a multifaceted collaboration.

Indeed, it's not really accurate to think of this simply as a coral, it's a conglomeration of disparate but interconnected partners working in concert. As well as the algae that shoulder so much of the partnership's work, a community of intrinsic microbes contributes to the smooth running of operations. The microbes assist with nutrient cycling, protect against disease and even provide vitamin supplements. In addition, the microbial community has a powerful role to play in supporting recovery from bleaching. And if the colony does recover, the microbes seem to protect against future heatwaves, making bleaching less likely the second time around in a phenomenon known as ecological memory. Unpicking exactly what features of the microbial community best support resilience would also provide us with a means of protecting reefs.

If we support it, or at least stop hampering it, the natural world is incredibly resilient. Coral reefs, however, are on the brink. There were celebratory headlines when in 2022, following a series of horrific mass bleaching events, it was reported that the Great Barrier Reef had more coral cover than at any time since monitoring began in the 1980s. Regrettably, this is akin to saying the rampant poppies that coated the battlefields of Flanders in 1915 represented a vegetative recovery. Just as poppies are no replacement for forests, so new corals are no replacement for the richness and complexity of the ancient growths that were lost, nor the number of other species that they can support. A recovering reef is a pale shadow of its former self.

Though there are grounds for optimism, especially in regard to the number of avenues that show promise for research into promoting coral resilience, it's important to recognise that the absolute best that we can hope for using these approaches is to provide a stay of execution for some coral species. If we don't address the underlying cause of global warming, then no amount of palliative treatments will ultimately save the situation.

Back at One Tree Island, my group heads to the edge of the reef. It's late afternoon and time and tide have aligned to provide us with a joyride. The shallow channel along whose banks we're walking will soon be a torrent because it's here that the flood tide gains entry to the lagoon. Twenty metres away, on the seaward side of the reef crest, the triangular fins of sharks show above the waterline. These nocturnal hunters are keen to cross into the lagoon and begin their prowling; they won't have long to wait. We lie on our stomachs on coral rubble covered by just an inch or two of water, and we get hold of anything we can to ballast us in the currents that will come. A large boulder is best. Gradually the waters rise, and the flow begins. Soon, it's just deep enough to duck our face masks under the water, to see the bubbles of the whitewater torrent rushing in and to hear its power in our ears. The moment it's deep enough, the sharks come firing past us like missiles. A few more minutes and we're fighting to hold our ground against the rising torrent. Eventually the sea's power wins out, as it must. We spin loose, tumbling in the rapids, but as the channel widens and deepens, the flow slackens and we're borne at a gentler pace through a corridor wonderland of corals and a spectacular array of impossibly beautiful animals. It's such a privilege just to drift effortlessly like this for minutes on end amid the serenity and the glory of the reef, and it's a rite we perform at every visit.

This year, 2024, we have no appetite for it. The spectre of global warming has visited the far south of the Great Barrier Reef and has hit One Tree for the first time. The coral here is dying or dead, almost all of it; a pall hangs across this place and the sense of loss is overpowering. Throughout the tropics, these priceless jewels are crumbling as we look on. What will it take to change our ways?

7.
Amid the Flashing and Feathery Foam

The open ocean is both highway and habitat for marine species

FAR FROM LAND, THE OCEAN TAKES ON A DIFFERENT character. The waves glide unhindered across the great expanse of salt water, each as it rises showing a face that's a deeper, richer blue than those that roil the shore far behind us. Two hours from the Azorean port of Madelena, we've long since swapped the turmoil of inshore waters for the timeless roll of the open ocean. The only sound out here is the rhythmic swash of water against the boat. To the north, the volcanic peak of the island of Pico disturbs the uniformity of the horizon, but aside from that small reminder of home, the immensity of the sea stretches in every direction. The effect of this is to make me feel small and insignificant, a mere speck on this enormous, undulating aquatic plain. This sense of humility will be compounded when I meet the creature that has drawn me here, the largest toothed predator on the planet, the sperm whale.

This oceanic behemoth has exerted a powerful hold on the psyche of mariners for as long as we have been seafarers, its mystery and its sheer size conjuring fear, especially since our ancestors lacked perspective sufficient to judge it by any other than their own standards. Notions of its power, its cunning, and even its capacity for deadly rage were placed firmly in the public consciousness by Herman Melville's classic novel *Moby Dick*, itself inspired by the ramming and sinking of the Nantucket whaler *Essex* by a sperm whale a generation earlier. The story of the *Essex* is powerful not only for the supposed calculated vengefulness of the whale's attacks but also for the incredible privations, extending to cannibalism, suffered by the crew in the lifeboats subsequently.[68] Yet the *Essex* is far from the only whaler to have been attacked by sperm whales.[69] The *Ann Alexander*, the *Parker Cook* and the *Kathleen*, to name just three, had their planks stoved in by sperm whales, typically while the hunt itself was reaching a climax and the animal was injured. Alongside these, dramatic accounts exist of whales, wounded by harpoons, lashing out with their powerful tail flukes at the open rowing boats that put out from the whaling ships to do the task of killing.

Among these stories are frequent references to the aggression of the whales, the crushing of boats, even of the chase boats being bitten in half, and of sailors carried off in the jaws of a whale. Legends of maddened sperm whales were rife among whaleship crews, huge bulls that carried the scars of many battles and who harboured a cold malice. Oftentimes, these individuals achieved a kind of fame, were given their own soubriquet and were celebrated as worthy adversaries. One such was Mocha Dick, an albino whale, named for the island of Mocha, off the coast of Chile, where he was first spotted. He is described posthumously by Jeremiah Reynolds in a New York periodical in 1838 as 'emphatically among fish, the Stout

68 The tale is brought vividly to life by Nathaniel Philbrick in his book *In the Heart of the Sea*.

69 Detailed accounts are given in A. Hyatt Verrill's 1916 book *The Real Story of the Whaler*.

Gentleman of his latitudes' and a 'renowned monster, who had come off victorious in a hundred fights with his pursuers'. Whalers rounding the Horn to enter the Pacific grounds were keen to test their mettle against this legendary animal, whose back was 'serried with irons' and who had 'fifty to a hundred yards of line trailing in his wake'. But though stories of the ferocity of Mocha Dick were many, and the plaudits given to his eventual killers were great, there is a contradictory motif to the story that coexists with the narrative of his spitefulness.

In Reynolds' account, Mocha Dick, for all his supposed ferocity, was not given to acts of aggression unless subjected to attack. 'On the contrary,' Reynolds writes, 'he would sometimes pass quietly round a vessel, and would occasionally swim lazily and harmlessly among the boats, when armed with full craft, for the destruction of his race.' Mocha Dick was killed 28 years after he was first attacked by whalers, touchingly while swimming to the aid of a female whose calf had been harpooned. While his size was scarcely exaggerated – Mocha Dick was measured at 21 metres – there's more than a little licence to all of these stories. For one thing, no sperm whale would be capable of biting a 10-metre rowing boat in two. For another, sperm whales very quickly adopted an ingenious defence against whalers in the era of sail. Their natural response to attack from their only marine enemy, the killer whale, is to close ranks in a wagon wheel, known as a Marguerite formation, with their most formidable weapons – their tails – pointing outwards. Such a defence is self-defeating against whalers, since by drawing together like this, the harpooners simply gain more targets. Instead, the whales quickly learned to swim upwind of an approaching ship, thereby stymying the efforts of the whalers to give chase. This and other defensive measures specific to human persecution seem to have caught on rapidly within sperm whale populations following their first encounters with whalers in the North Pacific. As a highly social, intelligent species, the facility exists for them to learn from one another with the effect that, although the frequency of whale sightings remained consistent, the successes of the whalers fell by well over half in just a few years.

The decline of wooden ships, and of whaling, seemed to spell the end to the tales of hostile sperm whales. Indeed, many of these historical accounts are looked on with scepticism because of both the legendary fondness of sailors for a yarn, and the likelihood that many so-called attacks were, in fact, accidental collisions. Inadvertent whale strikes were common and remain so; for a wooden vessel the damage could be considerable; historical accounts of such strikes describe the impact as being comparable to running aground. As the years passed with no further incidents reported, it appeared that sperm whales had called a truce. Then, in 2007, a bull sperm whale appeared to deliberately charge a ship. To add credence to events, the ship in question was a marine mammal research vessel and the whole thing was captured on video.[70] From the perspective of the people aboard, three males approached from distance, the largest of them accelerating before ramming into the ship. Fortunately, none of the people were hurt, and the whale appeared to be untroubled; nevertheless, the episode raises questions, chief among which is why might a whale do this?

A key detail is that the males present seem to have been fighting for the attentions of a nearby female and all of them bore rake-like wounds on their heads, likely inflicted by the other males' teeth. Moreover, we know bull sperm whales use their heads like biological battering rams, pairs of rivals charging headlong into one another

70 Fulling et al. (2017) *Sperm Whale Ramming*. Available at: https://www.aquaticmammalsjournal.org/fulling-et-al.

like male muskoxen or bighorn sheep do on land. It's possible that the combatants in this case, roused to fury, mistook the hull of the ship for a rival, or that this was a case of redirected aggression in the heat of the moment. The observers raised a third possibility, which is more troubling. Given the long lifespan of sperm whales – they live roughly as long as we do – the adults in question are old enough to have direct experience of persecution by whalers, which extended until late into the last century. Perhaps these whales still bear a grudge.

The idea of resentful whales did play on my mind while I was planning the first trip to the Azores, but by the time I slipped into the glassy waters of the Atlantic, my excitement at getting the chance to meet these intensely charismatic animals far outweighed any trepidation that I felt. Besides, I was by no means guaranteed to meet them. Any meeting with sperm whales is by necessity on their terms; notwithstanding their size, with an ocean of space to manoeuvre, a whale could easily avoid a floating researcher. Activities that involve interactions with whales in the Azores are, quite rightly, strictly regulated. Accordingly, the typical drill was to be dropped into the ocean from a boat several hundred metres ahead of the animal. This done, the boat then departs, and you just hang in the water and hope that the whale is feeling sociable. Happily, they always were.

As I bob around at the surface, I feel something like vertigo looking down into the crystalline water that descends kilometres beneath my fins. The ocean seems devoid of life until a faint shadow appears at the extremes of my vision, an almost imperceptible darkening of the azure canvas that grows and takes shape. By degrees, it resolves into the distinctively peculiar shape of a whale, heading straight for me. She produces a staccato commentary of clicks as she approaches, crisp sounds that I feel as much as hear. As she draws close, she rolls onto her side, the better to regard me with a huge, dark eye.

Her flanks, combed with small ridges like the sand patterns left on a beach by the ebbing tide, had seemed a uniform grey at distance. Now up close, I see that they are patchy and mottled, a greyscale pastiche, with swirls of white along her belly and lining her mouth. This mouth, supposedly fearsome enough to crush a boat, is long and thin, and seems almost delicate. Even her powerful tail as it cuts through the water is a study in elegance. She has an enormous presence, this animal, one that outstrips her 15 tonnes; she is quite simply the most magnificent animal I've ever seen.

It's tempting to imagine that I was in the whale's world, but the reality is that our encounter took place in the narrow intersection at the extremes of each other's existence. Sperm whales are indefatigable predators who spend the majority of their time in the crushing pressure of the deep sea. Exactly how deep they go depends on their prey; a dive of 45 minutes' duration down to 800 metres is fairly typical, though these incredible submariners have been known to descend beyond 2 kilometres and for two hours at a time. The first whales that I met in the Azores were enjoying a brief hiatus of perhaps ten minutes between dives, replenishing their oxygen supplies and gathering themselves for their next foraging expedition to the permanent night of the deep. Then, with a final snort to empty their lungs, they leave the sunlit upper echelons of the ocean once again, a stupendous tail raised briefly above the waves in a farewell gesture.

The whale's clicking as she loomed out of the blue is part of her sophisticated, inbuilt biosonar, a kind of echolocation that has much in common with that employed by hunting bats and indeed by some blind people. The huge head of the whale acts as an acoustic lens, training and focusing a beam of sound out into the distance. The returning echoes have a texture to them that is imparted by infinitesimal differences in the time they take to reach the whale.

We know from studies of people with this ability that the visual cortex is recruited to translate the incoming acoustic information. It's possible that the sperm whale's giant brain performs a similar role, rendering a kind of image of the environs in their mind's eye that emerges not from light, but from sound. Regardless of exactly how sperm whales perceive acoustic information, it's crucial to them when foraging. Their dives to half a kilometre or more below the waves take them into the heart of oceanic darkness where they use their sonar to seek out their prey.

Sperm whales also make use of the strange oceanographic phenomenon known as the oxygen minimum zone, or OMZ, the layer of suffocating water that marks the boundary between the oxygenated currents of the deep sea and the gas-saturated surface waters. Exactly where the OMZ sits depends on where you are in the world, but it typically ranges from about 500 to 1,000 metres depth. Life in the OMZ is challenging for its resident animals, and it's one that they cope with in two main ways. Firstly, they've adapted hyper-efficient respiratory systems to glean as much precious oxygen as possible. Secondly, they tend to live at a slow pace because, as mountaineers know, it's dangerous to exert yourself when you're struggling to breathe. Not that this troubles a diving mammal like a sperm whale; they take their own air supplies with them by using their bodies as a gigantic oxygen store. A whale's blood and muscle are packed, respectively, with haemoglobin and myoglobin, which cling ferociously to oxygen. The concentrations of myoglobin in whale muscle are 20 times greater than our own, lending it a dark, purplish-black appearance. The blood, meanwhile, is crammed with red blood cells, which make it exceptionally gloopy, almost like motor oil, and they have stupendous amounts of it. Blood makes up about 7% of our body weight, whereas for a sperm whale, the value is closer to 20%, so the female that I encountered on my first day was carrying something like 3,000 litres of the stuff.

It was once thought that foraging sperm whales followed a very basic strategy, operating like giant marine versions of flypaper by lingering motionless in the deep sea with their mouths open and waiting for their victims to blunder into them. Confidently though this was asserted, it's far from correct; sperm whales are relentlessly active hunters. Their inbuilt cetacean scuba apparatus, combined with the sluggish behaviour of their oxygen-starved prey in the deep, gives them a crucial edge, which is just as well because they have an appetite to match their size. To fuel herself, an adult female sperm whale has to catch around half a tonne of prey every day, and a large male might need twice as much. A former colleague of mine once remarked sniffily to me that he wasn't interested in whales because, despite their size and charisma, their relevance to marine ecology was negligible. The claim doesn't stand up to scrutiny. Sperm whales alone capture more food from the ocean than we do with all of our fishing fleets, and no one ever accused us of having a negligible impact on marine ecology.[71] This shouldn't imply that sperm whales are a problem, or that they're overfishing and thinning out stocks, for the simple reason that they hunt way down the food chain, eating plentiful animals that have short generation times. Though they do, occasionally, go off-menu to sample fish, squid are the sperm whale's staple.

Adventurous accounts of sperm whales wrestling with leviathans like the giant squid and its even heftier counterpart, the colossal squid, paint a somewhat inaccurate picture of their diet. Although it's true that large male sperm whales especially tackle these stupendous cephalopods, and have the circular sucker scars on their heads

71 According to Hal Whitehead, the world's foremost expert on the species, there are about 740,000 sperm whales alive today.

to show for it, the more typical prey for these mammals is much smaller. Researchers around the world who've performed autopsies on the fragrant carcasses of beached whales report that the average squid found inside the stomach of sperm whales is about 30 or 40 centimetres in length and a little under a kilogram in weight. At this kind of size, the whale has to eat several hundred squid every day, and possibly over a thousand in the case of a large bull.

Thanks to the wonders of multi-sensor tagging technology, we can now follow sperm whales as they descend to their deep-sea dining hall and, for the first time in our history, understand how the world's largest toothed predator hunts. A sperm whale departing the surface has some way to travel before it reaches its foraging ground. It spends the first part of its dive in relative silence, but as the depth increases and the surface light fades, the whale activates its sonar. For a squid, ensconced in the velvet blackness of the deep, the click-click-click of a hunting whale should represent its worst nightmare. Ironically, however, it seems as though they lack the auditory ability to perceive the ominous clicking of the loudest animal in the world. So, the whale is able to locate flotillas of its cephalopod prey at distances of up to 500 metres, while its quarry is unable to either see it or hear its approach. In any case, there's nowhere to hide from the penetrating sonic searchlight and, as the whale closes in, it ups the rate of clicking dramatically until it sounds to us like a continuous buzz or creak. What matters to the whale, of course, is not the sound it produces, but the faint echoes that report back from its target. From these, the whale can precisely locate its target and rush in for the kill.

I say rush, but the lack of athleticism shown by some of the squid in the OMZ makes them so obliging as prey that it scarcely seems fair; it's like a cat being confronted with a troop of elderly, asthmatic mice. In any case, the sperm whales hoover them up with little difficulty. The problem for the whales, however, is that the easiest kinds of squid to catch are also the least nutritious, so the hunters face a trade-off between these and the more rewarding but more elusive squid. As their stomach contents testify, sperm whales are up to the task.

The technology that has opened up the sperm whale's world to us also reveals their tactical flexibility in the hunt. Sometimes the whales glide stealthily towards their prey – although their intended victims may not hear the clicking sonar, they're highly sensitive to pressure waves that indicate movement. Humboldt squid, for instance, flee into the depths at the first inkling of sperm whale activity. At other times, the whales abandon caution, accelerating and turning sharply to capture fast-moving prey. Cameras mounted on the backs of the whales show the gory aftermath of such charges, a forlorn cloud of ink and fragments of dismembered calamari. Oddly for a predator, the sperm whale's teeth, which are conical and as long as a paperback is tall, seem to be little more than decorative. Though the whale might use them to savage its more recalcitrant victims, more usually it simply suctions the intact squid down its gullet. Indeed, they don't even grow their teeth until they're on the cusp of maturity at around ten years of age, and even superannuated individuals that have lost their teeth seem to be perfectly capable of hunting.

The relentless cycle of diving and surfacing continues through day and night with scarcely a break. The closest they get to rest is when they position themselves vertically in the water like giant candles – head up or head down seems to make little difference – and catch 15 minutes' snooze. Over the course of a full day, the whales might manage a series of these naps, adding up to a meagre hour or two's worth of sleep for each whale, even the teenagers.

Despite their relentless schedule, sperm whales – particularly females and calves – always make time for another essential activity.

In the late afternoons, the whales take a break from terrorising squid to socialise. I had the incredibly good fortune to experience one of these cetacean conventions, involving a family of five whales that spanned two or perhaps even three generations, and a bottlenose dolphin with a twisted spine that had apparently been co-opted into their circle. As I watched, the whales rolled at the surface, the smaller members, and even the dolphin, taking it in turns to swim lengthways into the mouth of the largest female, and to receive a gentle nibble for doing so.[72] The frolics went on for at least an hour, and floating alongside these giants, occasionally manoeuvring to stay clear of their enormous tails, I gained a wonderfully clear sense of the whales surrendering themselves to the joys of socialising.

As with so many mammals, sperm whales live in matrilineal societies, wherein a family group is built around a large and experienced adult female – the matriarch. Males leave their mother's side as they approach maturity, initially congregating in bachelor groups before becoming ever more solitary as they age. As a rule, the females and calves tend to keep to the warmer waters of the tropics and subtropics. The males, instead, head towards higher latitudes where they roam independently, returning to sunnier climes only to consort with the females for breeding.

While the males live on the margins of sperm whale society, the matriarchal family units are its foundations. These groups are close-knit, numbering up to ten individuals. Family members may spread out, especially during hunting, but they're seldom separate for long. They travel together, defend against common threats collectively and look after each other's calves, even to the point of suckling them, all of which forges powerful bonds that endure for a lifetime.

72 I describe this in detail in a previous book, *The Social Lives of Animals* (2022).

Although the family is the centre of the sperm whale's world, that's not to say that they shun wider connections. Families congregate from time to time, and dozens of whales may be seen together for periods lasting from a few hours to a few days. There are rules to this, however; a sperm whale family unit will happily mingle with some groups but not with others.

The clan is one of the most fascinating aspects of sperm whale society. It's a word that inevitably puts me in mind of the Scots, both because of its Gaelic etymology and because of the famous blood feuds between the Campbells and the MacDonalds, or between the Camerons and the Mackintoshes. Despite the imagery, however, the comparison isn't especially apt; for all that they keep to their own, the whales don't engage in inter-clan fighting. Nevertheless, each sperm whale clan, which might comprise thousands, if not tens of thousands, of members, has its own distinct cultural identity. Some are more sociable than others, and there are also differences between clans in how they forage and migrate. Babysitting is a carefully coordinated activity in some clans, whereas others take a less structured approach. What the whale clans do have in common with their historic Celtic equivalents is that they keep themselves to themselves. Even though clans may overlap geographically, females never fraternise with their sisters from other clans.

Perhaps surprisingly, there are no major genetic differences between the clans. That's because males don't have much truck with clannish propriety and are only too happy to make the sexual acquaintance of all and sundry, so there's plenty of gene flow throughout the entire sperm whale population. What keeps the clans distinct is their strictly enforced social isolation, which provides fertile ground for the development of clan-specific behaviours, a culture that is learned and shared among peers, and is passed down to calves by their mothers.

With little to differentiate the clans in terms of blood, the whales maintain their social boundaries by means of so-called codas, the sounds that the animals produce when they're in company. A coda is a series of clicks made by the whales, and each clan has a unique coda of its own. For instance, that of the EC1 clan in the Atlantic follows the pattern 'click-pause-click-pause-click-click-click', while the Mediterranean clan use 'click-click-click-pause-click'. On the basis of these stereotypical codas, we've identified to date seven sperm whale clans in the Pacific and four in the Atlantic, with a strong likelihood that more remain to be found elsewhere.

If maintaining clan-based divisions within sperm whale society was the only function of their codas, it might seem slightly underwhelming in such an intelligent animal. The rough equivalent would be for me to run around shouting, 'Ward! Ward!' to anyone and everyone I encountered, without ever imparting more information to them than my name. In fact, there's much more to their communication than mere badges of clan identity. Indeed, the majority of the codas that whales produce seem to have little to do with announcing their credentials. The whales' use of so-called 'non-identity' codas seems to transcend clan boundaries, potentially allowing them to converge on similar patterns of expression.

The wonder that I felt when mingling with the cavorting sperm whales wasn't solely based on the extraordinary visual spectacle that they presented; the soundtrack to their celebrations was also astonishing. The clicking was cacophonous, but what really struck me was the richness of their phonations, the whales' continual dialogue textured with subtle rhythms and shifts.[73] It could be, of course, that these tones convey only a very limited meaning, and indeed this

73 The term 'phonations' is preferred to 'vocalisations' because the whales don't produce these sounds by means of vocal cords, as other mammals do, but by structures in their giant heads known as monkey lips.

has been our prevailing view of animal calls throughout history. Yet the diversity of the sound patterns produced by the whales, and the way in which different individuals clearly took turns to pipe up, is intriguing precisely because it is so reminiscent of human conversation, albeit in a patois as alien to our own as could be imagined.

The idea of us actually listening to other animals has a much shorter pedigree than that of butchering them. It was in 1970 that the scientist Roger Payne released the groundbreaking album *Songs of the Humpback Whale*, from an archive of recordings given to him by a US Navy listening station in Bermuda. Payne was inspired, or perhaps more accurately, provoked, into action by an event some years before when, after hearing that a porpoise had washed up near his home in Massachusetts, he drove to inspect it. By the time he arrived, its fins had been hacked off, people had carved their initials into its flanks, and someone had put out a cigar in its blowhole. The desecration lit a fire under Payne. His elegant response, in the form of the whalesong album, brought about a re-evaluation of cetaceans and kindled the conservation movement 'Save the Whale'.

Even so, there's a long way between empathising with an animal and acknowledging it as an equal. Charles Darwin wrote that the differences between the minds of animals and of man were 'of degree not of kind', but in the 150 years since the publication of *The Descent of Man*, this remains a controversial idea for many. This is never more the case than with human language which, as many linguists are keen to point out, is both quantitatively and qualitatively different from animal utterances. We can refer to abstract concepts and generate new ideas using complex, and often novel, arrangements of discrete phonemes. We can combine our words in an almost infinite number of ways, based on a shared frame of reference, and using rules of grammar and syntax, to describe how we're feeling, or draw analogies, or to relate to other times and places, among much else. Our language is, to a large extent, what makes us, us, and it underpins our sense of human exceptionalism. Animals have travelled some way down the road towards language but, by our current understanding, even the most intelligent of our fellow species are a long way short of us in this respect.

There's always the possibility that this current understanding is imperfect. If we start from the principle that our language developed as a means to navigate our social world, and it has co-evolved with our brains in a positive feedback loop, then it follows that the likeliest animal candidates to emulate our linguistic skills would also be gregarious and brainy. Sperm whales fit the bill on both counts. The problem is that the sperm whale's Morse code of nose clicks seems, at face value, too simplistic to fit the criteria of a language. The more detailed examination permitted by new technologies this century, however, revealed subtle yet tantalising suggestions of complexity in sperm whale phonations. At the same time, we have begun to piece together evidence of consistent structures within the phonations, hinting at the possibility that meaning is embedded in these sounds.

Having uncovered the rudiments of sperm whale communication, it's an enormous challenge to work out whether this extends beyond basic utterances to meet the definition of a language. And even if it did qualify as such, how could we understand it? We lack the shared frame of reference with sperm whales that provides the foundational basis for human conversation. Nevertheless, if we're to claim a deep understanding of the animals that we share the planet with, we can't afford to shy at the challenge. As the proverb has it, a journey of a thousand miles begins with a single step. In the case of sperm whales, that first step is to identify the subtleties and intricacies of their phonations and to relate these to the contexts in which they made them. This necessitates trawling through a massive amount of data – billions of codas, most likely. Until recently, this would have been impossible, but no longer.

The Caribbean sperm whale hotspot of Dominica is now at the centre of the most ambitious attempt in history to decode the communication of another species. It's an effort coordinated under the banner of Project CETI (the Cetacean Translation Initiative), established in 2020, to bring together a diverse and multidisciplinary team of biologists, linguists, acousticians and experts in artificial intelligence, machine learning and robotics to delve into what they call whale-speak. It's entirely appropriate that Roger Payne was an adviser to Project CETI until his death in 2023. Fifty years on from *Songs of the Humpback Whale*, this new analysis of whale sounds promises to uncover the hidden currents of cetacean life.

Prior to the advanced analysis offered by Project CETI, the idea of sperm whales as conversationalists would have seemed a stretch. A total of only 21 discrete codas, based on the number and timing of clicks, had been identified in the Caribbean population, representing a somewhat limited repertoire for a language. Closer examination, however, revealed much greater nuance and layering within those codas. Different inflections and pacing, along with periodic additional flourishes to established codas, pushed that initial 21 up to at least 156. If the researchers are correct in their assertion that these represent a phonetic alphabet, then the sperm whales could potentially combine these units into complex sequences that encode information, exactly as we do in our own languages.

The advances already made by Project CETI represent some of the most exciting and novel of their kind in my lifetime, but there's a long way to go. For one thing, even if the sperm whales do have a language, it'll be some time before we can create a cetacean equivalent of the Rosetta Stone that would allow us to infer meaning from it. At the same time, it remains a distinct possibility that there is no profundity to the sperm whale's codas, or that their sounds represent something different to a language. Rather than conveying information, the sounds could instead be expressing something that the whales recognise and respond to, but which is ineffable, in the way that music conveys emotion. Regardless of what the future holds for this initiative, if the richness and the complexity already revealed by the deep analysis of the sperm whale's soundtrack leads, like Roger Payne's *Songs of the Humpback Whale* did half a century ago, to a step change in our appreciation of these animals, then Project CETI can only be counted a huge success.

* * *

From the window of an aircraft, the silvered skin of the open ocean stretches to a distant horizon, the altitude diminishing the waves until they seem no more than the contours of a fingertip. What seems like a uniform plain, however, is patterned by gravity to trace the topography of the seafloor below. The peaks of seamounts lie below the surface, yet each is crowned by a distinctive bulge of water rising above the surrounding sea. The gradients of these bulges are so slight that we can't detect them by eye, but precision instruments mounted on satellites can. What they tell us is that the oceans are strewn with tens of thousands of undersea mountains collectively covering an area greater than that of Europe.[74]

The pelagic, a vast area of free and untrammelled ocean far beyond the coasts, is the largest biome on Earth. It is bounded by the continental slope that descends from the shallow shelf waters that hug the land and extends across the ocean basins to cover more than 60% of the surface of the planet. Largely isolated from

74 Seamounts are typically (but not universally) defined as conical structures that stand 1 kilometre or more above the seabed. Below this height, the features might be referred to as guyots, knolls or abyssal hills depending on their height and topography.

the nutrients that flow from the land, and from the mineral riches of the cold, deepwater currents, much of the pelagic is a great, glassy desert. Compared to coastal zones, there is little food to be had here, and still less shelter, and typically, it's sparsely populated with life. Seamounts, however, are the exception to this rule. The rocky bulk of these geological features diverts deep currents, forcing them from their sunken course and creating upwellings of enriched water. This in turn fosters the development of teeming biological oases. Enormous shoals of fish encircle the seamounts, shimmers of energy against the untenanted stillness of the open sea beyond. Below them, on the basalt flanks of the mounts, sea stars slide and lobsters clamber through flourishing sunken forests of sponge and coral, while crinoids and sea whips sway in the flow. And perhaps most incredible of all, the foothills of some seamounts play host to dense aggregations of octopuses.

Davidson Seamount in the deep waters off the coast of California is home to what's become known as the Octopus Garden. Its slopes are strewn with pearl octopuses, with as many as 20 of the ghostly mauve-grey creatures crammed into each square metre, like sun worshippers congregating at the beach on a hot summer's day. There's no bustle of activity, however; most of the Garden's occupants lie near motionless for months on end. It's this inactivity that gives the clue as to why the octopus are here; they're engaged in one of the most remarkable displays of self-sacrifice in the animal kingdom. Most are females, and after laying the only clutch of eggs they will produce in their lives, they curl protectively around their brood and begin a lengthy vigil, which will end with a generational transition: the emergence of the young coinciding with the death of the exhausted mother. It takes almost two years for the eggs to hatch, and during this time the mother barely stirs, except to discourage threats to her developing offspring. Like most seamounts,

Davidson is volcanic, oozing gentle hydrothermal streams that raise the temperature in the Garden. It's thought that this is what brings female pearl octopuses here, the warmth nurturing their embryos and speeding their development even as it propels the mothers to an earlier end.

Seamounts are by definition submerged, but where active volcanoes build to climb beyond the waves, new land is created. The island groups of Hawai'i, the Galapagos and the Azores were all formed in this way, their exposed parts representing the upper flanks of stupendous mountains, which in the case of Hawai'i's Mauna Kea is taller than Everest from base to peak. Regardless of the distinction between seamount and island, their oceanographic and biological impacts are equally stunning.

The Galapagos Archipelago lies at the confluence of three major ocean currents. As the Panama Current brings warm water down from Central America, the Humboldt Current surges north from the Antarctic, and the Cromwell Current powers eastwards across the Pacific. While the islands are justly famous for their unique terrestrial fauna and flora, including two giants – the tortoises and the bizarre daisy trees – it was the marine animals that these currents support which drew me to the islands. For a biologist, visiting the Galapagos is a kind of pilgrimage and I was full of excitement as I stepped off the plane into the arid heat of Baltra. Before I'd even reached the terminal, I encountered one of the famous finches that inspired Darwin's thoughts on evolution; it felt like a good omen. By the time I reached my hotel, I'd stared wide-eyed at the groves of huge cacti that flourish in the dry, red earth of the lowlands, and the luxurious cloud forests of Santa Cruz's highlands. I was chafing to see another biological wonder of the Galapagos, the marine iguana, and this was top of my agenda when I spoke to the hotel manager, John.

'Do you think I could find one today?' I asked.

He looked at me as if I'd asked whether the rooms had beds before replying, 'Just head down to the harbour, you crazy foreigner.' He didn't actually say the last part, but his eyes did.

Minutes later, I was hurrying down to the docks of Puerto Ayora on a mission, arriving to find the locals studiously ignoring what, for them, were commonplace: marine iguanas basking on the rocks and jetties in the heat of midday. For me, it was how I imagine meeting a film star must feel for other people. Here's an animal whose distant ancestors ate leaves, fruits and flowers in the forests of South America, and whose slightly less distant ancestors became accidental tourists, rafting a thousand kilometres to the islands on floating vegetation, where they somehow carved an existence for themselves. There's precious little in the way of leaves and fruit to sustain an iguana here, and so by degrees they came to specialise on the one item of salad that's abundant around the Galapagos: seaweed.

The adaptability of those original iguana castaways was incredible, requiring them to cope with an influx of salt in their diet and the cold waters that, thanks to the Humboldt Current, are typical of the Galapagos. Still more remarkably, it required them to swim, and to cling to submerged rocks to crop away at algal turfs, a behaviour that is as far removed from their ancestral, arboreal habits as can be imagined. The following day, I joined these wonderful creatures as they foraged in the surf at Tortuga Bay, marvelling at their mastery of the conditions that swept me around helplessly. The chill of the water quickly made itself known to me, a homeotherm, but it's even more significant for the lizards. Their tenure underwater is limited by the low temperatures, which can slow them down into a torpor, potentially risking their lives. To avoid this, they have to return to land in order to top up their metabolic batteries by basking in the sun, and so, periodically, one would detach itself from its submerged rocky platter and swim for shore. En route, they would glide past me, utterly indifferent to my presence, with legs folded tight to their flanks and powering themselves along with their flattened paddle-like tails.

Snorkelling in the shallows with the iguanas gave me my first look at the Galapagonian seascape, which included a number of damselfish who were furious at my incursions and weren't afraid to show it by charging at me. The whole experience whetted my appetite to explore further beyond the shores and to dive in the blue waters around the Galapagos where oceanic megafauna come close to land. Just after dawn the next day, I boarded the dive boat to begin my adventures. I was headed to Gordon Rocks, a great underwater coliseum formed from the weathered remains of a volcanic crater that lies 3 kilometres off the east coast of Santa Cruz. Only the remnants of the crater rim remain above water, weathered and washed away by the relentlessness of the Pacific, but beneath the waves, steep walls descend into the deep cup of the crater floor. Blue-footed boobies, frigates and tropic birds wheeled in the bright morning skies above, while the barks of sea lions echoed among the rocks. It's always a good sign to see so many predators, not least because it means that there must be riches in the waters below. Time to take a look. I rolled off the gunwale and into the swell, descending a few metres with my buddy to escape the surge of the waves and then making more leisurely progress down the wall of the crater and into the most incredible underwater wonderland that I've ever experienced.

A school of Pacific creole fish faced into the currents that channelled through the broken remnants of the volcanic rim, snapping at morsels borne by the flow. The fish in turn were being chivvied by one of the sea lions, whose graceful arcs choreographed the shoal into ever-closer formation. The sea lion seemed to flow through the scene, a seal-shaped liquid within a liquid, dipping, sliding and banking in an effortlessly masterful display. At one point, the curious mammal broke from harrying the fish and glided across to me, somersaulting as it arrived to peer into the glass of my mask. Conscious of my own comparative awkwardness, I held my breath as the bewhiskered face filled my vision. After a few moments, curiosity satisfied, the sea lion soared away to resume its assault on the fish. The shoal parted on its approach and coalesced behind it, enveloping the sea lion in its mass.

We continued our descent and, as we did so, large shapes resolved out of the blue ahead. They belonged to one of the most instantly recognisable of animals, hammerhead sharks. There were dozens of these wonderful animals, each of which approached 3 metres in length, swimming languidly against the currents. Their peculiar and distinctive head is a multifaceted supersensor. It's covered in thousands of electrosensitive pores which aid them in detecting the electrical impulses emitted by their prey, whether they're juveniles foraging among the leaf litter of their mangrove nurseries, or adults hunting at night, or in the sunless waters of the mesopelagic, hundreds of metres below the surface. The head also facilitates a wide separation of their eyes and nares (or nostrils), promoting respectively an uninterrupted, binocular field of vision and an enhanced capacity to locate and follow scent trails.

Unusually among larger sharks, scalloped hammerheads form into shoals and even more unusually they often segregate by sex. The group I was watching now were almost all large females, cruising along together companionably, though it has to be said that a strict etiquette is observed within the groups. At the top of the hierarchy, the largest, most dominant individuals occupy the most coveted positions in the group, which for them is the middle. They're quick to put inferiors in their place should they transgress, initially with an unmistakable head-shaking display that can escalate into a peculiar corkscrew swimming pattern that might culminate with a disciplinary bite to restore the status quo. They gave no sign of being interested in us as they cruised along, their indifference contrasting with the behaviour of their pelagic cousin, the blue shark. Blues are incredibly beautiful creatures, their elegant, sinuous bodies being a model of hydrodynamic efficiency, and they're intensely curious when they detect novel objects in the great expanse of the open ocean. A colleague of mine, Jens Krause, travelled to South Africa to study these creatures, only to find himself being the object of study. The 2-metre sharks approached to investigate Jens in the only way that they can, by bumping and biting. At different points he had the top of his head in a blue shark's mouth, as well as one on each of his legs, and another even tangled itself in his regulator hose. None of this is aggressive, though; Jens describes it as like being surrounded by puppies.

The hammerheads might be less forward, but that was more than compensated for by their numbers. There were perhaps 60 of them when I was there, though my dive leader, Desi, told me that on a good day there can be hundreds gathered at Gordon Rocks. Few places in the world now host such aggregations since all of the larger hammerhead species have been in steep decline for decades. In many parts of their range, they're all but gone. As so often, man is the agent of destruction, and countless millions of sharks are lost to the fishing industry every year, many to supply the thriving trade in shark fin. Even though the Galapagos marine reserve is closed to shark fishing, Ecuador lacks the resources to patrol the area – which is roughly the size of England – effectively. The result is a cordon of massive international vessels that sit outside the reserve's margins, playing a cat and mouse game with the authorities. More often than not, sadly, the illegal fishers get away with it. Sharks, along with tonnes of other protected species, are stolen in the reserve by small, highly manoeuvrable boats. These are then deftly shuttled to a larger factory ship outside the reserve, which is beyond the jurisdiction of the authorities. This is where the processing takes place for the majority of the catch. Sharks, however, are often dealt with immediately, and barbarically, after landing. Their fins are cut off and the mutilated animals are thrown back into the water, alive but unable to move, and condemned to end their days in unimaginable suffering. Although Gordon Rocks is a long way from the boundary of the reserve, hammerheads are long-distance travellers, and their sociability makes them a poacher's jackpot. I could only hope as I watched them that, somehow, they'd dodge the fate that awaited them just beyond the horizon.

It's little wonder that an international poaching operation is drawn to the Galapagos, since the wealth of marine life that I experienced there is beyond anything I've ever seen. A densely packed, city block-sized school of striped salima stretched beyond the limits of vision, the mass shape-shifting, streaming and pooling like the cytoplasm of some colossal amoeba. Immense, metallic tuna sliced through the water, quartering the school. Flocks of eagle rays flying through the blue in formation, their wingtips almost touching as they soared on their mission to who knows where. Galapagos sharks cruising in the deeps, while whitetips made log pile dormitories of their bodies and slept the day away on the seabed 30 metres down.

The Galapagos and the sprinkling of other places where the pelagic collides directly with the shore are excellent places to meet oceanic oddities, and there were still two that I was greedily hoping to see. One of these I'd encountered fleetingly once before, as a small boy on an open fishing coble in the North Sea off Flamborough Head when, training my gaze across the sea surface, I glimpsed something that looked like a giant, silvery hub cap floating among the waves. As we approached, I was able to make out an eye, and then a mouth and fins, but… no body. It was clearly a fish, but not one that resembled anything in my conception of the term. To my horror, the skipper tried to gaff it with a boathook. He was too slow; the peculiar fish dived out of sight, leaving me with a profound sense of wonder. How could a fish live without a body? Back on land, I discovered that the mysterious creature was a sunfish, an oddity that only grows stranger as you learn about it. This is a fish that is on a par, weight-wise, with an Asian elephant, making it the heftiest bony fish in the world.[75] And when it breeds, it produces more eggs than there are people in the United States. Alongside its prodigious size, the appearance of sunfish is unforgettable. It's a greyish bust of a fish's head, adorned with two paddle-like fins, one below and the other on top. It lives a mysterious, roaming existence subsisting on a diet of jellyfish and even though it was now decades since I'd seen one, I retained my childlike enthusiasm for them.

True to form, when I did finally see the first sunfish of my adulthood it was doing something weird. A sudden loud splash behind the boat seemed to suggest that someone had fallen in. Then it happened again, and I watched as a group of three sunfish took turns jumping out of the water. If someone had told me they'd seen sunfish breaching, I'd have thought they'd lost their minds. Sunfish are slow swimmers, and heavy, yet here they were leaping like salmon. Perhaps they were spooked by something below, or maybe trying to shed parasites. Possibly they were just having fun, but in any case, I scrambled to get my gear on so that I could meet them. Dive kit resists haste, however, and they'd gone by the time I toppled over the side. Far from being disheartened, I was

75 A specimen of the giant sunfish, *Mola alexandrini*, washed up in the Azores in 2021 measuring more than 3 metres in diameter and weighing 2,744 kilograms.

just thrilled to have seen them, and the dive soon got even better. Silhouetted against the bright sky, a posse of kite-shaped devil rays cruised overhead with languorous flaps of their fleshy wings. As I admired their grace, I wondered if I'd ever see their close relative, the giant, oceanic manta, an animal I'd been hoping to encounter for more years than I cared to remember. I was just wondering whether, from a distance, I'd know a manta from a devil ray when events supplied a very clear answer.

I turned my head to see the largest fish I've ever encountered. A magnificent oceanic manta, an utterly unmistakable icon of the sea. I saw her first head-on, enormous mouth agape, using a pair of fleshy lobes positioned to either side of her head to funnel water, and the plankton it contained, into her mouth. I had just a moment to admire this view before the manta swept over me, this great sail of a creature awing me with her size and the beautiful purity of her markings, the luminous alabaster of her white belly and wingtips contrasting with the gunmetal of her back.

Plankton feeders are not typically known for their intelligence; paradoxically, however, the manta has the largest brain of any fish, even correcting for its body size. This might be connected to their sociability, since reef mantas are known to fraternise with preferred partners for weeks on end, and they aggregate at cleaning stations to await the parasite removal services of cleaner wrasse in some locales. There are even suggestions that they use their highly manoeuvrable cephalic lobes as a means of communication with other mantas since they seem to gesture with them, the lobes adopting specific conformations in different social contexts. But like so many marine animals, what we don't know about the enigmatic manta far outweighs what we do. As the manta faded from view, I looked across to see Desi, who was punching the water in delight. I found that I was doing the same with sheer elation and the realisation of just how fortunate we are to share our planet with these extraordinary creatures.

* * *

Charismatic and majestic as animals like the sunfish and the manta are, it'd be remiss to ignore the massive contribution of other, much smaller kinds of organisms to the functioning of the marine environment. If you were to take a glass jug and scoop a litre of water from the surface of the ocean and then hold that jug to a light, what would you see within? The answer, most likely, is not a lot. Although the water would sparkle and perhaps play the rays of sunshine into rainbows on your peering face, it would seem clear. But this apparent emptiness is an illusion. Every litre of water teems with life. We don't perceive it as such because the creatures in question are individually smaller than we can make out. Your jugful is a world within a world, and in the litre of water before you there are probably ten times more microbes than there are people on our planet.

Most of these microbes are viruses, tens of billions of them. If this is a troubling thought, take heart from the knowledge that the overwhelming majority aren't coming for you, but for easier prey like bacteria.[76] Whether viruses count as living organisms is a moot point since they lack a cell of their own to call home, or the means to produce their own proteins or energy. Some biologists contend fiercely that they are alive; others – just as vehemently – insist that they're not. Caught in the middle, many settle for describing them enigmatically as 'entities'. Not that the furore makes any difference

76 There are viruses in the marine environment that can cause significant human health problems, however: enterovirus, norovirus, rotavirus, adenovirus and hepatitis A, to name five, all of which are prevalent in coastal waters, particularly those that are polluted by sewage.

to the viruses themselves, of course. They simply get on with doing what viruses do, which is to hijack living cells and turn them into factories to make more viruses. Phages, which make up a high proportion of marine viruses, make their living by alighting on the surface of a bacterial cell in a manner reminiscent of the lunar landers that they resemble, and then injecting genetic material through the membrane. This ultimately kills the bacterium, but not until its body has been used to make 100 or more new phages.

The bacteria that are the viruses' meat and drink are the next most numerous microbes, with perhaps a billion or so partying in your litre of water. Just like the viruses, most marine bacteria have better things to do than infect us, such as to photosynthesise. Take, for instance, the 'primitive green berry', a cyanobacterium called *Prochlorococcus*. Chances are that you've never heard of it. In fact, it was only discovered late in the twentieth century, yet it's possibly the most abundant microorganism on Earth and contributes close to half of the oxygen produced in some regions of the sea.

Much of the rest of that oxygen comes from the photosynthesisers in another group of microorganisms, the protists.[77] If, for comparison's sake, we were to magnify the average marine virus several thousand times until it was the size of a full stop on this page, then a typical marine bacterium given the same treatment would be roughly between the size of a lychee and a lemon. By this scheme, the dimensions of a protist could range from a yoga ball to a horse. Among microbes, size and complexity go hand in hand, and indeed

77 Protists as a group, however, are a loose conglomeration of creatures, each encapsulated within a single cell and bundled together in the taxonomic equivalent of a kitchen drawer. Beyond their unicellularity, there often seems little to unite protists. Attempts to redraw the tree of life are ongoing, but there's a great deal of complexity and controversy. While acknowledging the flawed nature of protists as a taxonomic group, I'm going to let the taxonomists duke it out among themselves as we, instead, explore the wonderful world of these creatures.

protists are much more elaborate than bacteria. For one thing, protists are eukaryotes, which is something they have in common with you and me, as well as lychees, lemons and horses (but not yoga balls).[78] Yet while they dwarf viruses and bacteria alike, they still amount to just a single cell and most are so small as to be invisible to the naked eye; consequently, there could be more than a million – some say as many as 100 million – drifting around undetected in the jug. Many of these will be diatoms, microalgal drifters that, alongside other photosynthetic protists and bacteria like *Prochlorococcus*, make up the phytoplankton, the foundation of the marine food web.

Wherever there's sunlight and nutrients in the ocean, you'll find diatoms. Perhaps a quarter of a million different kinds of them exist in stupefying abundance among the surface layers. Under the microscope, this cornucopia of diversity emerges in all its glory as each individual diatom is revealed in its ornate and beautiful symmetry. Among much else, there are wagon wheels, five-pointed stars, triangles, diamonds and hexagons. What all of these configurations have in common is that each is formed from a pair of closely matched shells, known as frustules, made of transparent opaline silica. One of the frustules is very slightly larger than the other, allowing the two to fit together snugly, like a shoebox, to encase the delicate interior of the diatom. To multiply, the diatom divides into two, one daughter cell taking the smaller of the two frustules and creating a new frustule to fit within it, so that with each passing generation the diatoms grow progressively smaller. Inevitably, this matryoshka doll-like shrinkage must have an end, and it occurs when the diatom gathers itself into a spore and puts

78 Eukaryotes – 'good nuts' – are distinguished from bacteria (prokaryotes, or 'before nuts') by their possession of advanced, compartmentalised internal structures, including the aforementioned nut, a reference to the nucleus, a cellular vault that holds DNA.

on a growth spurt to restore itself back to its ancestors' size before the process starts again.

Aesthetics aside, the value of diatoms is beyond calculation. They're major players in the phytoplanktonic community that collectively sucks up massive amounts of carbon dioxide and produces around half of all the world's oxygen. As if being gas factories and supporting the world's biggest food chain wasn't enough, diatoms have a couple of other tricks up their silica sleeves. One is that they sink. I know, it sounds underwhelming, and possibly why they get few bookings for children's birthday parties, but it's crucial. Their gorgeous frustules make them heavier than water, and though they can trim their cell contents to stem their descent, their fate is to slide inexorably into the deep ocean. The reason it's so important is that the diatoms take with them a massive amount of carbon, which ends up being stored away safely in the abyss where it can't interact directly with the atmosphere. The other thing that diatoms do, in conjunction with other plant-like protists, is a bit more impressive than sinking: they make clouds. They don't do this deliberately, but through the production of dimethylsulphide, or DMS, that, when mixed into an aerosol by wave action and released into the atmosphere, catalyses the formation of clouds.[79] These clouds then reflect solar radiation, thereby providing a powerful cooling effect. One recent estimate of this suggested that the cooling effect of clouds formed through the efforts of phytoplankton is one of the most important factors in holding back the rate of global warming.

Mingling with the diatoms, other protists such as coccolithophores and dinoflagellates shoulder a share of the photosynthetic work. Coccolithophores protect their spherical bodies with a series

79 DMS is also the major chemical in the mix that gives the sea its characteristic, refreshing smell, which people sometimes mistake for ozone. It's also the compound that birds use to home in on plankton blooms, as described earlier in the book.

of overlapping chalky plates. Individually, each plate looks like a miniature button; arrayed together, they resemble a Viking shield wall. Like diatoms and dinoflagellates, they're minuscule yet their sheer abundance in the sea is dazzling. The famous white cliffs of Dover are testament to this; the towering escarpments that rise more than 100 metres above the sea are formed by the accretion of innumerable microscopic limestone plates bequeathed by ancient coccolithophores. Dinoflagellates, however, differ from the more conspicuously armoured diatoms and coccolithophores by their possession of a pair of eponymous flagella, the outboard motors of the plankton world, meaning that they're equipped to travel, albeit in a very limited way. They're responsible for the hauntingly beautiful shimmering blue phosphorescence that lights up the sea surface at night, as well as for some of the sporadic and deadly so-called red tides that form when excess nutrients are washed into the sea from rivers.

Dinoflagellates are emblematic of the challenges involved in categorising the smaller members of the marine plankton community. It'd be helpful if they confined themselves to photosynthesising; then we could put them firmly in the phytoplankton camp, but they have other ideas. A sizeable proportion of them have given up on the idea of making sugar and oxygen out of sunlight and have instead become predators. Thus, while some dinoflagellates merrily photosynthesise alongside diatoms, others simply eat them, and still more are undecided, so do both. Meanwhile, plenty of protist zooplankton, organisms that were long thought of as predators, turn out to have a lucrative sideline in photosynthesis. The main groups of microzooplankton – ciliates, radiolarians and foraminiferans – all have members that have a dual role as part-time hunters and part-time plants. Some of them adopt an ingeniously barbarous approach; having captured an algal cell, they consume it all bar its precious chloroplasts, which they kidnap and put to work, like solar panels within their bodies. Ultimately, all of this nonconformity means that distinguishing among microbes on the basis of their ecological role turns out to be far trickier than we'd thought, and this represents a problem because it complicates our ability to understand the processes of marine biology at a foundational level.

At the same time, life in the microbial community is one of endless flux, where lifespans are measured in hours and death is always just around the corner. Consequently, the billions of microbes in the jug of water aren't merely existing and getting on with their diminutive days; they're interacting, often savagely. In every drop of seawater, it's Gettysburg, or Waterloo, every day. This relentless ferocity, though unseen by us, is critical because it's the wellspring of the entire marine food web.

There are many reasons why you wouldn't want to be a marine bacterium, but on top of everything else, more than a third of them are killed every day by viruses. Given that a similar proportion of them are consumed by protists and by others of their own kind in the same period, it'd be an optimistic bacterium that made plans for the weekend. You'd scarcely be better off as a protist – viruses and predators are after them, too. In some parts of the world, as much as half of the phytoplankton is eaten every day by zooplankton alone. To counteract this, phytoplankton have to breed, and fast. Under the right circumstances, these miniature photosynthesisers are capable of dividing four times per day; what holds them back is generally a lack of nutrients. This realisation spawned the idea of fertilising the ocean, broadcasting vital trace elements at sea to release the phytoplankton's shackles. By doing this, its proponents argue, we could encourage mass blooms of phytoplankton with the ultimate goal of taking carbon dioxide out of the atmosphere. It sounds great, but it carries risks in the shape of destabilising the local ecosystem and depleting oxygen once the population explosion begins to die off and decay. And it also reckons without the intervention of other microorganisms. In 2009, one such project dumped 6 tonnes of dissolved iron into the Southern Ocean with the goal of artificially creating an algal bloom. No sooner had the phytoplankton begun to exploit the opportunity than a swarm of delighted zooplankters descended on them and ate the lot.

Zooplankton are spectacularly messy eaters, albeit not as messy as the effects of viruses, which cause their host cells to burst as the next generation of infectors makes a bow. The result of all this microscopic brutishness is the release of huge amounts of organic matter into the environment and it's for this reason that the stupendous populations of microorganisms in seawater are so important. Rather than being lost, this stray organic matter is eagerly gobbled up by bacteria and protists in a massive nutrient recycling scheme known as the microbial loop that is fundamental to the health of the oceans.

Within the glass walls of the jug, and beneath our notice, among the most significant biological processes on the planet are taking place. The transparency of the water, its apparent emptiness, is a lie told to us by our imperfect eyes. This microbe megalopolis exists in every one of the sextillion other litres of water in the ocean, where the populations of microbes are so great as to defy our imagination. Rather than counting, we can perhaps get a clearer idea by weighing them. Take for instance *Prochlorococcus*; it's one of the most abundant microorganisms in the ocean yet it's so small that ten billion of them weigh only as much as a grain of sugar. Despite their minuscule proportions, the sea's microbes collectively outweigh marine animals – everything from worms to whales – by at least two to one. It's not just the quantity of marine microbes that matters, though, but their sheer industriousness. About half the world's primary production is done by plants on land, and the other half by phytoplankton in the sea.[80] On land, it takes a combined 450 billion tonnes of shrubs, grasses and trees to do the job. The same workload is performed by less than 1 billion tonnes of phytoplankton in the ocean. Their brief but brilliant lives sustain the sea today, just as they transformed the atmosphere hundreds of millions of years ago with their waste product, oxygen. And the fossilised bodies of microbes, rich with carbon and deposited on the ancient seabed, gave us oil, which transformed human existence but now, through overuse, threatens to extinguish it.

* * *

Look a little closer into the jug, and right on the edge of vision, you might see movement. A mote of life cruising or hopping along in its makeshift aquarium. By the time marine life becomes visible to us, we might already be two or three steps along the food chain and passing into the realm of the microanimals. The swimming speck may be a copepod, a crustacean relative of crabs, that's typically about the size and shape of a comma. Diminutive though they are, copepods are ferocious and highly successful predators that take on anything from phytoplankton to larval fish that are larger than themselves. At the smaller end of its prey scale, each copepod might claim 100,000 victims every day, so the copepod in the jug is feasting tirelessly as you watch it. When copepods are abundant, their diatom prey respond either by bolstering their frustule armour, or by becoming sticky. In the latter case, they use their tackiness to glue themselves into groups, presenting a united front against these planktonic piranhas.

---

80 Primary production is a measure of the amount of work done by photosynthesising organisms using sunlight to power the conversion of carbon dioxide to sugar and oxygen.

Copepods, in turn, are on the menu of much larger organisms, including fish and, incredibly, whales. Not that they're easy to catch; copepods are capable of reacting to impending danger with surprising alacrity. Just 2 or 3 milliseconds is enough to set their escape response in train, and when they move, they do so at extraordinary speeds, covering more than 500 body lengths per second, a velocity that, when scaled, makes jet fighters seem sluggish. They're even capable of leaping clear of the water surface when under duress, and while you won't get boats of tourists turning up to admire the breaching copepods, breaking the surface tension is so hard for such minuscule creatures that it's arguably more impressive than the cetacean equivalent.

Joining the copepod at the banquet are a host of other hunters. Some are the larvae of much larger organisms: sea stars, urchins, sea snails, crabs, lobsters, octopus and fish, all gorging themselves on the plentiful food until they graduate from the plankton to follow in their parents' finsteps. Arrow worms, which look like a fish whose head has been replaced with a bear trap, are voracious predators that latch onto more or less anything that crosses their path and back up their ferocity with toxins. Shrimps and mysids either scoop armfuls of phytoplankton or feed in the manner of miniature whale sharks, filtering their food from the water. Salps and their tunicate cousins, the larvaceans, operate in this way too, chowing their way through clouds of microbes. Salps, which resemble little more inspiring than a gelatinous funnel, are nonetheless hugely important members of this community. Their efforts in capturing vast quantities of algae results in correspondingly large amounts of carbon-rich crap, which sinks into the deep ocean. They're capable of responding to phytoplankton blooms with an explosive boom in their own population, one that they achieve by frantically cloning themselves. The new clone attaches to the side of the old, resulting in a colonial chain of gooey tubes that looks like a discarded ammunition belt.

Larvaceans, meanwhile, eschew the colonial life, preferring instead to boost their foraging activity by constructing a house. The term, regrettably, makes the structure seem commonplace, which the larvacean's creation absolutely is not. Uniquely among animals, tunicates are able to produce that constructive cornerstone of plant architecture, cellulose, and they use this and a soupcon of mucus to build what appears like a gummy balloon surrounding their tadpole-like bodies. What this amounts to is less a house than an incredibly sophisticated plankton filter, complete with meshes, channels and valves. The larvacean powers this contraption by generating a current with its slender tail, and its reward is to snare bagfuls of delicious plankton. So effective is the house at trapping particles that it often gets clogged, an eventuality that the larvacean responds to by shedding its house and starting again. Just as salps make a crucial contribution to the carbon pump, exiling excess carbon to the safety of the deep sea, discarded larvacean houses play a similarly impressive role in oceanic nutrient flows.

Larvaceans, along with salps, are often mistaken for jellyfish, lumped together into the club of gelatinous animals. Jellyfish, however, have a much longer evolutionary pedigree than their imitators; indeed they belong to what may be the most ancient animal group on Earth, the Cnidarians.[81] The term 'jellyfish' is an umbrella that unites thousands of species, from true jellyfish to the colonial siphonophores whose bodies are a collaboration of many distinct but interdependent animals. What they all have in common is simplicity. True jellies may be as much as 96% water, have no brain and no systems for respiration or circulation, yet they're full of surprises. For instance, brainless though they are, they can learn

81 The phylum Cnidaria, comprising anemones and corals, hydrozoans and scyphozoans, is rivalled by Ctenophora, the comb jellies, for the title of oldest animal group.

through experience, and a few species have evolved a supercharged ability to protect and repair DNA, which in the case of the immortal jellyfish means that it can cheat death more or less indefinitely. Their fundamental squidginess, held together by a matrix of proteins, means that they're perfectly adapted to their pelagic lifestyle. There are no hard surfaces to cause damage in open water, while their lack of substance makes then neutrally buoyant and saves the animal energy. Little wonder, then, that tunicates in the open sea have adopted gumminess as a strategy. There's one other unique feature that has underwritten the success of jellyfish; the nematocyst. It's an evolutionary wonder, an explosive, barbed stinger, triggered by fine hairs and loaded with toxins.[82] Jellyfish might have one of the simplest body plans in the animal kingdom, but the nematocyst is a miraculously complex cell, and it's been at the cutting edge of hunting for more than 500 million years.

The familiar jellyfish that glides through open water with rhythmic pulses of its bell is the adult phase, or medusa. Before reaching this stage, it has passed a chapter of its life as a wandering larva, and a subsequent one as a polyp, when the animal attaches itself to a substrate and does an excellent impression of an anemone, albeit one that is only a few millimetres tall. Only when it casts off from its mooring to become a drifter does it attain the full glory of jellyhood. The adults of many species trail long tendrils, festooned with nematocysts, whose embrace means death for any one of a multitude of marine creatures. When prey are snagged, the barb holds the victim in place as a hypodermic needle injects a killing venom and, in some cases, enzymes to get digestion under way. Enmeshed in this deadly grip, the hapless quarry is drawn into the mouth on the jelly's underside.

---

82 You can get a sense of the barbs if you touch the tentacle of another cnidarian, an anemone. It feels tacky, as a result of the barbs penetrating your epidermis.

Some jellyfish trail enormous locks of tentacles behind them, extending for more than 30 metres in some cases, but some have barely anything that could dignify the term tendril. One, the upside-down jellyfish, often rests on the seabed, head down and revealing an underside that looks like a garden of spoiled cauliflower. The colour – a kind of sickly taupe – comes from an internal algae farm, and the posture is designed to maximise the sun that the lodgers obtain. Rather than trouble itself by deploying stinging tentacles, the upside-down jellyfish relies for its defence on grenades of mucus-coated stinging cells, which it flings at its persecutors.

The stinging wizardry of jellyfish is exploited by other animals for their own uses. The young of dozens of fish species seek the shelter of a jelly, swimming among its deadly tendrils in a manner reminiscent of anemonefish. Though this shields them from the threat posed by many predators, jellies themselves are avid consumers of small fry. So, it's a dangerous game for the fish especially since, unlike anemonefish, most have no resistance to their host's venom. Few take the relationship as far as the bluebottlefish or the spectacular nudibranch known as the blue sea dragon. Both associate with the fearsome Portuguese man o' war, which though not a true jellyfish, certainly acts like it. The bluebottlefish is claimed to have some resistance to the stings of the man o' war, but it chiefly relies on being an exemplary tentacle dodger. The blue sea dragon, meanwhile, has cracked the man o' war's code and is immune to its stings. In fact, it adds insult to injury by not only ignoring the man o' war's armoury but actually eating parts of it and sequestering its stings. Most impressive of the lot, however, is arguably the blanket octopus, which dextrously pulls the tentacles off the man o' war and carries the dismembered tendrils around in its arms to use as bespoke whips with which it can lash out at its prey and predators alike.

I vividly remember one seaside holiday as a child, watching a gang of small boys furiously smashing a stranded jellyfish with stones. Many people feel similarly hostile, and it's understandable when you see, as you often do, the vivid purple welts left on bathers' skin when bluebottles are blown inshore towards Sydney's coast, or hear of people succumbing to the most venomous animal in the sea, the diminutive box jellyfish.[83] The negative press that such creatures attract has only intensified in recent times as huge blooms, sometimes kilometres across, appear at our shores. For the most part, the jellies in question tend to be about the size of a soup bowl, but there are exceptions. Swarms of Nomura's jellyfish off Japan are astonishing for their size as well as their numerosity. Each is a vast ball, 2 metres across and weighing as much as a piano, resembling a haystack with a gloopy hat. Aside from their unpleasant sting, swarms of Nomura's jellyfish blight fisheries in the East China Sea, in one case even sinking a fishing vessel when they inundated its nets. The Japanese, ever resourceful, make ice cream out of them, but this scarcely makes a dent in their populations.

The jelly headache isn't restricted to Japan; fisheries have been hit around the world, while power plants in the US, the UK, Sweden and China have been forced to shut down as clouds of these animals clogged their intakes. Nevertheless, for most people, the immediate impact is the closing of beaches by the blooms; no one wants to swim in a sea of stinging bubble tea. There's a sense of the apocalyptic as these plagues descend without warning to sabotage our leisure time.

The causes of such outbreaks aren't well understood, but a host of culprits have been identified. Overfishing has driven a reduction in the jellyfishes' competitors and predators, taking the brakes

83 Bluebottle is the local name for the Portuguese man o' war, and despite the popularity of peeing on such stings as a cure, it's a completely worthless approach, and might even make things worse.

off their gelatinous fecundity. Implicated, too, are the floods of nutrients that pour off the land and into the sea, triggering both algal blooms for the jellies to feast on, and precipitating hypoxic conditions in coastal waters. As other organisms gasp for oxygen, the jellyfish march on, almost uniquely adapted to cope by dint of their bloodless simplicity. Jellies also favour warmer waters, so climate change is a boon to them.

Whatever the cause, the blooms of jellies compound existing problems. The swarms are relentless killing machines that strip the life out of the sea, wiping out cohorts of larval fish and removing zooplankton. The effect of this is to fracture the food chain, and the impacts reverberate through marine ecosystems. Their ruthless efficiency, however, carries a self-inflicted cost. As the bloom runs out of food to fuel it, the jellies – those without symbiotic algae to sustain them, at least – begin to die. The sinking carcasses decay and release nutrients, promoting an upsurge in bacterial activity that, in turn, sucks up the water's precious oxygen supplies.

Despite the depiction of jellyfish as bellwethers of collapsing ocean ecosystems, their swarms, as far as we can tell, have always occurred. The question is, are the outbreaks happening more frequently now? The answer seems to be yes, but the data are noisy. In a pristine ocean, sporadic blooms can be a good thing, representing opportunities for other marine life. A mouthful of jellyfish might be 30 times less nutritious than the same quantity of fish, yet there are many animals that seek them as food. Ocean sunfish and leatherback turtles are jelly specialists, but countless others, including tuna, swordfish and even penguins enjoy an occasional gummy supplement to their diet. Furthermore, dead jellies that reach the ocean floor are manna for a host of benthic creatures. The worry, however, is that depauperate seas, those where we've driven the biota into mass retreat, can't cope with the additional burden of jellyfish blooms. In the worst-case scenario, ever increasing blooms could trigger a phase change in marine ecosystems. Instead of being rich in diversity, they could become dominated by suffocating jellyfish. If this is indeed the canary in the oceanic coal mine, we'd better start to listen.

* * *

From the periodic appearance of swarms of jellyfish to the swash of waves, life in the sea is characterised by rhythms. Even in the invariant darkness and cold of the deep sea, pulses of marine snow, the remnant remains of seasonal plankton blooms at the surface, dictate the cadence of life. While seashore animals live in tune with the tide cycle, those of the open ocean attend more closely to the alternation of day and night. In the early 1940s, US Navy technicians using newly developed sonar equipment to ping the depths reported an anomaly. Something was amiss with their readings far from the shallow coastal shelf. The data were telling them that the seabed lay just a few hundred metres down, something they already knew to be impossible. Were the instruments faulty? It didn't seem so, and, this being wartime, there was initial concern that the phenomenon might be related to enemy submarines. Yet this deep scattering layer, as it came to be called, was mobile, rising at night and sinking during the day. It took some years to resolve the mystery of the ocean's false bottom when the culprit, or culprits, were identified. The sonar was reporting an unimagined aggregation of animals.

In the early nineteenth century, the pioneering naturalist Georges Cuvier noted that tiny crustaceans known as *Daphnia* appeared and then vanished from the surface layers of lakes on a daily basis. In the decades that followed, a similar phenomenon was documented in the marine environment, of planktonic animals descending to the depths at dawn and returning to the surface at dusk. Even so, it wasn't until the advent of sonar that the scale of this movement became apparent. We now know that this twice-daily commute of animals, known as the diel vertical migration, in the ocean represents by far the largest translocation of biomass on the planet. Gigatonnes of animals, from millimetre-long copepods to 5-metre megamouth sharks, undertake the journey. The significance of this mass movement isn't solely biological. The inevitable consequence of eating is pooping; animals that have filled up in the shallows carry a gutful of food to the depths where, a few hours later, they expel huge quantities of excrement. This conveyor belt of crap both fertilises the deep sea and locks tonnes of carbon out of harm's way.

From the perspective of tiny animals like copepods, the distances travelled – hundreds of metres up and down every day – are stupefying. The reasons must be compelling, and they are. Leaving the surface waters before dawn breaks means the smaller migrators must abandon their phytoplankton-rich feeding grounds, but to stay at the surface after sunrise would expose them to the danger posed by legions of sharp-eyed predators. Caution is the copepods' watchword, and they wait out the perils of the day in the comparative safety of the deep. The return journey at dusk, meanwhile, is a race driven by hunger since small animals typically can't sustain a fast. In turn, the migration of the smaller animals is the driver for many of those that prey on them, so the 12-hourly commute represents the mass movement of much of the open ocean's food chain. Squid, shrimps, siphonophores and countless fish species, make the trip. Since so many predators echo their vertical movements, the long trek might seem like a pyrrhic gesture on the part of the smaller zooplankton, yet the ubiquity of this pattern throughout the world's oceans suggests that whatever the costs of moving, those associated with staying put are greater. And so, the metronomic sequence continues, even in the dead of polar winters, the DVM, a great oceanic pulse, breathes life and nutrients throughout the seas.

* * *

The daily migration of oceanic creatures from the depths to the surface, and back again, is the largest of its kind on Earth, but as these animals shuttle up and down, others traverse the ocean. The helter-skelter charge of hatchling turtles down a beach to the surf is the beginning of an odyssey that few will survive. Many fail even to taste salt, their lives abruptly curtailed before they even reach the waves' embrace, intercepted by the predators that gather in anticipation of their emergence. Notwithstanding the imperative to avoid interfering with the course of nature, the turtles seem so ill-prepared for the harshness of the world above the sand that it's hard not to empathise. For that reason, I've stood as part of organised cordons in Queensland, a human levee entraining streams of matchbox-sized turtles as they flow from nest to sea. Regardless of our efforts, we all knew that another expectant welcoming committee loitered beyond the shore. The turtles, too, seem to have some grasp of this; they arrow out into the blue, in a hurry to be elsewhere.

The elsewhere that they're heading for has long been a mystery to us. We see the Lilliputian and Brobdingnagian versions of turtles when they hatch and when they return to lay eggs, but between times, where do they go? It's a question that challenged turtle researchers for years, giving rise to the description of this period of their lives as the 'lost years'. Now, at last, the puzzle is being resolved, thanks to meticulous research and satellite tracking. For the first time in our history, we can now claim an understanding of the lives of these charismatic ocean wanderers.

The coast of Florida is a mainstay of the world's loggerhead turtles and attracts tens of thousands of expectant mothers to nest on its beaches each year. A little under two months after they were laid, some five million hatchlings, each weighing about as much as an AA battery, emerge from their eggs beneath the sand. They're genetically preprogrammed to dig upwards to the surface, but tunnelling through perhaps a foot of compacted sand is exhausting work. If a lone hatchling were to shoulder the entire burden of the excavation, it'd burn through something like two thirds of its energy just to escape the nest, leaving precious little in reserve for the journey ahead. This is one major benefit to the near synchronous hatching of the whole clutch of perhaps a hundred youngsters; simply put, many flippers make light work. They don't dig indiscriminately though. If the surface layers are warm, the miniature miners slow to a standstill. Their goal is to emerge into the comparative safety of the night, so they wait until things cool down.

If they've timed it right, the hatchlings break free of their sandy interment under cover of darkness. Even so, topside is a killing ground for the neonates, their race to the sea a kind of reversed Normandy landing. Each individual hurries towards the brightest light on the horizon, which for most of their 200-million-year evolutionary history meant that they would head seaward, guided by the glimmering reflection of moonlight on the waters.[84] The simultaneous emergence of the entire clutch also serves to limit their losses by swamping the appetites of opportunist predators. Once beyond the beach and into the blue, the turtles scatter to deeper waters where their predators are far less concentrated.

Their path is set initially by the sea itself, heading against the direction of the waves as a simple rule of thumb that invariably takes

84 We know that this is visual navigation because hatchlings fitted temporarily with blindfolds are unable to navigate to the sea, and, regrettably, because all too often the glare of artificial lights draw them in the wrong direction to certain death..

them offshore. Nevertheless, the turtles are born with sophisticated navigational abilities, thought to be founded principally on a kind of mental map, whose coordinates are set by Earth's magnetic field. A layperson's experience of this likely extends no further than using a compass to find due north. Many animals, however, are sensitive to the richer information contained in this magnetic flux, using their perception of local patterns of field strength as well as idiosyncratic features within the field that allow them to pinpoint their exact location anywhere on the planet. It's this remarkable facility that will guide the turtle's wanderings for the rest of its life.

As they progress out from the Florida coast, the turtles ride the great, clockwise swirl of the North Atlantic Gyre, which will be their home for more than a decade. Bounded to the west by the Gulf Stream as it flows north, and by the southerly Canary Current that sweeps down the Atlantic coasts of Europe and Africa to the east, the gyre is a huge, steadily circulating mass of water that encompasses almost the entire width of the ocean and stretches from the equator up to the waters off Newfoundland. Borne along at a steady pace of around 3 kilometres an hour, the turtles have the opportunity to conserve energy, though they need to be alert to the risk of spinning out of the gyre into the colder waters of the north, where the drop in temperature could quickly kill them. Happily, the turtles, equipped with a keen sense of location, are more than capable of trimming their course to avoid this fate. Their orienteering is all the more remarkable for the fact that it seems to be coded in their genes. Hatchlings that have been experimentally translocated over thousands of kilometres to different parts of the gyre, and that are entirely lacking in experience of navigating at sea, are innately adept at plotting the right course to join up with the main current. Not that turtle hatchlings exclusively go with the flow: many turn inwards, leaving the great circular conveyor belt of the gyre. Here, in the comparative stillness of the waters at the gyre's heart, they can enjoy a whole ecosystem's worth of support.

The uppermost layers of the open ocean are often barren places, devoid of structure and starved of nutrients. Yet the interior of the North Atlantic Gyre supports a thriving, floating forest of seaweed. This is *Sargassum*, a plant buoyed by air bladders and content to live an unattached, peripatetic life on the surface of the tropical Atlantic. So prevalent is this weed – there are at least 10 million tonnes of it here – that it gives its name to the Sargasso Sea, a uniquely coastless 'sea' region in the western part of the gyre's interior. Exactly how *Sargassum* manages to sustain such luxuriant growth despite the apparent paucity of the waters puzzled oceanographers for decades, becoming known as Ryther's Paradox. The answer to this biological conundrum, like many others, lies in the interrelationship between a host of different organisms. A huge number of marine animals find a home within the floating glades of *Sargassum*. It's been estimated that each kilogram of seaweed supports over a hundred fish and countless invertebrates. This includes everything from species that as juveniles seek the temporary shelter of the *Sargassum* to the specialists that live their entire lives in it. This latter group includes the aptly named sargassum fish, a predatory anglerfish whose outward appearance makes it near impossible to distinguish from the plant, and the flying fish, whose sticky eggs can be laid in such numbers on the plant that they cause it to sink.[85] Collectively, this rich community of animals in turn sustain the *Sargassum*, providing a reliable and abundant source of delicious excrement that provides the plant with all of the nitrogen it could desire.

The dense thickets of this buoyant pelagic mat provide a home

85 The eggs of flying fish are perhaps most familiar to people as the brightly coloured *tobiko* that is used to garnish sushi rolls.

for the young loggerheads. There's plenty to eat in the form of shrimps and molluscs, and there's shelter from would-be predators, especially since the smaller turtles are still light enough to climb on top of the vegetation as a means of escaping aquatic hunters. In addition to the food and cover provided by the Sargassum, the fronds of the plant constrain the movement of water, allowing it to heat up to 6°C warmer than the surrounding ocean. Pleasant as this must be to experience, it's of crucial importance to the turtles since as ectotherms the warmth turbocharges their growth and brings ever closer the day when they can abandon their exile in the open ocean and join the grown-ups in coastal waters.

The residence of the young loggerheads in the streams and eddies of the gyre carries them far from their natal beach, but they're never in any sense lost. As they approach sexual maturity at around half a metre in length, and perhaps 15 years of age, their thoughts turn to home and the moment they're ready, they make their move. There are many examples of the iron-willed homing instincts of turtles, not least the Japanese-born loggerheads that mature in Mexico's Baja California, before making the 12,000-kilometre journey across the Pacific, back to the Land of the Rising Sun as adults. But the story that caught the public attention recently was that of an indefatigable turtle which embarked on an odyssey that was impressive even by loggerhead standards.

In July 1997, a Japanese fishing boat put into Cape Town carrying an unusual passenger, one that they'd scooped from the ocean after seeing it in distress. The passenger was a loggerhead, approximately the size of an iPad and 2 kilograms in weight but suffering from a badly damaged shell. The crew had grown fond of the turtle as a shipmate and they'd named her Yoshitaro, in honour of the boat's diminutive chef. Nevertheless, she needed urgent care, so she was transferred to convalesce at the city's newly opened Two

Oceans Aquarium. She adapted well to her new circumstances and her shell healed, though it still bore the scars of her early misadventures. Now known by the shortened name, Yoshi, she rapidly became one of the most popular exhibits at the facility. There remained, however, a dilemma over what to do with her, a dilemma that intensified as the years passed and she approached maturity. Loggerheads are an endangered species, and every extra adult in the population counts, so the aquarium took the laudable decision to return her to the ocean. Before this could happen, however, she needed a little training to build up her strength for the challenges ahead. The staff spent a full 18 months in the run-up to her release improving her stamina by coaxing her to perform a series of shuttle runs each day up and down the 20-metre length of her tank.

In December 2017, just over two decades since she'd arrived at the Cape Town aquarium, the now 183-kilogram Yoshi was released from a vessel around 50 kilometres off the coast of South Africa. Back in Cape Town, her erstwhile carers followed Yoshi's progress by means of a satellite tag attached to her carapace. Despite her extended stay in captivity, Yoshi seemed to embrace her freedom, heading west and then north all the way up to the warmer seas of Angola. She kept largely to inshore waters, just as her wild counterparts instinctively do once they reach adult size, reflecting the change in their foraging patterns as they grow to specialise on bottom-dwelling invertebrates, such as urchins, crabs and molluscs. After several months fossicking around along the coast of west Africa, Yoshi doubled back and returned to the waters off the Horn before embarking on an epic crossing of the Indian Ocean. Two years after she was first released, Yoshi showed up among the loggerhead population of Western Australia, having covered 40,000 kilometres – the equivalent of swimming all the way around the world – during her oceanic peregrinations. The likelihood is that this journey represents Yoshi's return to her roots, to the place that she retained some memory of as a hatchling in the mid-1990s. Her tag has now failed, but it's hoped that in the not-too-distant future, Yoshi will ascend the beaches in this remote part of Australia to lay her own eggs and to begin the cycle once more.

The wanderlust that forms such a vital part of a young turtle's life might fade as the animal ages, but it never quite disappears. Adult females that nest on Florida's east coast disperse widely after egg laying, travelling up as far as Maryland and Delaware, out to the Bahamas, or into the Gulf of Mexico to the Yucatan Peninsula. Mostly, they keep to inshore waters, where there's plenty of food, but during their migration to their foraging sites they might travel for 30 kilometres a day. Even when they arrive, the turtles stay pretty active; a typical foraging site for a loggerhead encompasses around 200 square kilometres of sea. Roaming around as much as this is an excellent strategy for maintaining a diverse diet, but it's one that, thanks to us, carries considerable risk, especially in the shallows as I once witnessed in my time at One Tree Island on Australia's Great Barrier Reef.

Discarded fishing gear accounts for the lives of an astonishing number of turtles. A survey of the Timor Sea and Gulf of Carpentaria regions off Australia's north coast estimated that in this region alone, up to 2,000 turtles a year become entangled and die. It isn't just the nets that snag turtles, however, even something as simple as the line holding a marker buoy can become entwined. On a steaming hot morning a few years ago, I walked out to the front of the research station with my group to begin the day's work only to be greeted by the strange sight of one of the temporary buoys commuting around the lagoon. Clearly, something had been snagged, but what? When we got into the water to try and find out, the football-sized buoy disappeared under the surface, eventually popping up again a hundred metres distant. The last thing we wanted was for this animal, whatever it was, to disappear into the furthest reaches of the lagoon, or worse, out to sea, so we fanned out to try and keep it within the bay. After some patient stalking, we eventually managed to get a look at the animal; it was a big loggerhead, hopelessly snarled up around its head and one front flipper with the buoy trailing a couple of metres behind. The situation was worse than we'd first imagined because there seemed little likelihood that the turtle could free itself and in its predicament, the animal was clearly agitated.

Our first scheme to free the loggerhead failed utterly. We managed to corral the creature into a small, natural alcove of water bounded by reef, but when it realised that it was about to be trapped by us, it came barrelling out of the enclave in a panic. Much as we might have liked to imagine that we could impede a 200-kilogram animal travelling at a rate of knots, the reality was somewhat different. The turtle shot past us like a bullet. By now we'd spent more than an hour tailing the animal to no effect and we were running out of ideas. It was then that the station manager, Mark, recalled a documentary he'd once seen, which showed how Indigenous Australians used to catch turtles.[86] He stationed each of us around the shore of the lagoon to keep a lookout for the reappearance of the buoy while he and a member of my team took the tinny out to get into position.[87] I had an idea about what he was going to try, but in 5 metres of water I gave him roughly a 0% chance of succeeding.

Sure enough, the buoy popped up and the tinny puttered over, slowing to a crawl as it drew closer. Mark dived in and, after a huge amount of splashing, resurfaced empty-handed. We

86 Not Mark's real name - he swore me to secrecy.

87 A tinny is a small, open, aluminium motorboat.

were feeling completely disconsolate, but we resumed our watch, and a couple of minutes later, the buoy surfaced a few tens of metres further on. Once more, the tinny crept closer and again Mark swooped, this time astonishingly returning with the turtle clasped to his chest, its pale underside pointing skywards and its legs hanging limp. Just as with some sharks, being flipped upside down sends turtles into a kind of static reverie known as tonic immobility, and the reptile, which had been charging around the lagoon moments before, was now in a trance-like state that allowed it to be cut free from its tangles. I've never been so happy to be wrong, yet I still find it incredible that a novice armed with no more than a sketchy idea could manage to snare a fully grown wild turtle!

* * *

Spring is the time of year when a turtle's thoughts turn to love, and across the world, thousands of them convene off their breeding beaches. The males, ever eager, are first to arrive and form a bachelor gathering to await the appearance of the females. When they show, the result is something of a free-for-all, though given the males' persistence, it's not necessarily one that the females appreciate. In Lombok's Gili Islands, I watched as a male green turtle descended through the water column as gently as a piece of down, every delicate and breathless moment bringing him a fraction closer to the object of his desire. The female, meanwhile, straddled the reef top and was busily tucking into a salad of algae, blissfully unaware of his subtle approach. He alighted daintily impeccably on the sward alongside his beloved and turned his head to regard her. Suddenly aware of his arrival, she met his gaze with strands of her lunch trailing alluringly from her beaklike mouth and, after a moment's pause, bit him on the face. The male fled, chastened by his rejection; unrequited love is a terrible thing, even if you're a turtle.

The males are relentless, however, and it's a common sight to see pairs of turtles at the surface at this time of year, the males riding on their mates' backs. Neither looks to be enjoying it much, but then turtles' faces lack the means to express joy. As a result of the males' caddish behaviour, a large number of turtle clutches show multiple paternity. This might have some benefit to the female in terms of spreading the genetic risk associated with mating monogamously with a duffer; nevertheless, it seems that the ardour of the males takes choice out of the equation for some females.

Soon after mating, the males depart, leaving the females to face the arduous task of laying. The archipelago nation of Cabo Verde is, like Florida, a hugely important northern hemisphere nesting destination for loggerheads. The females come ashore mostly at night and throughout the laying season on the island of Sal, locals patrol the beaches at all hours to safeguard the turtles at their most vulnerable. Even as they heave their heavy bodies through the surf, it seems impossible that they'll be able to ascend the slope of the sands, yet I watched by the dim red light of a volunteer's torch as one female after another strained with the enormous effort of regaining the land. Every so often, the mothers-to-be let out a great snort with the exertion, until they were a few metres clear of the strandline and onto the dry sand where the process of nesting could begin.

In most respects, turtles represent a triumph of evolution. In the millions of years after their ancestors departed land to reconquer the waters, turtles have become superbly adapted to a marine existence. Loggerheads, in particular, can remain submerged for more than 10 hours at a time, and can descend on occasion to 200 metres. They even sleep underwater, and I've often seen napping turtles while scuba diving, wedged into a convenient cranny to stabilise themselves against the currents. Even so, their terrestrial heritage remains apparent in their need to surface to breathe and in the porous, amniotic eggs they lay. It's this tether to a life history that they forsook long ago that compels the females to come ashore now.

Having chosen a likely site to excavate her egg chamber, the female begins the rite by clearing sand away with all of her limbs, looking for all the world as though she's intent on creating a beach version of a snow angel. The point of this is to clear the loose substrate and to expose the harder packed sand beneath. Then she excavates a pit with her hind flippers, digging half a metre down into the sand to create a basin that will receive her eggs. This completed, egg laying begins. At a respectful distance, I watched as a large female dropped glistening white, ping-pong ball eggs, one at a time, into their seashore cradle. With each laying, her great body tensed momentarily, her front flipper rising a few millimetres above the sand. After half an hour of this, she was finished, and she could begin the process of burying them for safety. This is critical since there are any number of predators only too keen to exploit this bounty. In Cabo Verde, as in many parts of the world, feral dogs represent a major problem. I saw several shocking infrared images taken from drones of laying females with an escort of two or three dogs sitting alert at their sides, waiting for them to leave so that they could wolf down her brood. The strays still pose a threat, but this danger is partly what has mobilised the volunteers on Sal. With them in place, the turtles have a fighting chance.

Turtle numbers have climbed sharply in Cabo Verde in recent times, the ascent of so many females to the beaches rewarding the efforts of the local communities. Worldwide, however, the picture isn't so rosy. All seven of our marine turtle species are in trouble with three listed as vulnerable (olive ridleys, loggerheads and leatherbacks), one as endangered (greens) and two as critically endangered (Kemp's ridleys and hawksbills). We don't have enough data on the seventh species, the flatback, to be certain, but this, too, is a rarity. It wasn't always this way. Accounts of the voyages of Christopher Columbus to the Caribbean include references not only to unimaginable abundances – 'They saw tortoises of a vast bigness, and in such numbers that they covered the sea' – but also the fate of the unfortunate creatures – 'In the presence of our people they [indigenous fishermen] took four large turtles which they gave our people for a very delicate food.' It's thought that there may have been as many as 90 million turtles in the Caribbean alone at that time, in such densities that, according to legend, the sailors would use the sharp hiss of breath from a surfacing turtle as a means to navigate in poor visibility. Today there are an estimated 300,000 turtles in the Caribbean, less than 0.5% of their former numbers.

The rot started when Europeans began to realise the potential of turtles as a supposedly inexhaustible source of food in the New World. They were simple to catch and, best of all, they could be stored, conveniently if callously, on board ship, stacked alive (and thus fresh) in piles until such time as they were needed for their meat. Countless numbers were shipped to Europe and around the world, where turtle soup became a mainstay of banquets for the wealthy. When the inevitable happened and turtle populations began to collapse, the scarcity created more mystique and drove the prices up. The more parsimonious sought an alternative in the form of a broth made of boiled calf's head, calling it mock turtle soup, but even so, huge brands such as Heinz and Campbell's offered tinned turtle soup well into the twentieth century.[88] The global

88 Given that the taxonomy wasn't at the forefront of the manufacturers' minds, turtle could refer either to marine turtles or to freshwater ones.

catch of turtles peaked as late as 1968; Mexico alone caught 380,000 turtles that year, more than currently exist in the entire Caribbean.

Nowadays, turtle is firmly off the menu in most parts of the world. First and foremost, this is because populations fell to the point that commercial operations were no longer feasible, a situation that was later reinforced by legal protection and more enlightened views on the subject of eating desperately rare animals. Nevertheless, turtle fishing remains legal in some jurisdictions, accounting for, at a minimum, tens of thousands of animals a year, and unregulated killing to satisfy an extensive black market demand for turtle products is rife. Poaching of eggs, too, is epidemic in some parts of the world. If this were all that turtles faced, that would be bad enough, but in the modern era we've conjured a hellish farrago of other, worse, threats.

The commercial fishing industry exacts a terrible cost on turtle populations through bycatch. It's hard to quantify since, for the most part, it relies on accurate data emerging from what is effectively an honesty box system. Nonetheless, one academic study crunched the available numbers and came up with a figure approaching half a million turtles per year. Added to that, global warming imperils turtles. A strange feature of turtle biology is that their sex is determined in large part by the temperature of their nest as they develop; the hotter it gets, the more females are produced. That in itself isn't so much of a problem because females, which might nest only every two or three years, are always outnumbered by males near the breeding beaches, so an increase in the number of females would even this out in the short run. The greater danger is that the warmer it gets, the more nests fail. At 30°C, half the hatchlings survive, whereas at 31.5°C that number decreases to a third. Overarching all of these man-made challenges to turtles is a new problem, one that's emblematic of our times: plastic.

* * *

It's easy to forget just how recent the plastic problem is. Mass manufacture of the stuff only began in the 1940s, and it bumped along at a relatively low level until it took off in the 1970s and '80s. The industry has been accelerating ever since. We now produce 400 million tonnes of plastic each year, quadruple the amount in the late 1980s. Indeed, it's estimated that more than half the world's plastic has been manufactured in the last 20 years, and there's still no sign of a slowdown. By 2060, it's claimed, we'll be producing well over a billion tonnes a year. To put that into context, that's about three times the combined weight of all the people in the world today.

Like so many innovations, at the outset, plastic was a godsend. A book released in 1945, at the dawn of the plastic age, proclaimed: 'A world free from moth and rust, and full of colour, a world largely built of synthetic materials … a world in which man, like a magician, makes what he wants for almost every need.' Plastic provided the foundations for advances in electronics, medicine and food production. It drove down costs and cut down housework. In short, it made our lives better. But we've become addicted to plastic and to the convenience it provides, the exemplar of which is the 160 million tonnes of single-use plastic that we use and casually discard globally each year. To carry that away, you'd need a line of council bin lorries, bumper to bumper and some of them floating atop the sea, stretching to encircle the world more than three times.

The thing is, we don't carry it away or dispose of it in some neat fashion. We might diligently and responsibly put it in our recycling bins and consider ourselves good citizens, imagining, not unreasonably, that it is actually going to be recycled. It isn't. Why? Because we don't have the means to recycle this avalanche of plastic, and nor are we prepared to pay what it would cost to do the job properly. Instead, the vast majority, 80 to 90% of it, will ultimately be dumped in landfill or incinerated. The UK, for instance, has something like 57 incinerators, and Europe and North America each host hundreds. Some of these places burn household rubbish – including recyclable plastic – to create electricity. This is sometimes marketed as renewable energy, which stretches the definition to breaking point. Worse, the emissions from such incinerators comprise a hideous cocktail of dioxins, mercury and cyanide, and, by some reckonings, about twice as much carbon dioxide as a traditional gas or coal power station.

Even allowing for incineration and landfill, wealthy countries produce far more plastic garbage than they can handle. The fate of the remainder is to be shipped overseas, when it becomes someone else's problem. A substantial amount of waste used to be sent to China for processing, much of it poorly sorted and contaminated, making the already tough job of recycling next to impossible. Then, in 2018, China decided that it'd had enough and closed its doors to further imports. The effect of that was to redirect the torrent of trash to other countries, notably Thailand, India, Bangladesh, the Philippines, Indonesia, Vietnam and Malaysia. What these countries have in common is that they lack the infrastructure to cope with the massive influx of materials. Slag heaps of stinking, imported refuse grow like cancerous hills beside towns and villages, the task of sorting our heedless waste falling to workers being paid a pittance. As they toil, their children play among the carcinogenic grime. What can't be sorted is burned in open fires, or simply escapes into the environment, and still more arrives, each ship dumping the wages of our profligacy, blighting their lives and poisoning the environment.

Recently, a powerful infographic did the rounds on social media. It was a pie chart, showing the contribution to plastic waste in the ocean by country. If you're wondering which countries came out worst, I'd respectfully redirect you to the list in the previous paragraph. You could take this to mean, as hundreds of commenters did, that these countries are the world's polluters. 'We're not even on the map and I have to live with plastic straw drama,' wrote one enlightened analyst, before closing with an indignant interrobang and a clown emoji. Well, no. It's symptomatic of the fact that high-income countries have simply translocated the headache of what to do with garbage, the problem child of our throwaway societies. The average person in the US produces 130 kilograms of plastic waste per year, four times as much as, for example, a typical Indonesian. As a result, a large part of the pollution that spills out into the ocean from these countries originates in the wealthiest nations.

When a flock of Walmart bags blow into the sea off Manila, or a fleet of plastic Tesco bottles sails down the Mekong delta and into the Pacific, should we point the finger at Filipinos and Vietnamese people, or should we take a look at our own habits a little more carefully? It's true that some of the plastic debris that enters the ocean from these parts is local, but even in this case, we're aiding and abetting. Thanks to tightening regulations, the plastic exported by wealthy nations for processing in Southeast Asia tends to be better sorted than its local equivalent, and since the recycling capacity in such countries is swamped, the more lucrative imported plastic jumps to the front of the recycling queue, leaving domestic trash accumulating. Let's be honest: we aren't sending plastic to these countries because we know that they'll deal with it effectively, we're doing it because it's cheap. In any case, this isn't a cause for bellicose nationalism because this isn't a problem that's restricted to national jurisdictions. Besides, the plastic doesn't care where it came from. The ocean doesn't discriminate. A turtle dying because of the plastic it's ingested is indifferent to the provenance of its killer.

Over the course of my career, I've been fortunate to travel to distant places. My every journey has been shadowed by plastic. I've watched marine iguanas clambering over empty bottles in the Galapagos and I've seen plastic litter on the seabed, catching the lights of a sub at 600 metres. Even in the remote High Arctic, where walruses wallowed a thousand kilometres from the nearest city, an empty jar of Nutella on the shoreline was a reminder that nowhere is untouched. But nothing brought home the devastation that we're wreaking more than Sri Lanka's turtles.

The Mahamodara Sea Turtle Hatchery in Galle, on the southwest coast of Sri Lanka, is a hostel for the five species of turtle that visit this part of the Indian Ocean. It's a breeze-block building with wooden posts supporting a roof of corrugated iron, giving it a slightly ramshackle appearance. But though it lacks the sleek glamour of comparable marine exhibits in Europe or North America, this is proper, frontline conservation at work. The centre is perched, appropriately enough, at the edge of the sea. And where the buildings give way to an expanse of beach, they've enclosed a patch of sand that looks for all the world like a cemetery with row after row of small tombstones, each marked with names and dates. A closer inspection shows that what I'd taken for graves are, in fact, nurseries, and the labels announce the species and date. Beneath each mound, a clutch of eggs is developing, a process that takes a little under two months. When I was there, in mid-February, the nesting season was drawing to a close, but already over 10,000 hatchlings had been shepherded through this vulnerable stage.

Inside the building, a series of large concrete pools housed turtles of all sizes, and nearly all species. Some had been given over to a kind of chelonian kindergarten, with newly hatched youngsters being fed and experiencing seawater for the first time in their lives, prior to release. Under natural conditions, perhaps one in a thousand of these would live to adulthood, but this early-life intervention shifts the odds in these turtles' favour. The occupants of other, larger pools represented a sad litany of turtle misadventure. Cracked shells and missing limbs abounded, the consequence of straying too close to churning boat propellers. Despite their injuries, the wounded turtles – some now with epoxy-reinforced carapaces – seemed to be making excellent recoveries and, for most, their stay at the centre would be short.

The scarred turtles were in a minority here. Alongside them, a large number of others were far less perky despite bearing no obvious injuries. In many cases, they floated listlessly at the surface of their pools. They looked somehow as though they'd been inflated, their limbs, heads and their shells distended by an internal pressure. To demonstrate, one of the guides pressed gently on the back of one of these living dirigibles. When he released his hold, the animal popped to the surface like a float. All of the turtles in this predicament were the victims of plastic. When turtles ingest plastic, and all species do, the result can be a blocked digestive tract, causing gas to build up and for the turtle to become super buoyant. Such individuals are hostages to fortune; they can't forage, and nor can they escape the interest of predators or indeed collisions with outboard motors. In time, the blockage may pass naturally, but this can take weeks of pain and jeopardy in the open sea. The bloated turtles here were brought in by local fishers who scoop the hapless creatures from the water as they pass. These are the lucky ones; most will die at sea as a result of their ersatz intestinal cargo.

Eating plastic might seem a strange thing to do, and it's long been thought that the driver of this behaviour was the similarity in appearance between, say, a floating shopping bag and the jellyfish that species such as leatherbacks specialise on. Recently, however, a new, more insidious mechanism has come to light. Plastic in the marine environment rapidly becomes coated with a biofilm of algae, microorganisms and encrusting animals. This biofilm seems to act like a kind of olfactory disguise, causing the turtles to miscategorise it as food. The extent of the problem was revealed by a 30-year study, which reported that more than two thirds of turtles now have substantial amounts of plastic inside their bodies, most of it the single-use items that plague the modern world. With every passing year, the deadly smorgasbord of plastic grows, its offerings available in every shape, size and colour imaginable. The youngest turtles, those in the wandering, oceanic 'lost years', are particularly hard hit by this, drawn as they are to the interior of the great marine gyres, which, far from land and apparently beyond our concern, accumulate our societies' debris on an unprecedented scale.

The North Atlantic Gyre, which transports the hatchlings of Floridian and Cabo Verdean nests alike, is a colossal oceanographic feature, yet it's the smallest of its kind. There are five great gyres on Earth, two each in the Atlantic and Pacific, and one in the Indian Ocean. They're fundamental cogs in the great circulatory machine of the ocean, yet they leave little sign at the surface of their presence. Or, at least, that used to be the case. Now, the currents draw plastic into their orbit, ultimately concentrating huge rafts of debris at the centre of each gyre. The most notorious of these is the Great Pacific Garbage Patch, an aggregation of floating litter that's already larger than 90% of countries and it's still growing, like a melanoma on the face of the ocean.

The Great Pacific Garbage Patch is fed by an annual influx of some 10 million tonnes of plastic to the sea each year. Imagine watching a truck dumping a full load of plastic into the ocean; that's how much of the stuff we add to the marine environment every *minute*. Most people will never see one of these garbage patches or experience their scale, and so for most they remain almost illusory.

The same isn't true for marine life, however, whose interactions with plastic are all too real. Getting on for a million turtles and cetaceans die each year through entanglement in discarded fishing nets, but the numbers of marine animals that die after ingesting plastic are beyond measurement. The ocean's emblematic and charismatic animals are among the most obviously impacted by this insidious killer. The overwhelming majority of seabirds have plastic lodged in their insides; it kills an estimated one million of them every year and compromises countless others. Whales wash up with increasing regularity, choked to death by toxic plastic burdens. A Cuvier's beaked whale that washed up in the Philippines in 2019 had 40 kilograms of shopping bags in its stomach. The same year, an autopsy on a sperm whale in Scotland uncovered a plastic litter ball weighing as much as a fridge-freezer. We're killing these animals as surely as if we still pursued them with explosive harpoons, and on a far greater scale, and yet our addiction to plastic grows stronger by the year.

To some animals, floating plastic represents an opportunity. In the Azores a few years ago, the skipper of the vessel I was on paused to collect a tangle of floating plastic tubes. As he pulled the mass aboard, dozens of pilot fish painted with vivid black-and-white bands disappeared, the spring cleaning having robbed them of their hang out. Not so lucky were the goose-neck barnacles, their stony heads supported by fleshy crimson-black stems that quickly began to wilt in the sun, their attachment to the plastic proving their undoing. A small crab scuttled from amid the trunks of this crustacean forest, perhaps perplexed at the sudden change to its universe. There wasn't much I could do for the barnacles, but I intervened on the crab's behalf. I flicked him into the sea and watched as he swam away, as nimbly as any fish, in search of another floating island to call home.

The crab and the barnacles aren't the only denizens of the open ocean's plastic New Forest. In the garbage patches, an entire community of animals has sprung up, quick to exploit a novel opportunity. The plastic offers a foothold for creatures normally associated with coastal habitats. They likely arrived in the garbage patches like castaways, borne by currents from their inshore habitats on plastic flotsam. There's nothing new in this rafting, of course; marine animals have always been quick to colonise floating objects such as floating vegetation in their habitat, that has been carried across the oceans, sometimes to settle in new territories. Even so, their tenure in the open sea was transient, determined by how long their life raft remained afloat, which for a tree would typically be months at most. Plastic, however, is built to last, and where it congregates, so do the colonists, hundreds of species of corals, oysters, shrimps and crabs. Biology always finds a way and normally this is cause for celebration, but the conquest of the pelagic frontier by these coastal opportunists represents, to me at least, a sour cherry atop a garbage cake.

Durability is the hallmark of plastic; it's engineered to stick around. That which finds itself in the sea might take decades to centuries to degrade. Even as it breaks down, plastic doesn't magically disappear, it simply gets smaller, and more pernicious. It's estimated that there may already be as many as 350 trillion tiny fragments, known as microplastics, in the surface waters of the ocean alone.[89] In less than three generations, the marine environment has transitioned from its largely plastic-free past to today's slurry of synthetics. Sea creatures now live amid a blizzard of microplastics, the majority far smaller than can be seen with the naked eye. With so many different particle sizes drifting loose in the sea, there are pieces of plastic to suit every organism.

89 This catch-all term refers to pieces of plastic ranging from 5 millimetres down to 0.001 millimetres across. Particles smaller than this are referred to as nanoplastics.

Microplastics can be found in the bodies of virtually every individual of every species of marine animal. Nowhere is untouched. A new species of amphipod retrieved from the Mariana Trench and recorded in 2019 was named *Eurythenes plasticus* for the material it contained in its body. The initial route into animals might be via the gills as they breathe or their diet as they eat. The smallest particles even enter directly through the skin. Grazers find their diet of algal turfs supplemented by a sprinkling of synthetic particles. Microplastic specks the size of a diatom draw the attention of hunters like copepods, which readily ingest them. Filter feeders, such as mussels and oysters, draw in microplastics along with the water that they suck up. These in turn get passed on to the predators of these animals, so that prey bequeaths their microplastic burden to their predators, as happens when crabs eat contaminated mussels.

The fate of most ingested microplastics is to be ejected by the animals as unwanted guests. Bivalves coat the particles in mucus and vomit them out, while those ingested by other animals make their way through the digestive tract and are expelled in their faeces. The remainder, however, aren't always so easily dealt with. Some aggregate in the gut, where they can fatally clog. Smaller microplastics in particular are translocated within the body, insinuating themselves into the blood and tissues, where they interfere with crucial physiological functions.

Regardless of whether the microplastics are fleeting visitors to the body, or settle in for an extended stay, they bring with them a payload of toxins. Plastics are a diverse family of materials, incorporating more than 10,000 different chemicals. These might be integral to the substance itself, or additives that lend a particular quality to the finished product, such as fire retardant, or plasticiser that increases flexibility, among many others. A substantial proportion of these chemicals are incredibly toxic, and the more plastic weathers, the more easily it releases its baneful cargo into animals' bloodstreams. This is bad enough on its own, yet these substances are merely the tip of a noxious microplastic iceberg. The properties of plastics mean that they attract other man-made pollutants in the ocean like magnets. Forever chemicals and heavy metals are just two of the substances that are drawn to coat microplastic particles, augmenting the existing toxins borne by the plastic itself. Each fragment of microplastic thus becomes a vector, carrying a deadly arsenal of poisons straight into the bodies of marine organisms.

Worse still, microplastics gain a kind of chemical disguise by virtue of marine microbes, including microalgae, which colonise its surface. Within a month of floating around in the sea, these algae start to emit their signature smell of dimethylsulphide, or DMS, which, as mentioned previously, is powerfully attractive to many marine animals. Fish, turtles, seabirds, including penguins, and marine mammals are drawn by the alluring scent to feast on a banquet of plastic, masquerading as food.

The damage wrought by microplastics on marine organisms is devastating. The particles injure and contaminate the animals, compromising their immune systems and their ability to breed. Many are killed outright. The harm isn't confined just to individuals; it resonates through the ecosystem. Perhaps most concerningly, it interferes with the shuttling of nutrients throughout the ocean, including the biological pump that transports carbon from surface waters into the deep sea. Microplastics do this by compromising the ability of zooplankton to feed, clogging them with fake food and cutting their intake by almost half. If the microplastics transit through the gut, they end up in the faecal pellets that the organisms produce. The biological pump is dependent on the gentle fall of dainty, diminutive particles of carbon-rich crap from the surface, yet the amalgamation of buoyant plastic into the pellets hampers their ability to sink. Compromising the biological pump in this way could ultimately mean that instead of reaching cold storage in the abyss, billions of tonnes of carbon dioxide remain in the atmosphere each year. The consequences of that would be terminal for our species.

The price we pay for plastic at the shops bears little relation to its true cost. The difference between the two is a classic case of what economists refer to as an externality.[90] The ocean has long been subsidising our plastic lifestyles. There's a similar disconnect between the longevity of plastic and the fleeting uses that we often put it to. A soda bottle exists for a few days or weeks in the supply chain, but it has a lifespan of decades to centuries in the ocean. As it degrades, it bursts like a grenade into millions of microplastic particles, each with the potential to choke or to poison marine life. Until now, we've largely disregarded the costs that accompany this because they're not levied on us. A reckoning is coming, however. If we continue along our current trajectory with plastics and end up sabotaging the biological pump, no amount of money could fix it.

---

90 The upshot of the post-war decades of convenience is that we now have a world where microplastics are ubiquitous. Microplastics and nanoplastics are not just in the sea, they're everywhere and they're unavoidable. Each of us is part plastic. You are, I am. The food that you eat, the water that you drink and the air that you breathe all contain microplastics. A recent study found that a typical 1-litre plastic bottle of water contains almost a quarter of a million plastic particles. We don't as yet know what the impacts will be on human health. However, a 2024 review published by the *Lancet* concluded that human health risks associated with exposure to micro- and nanoplastics included 'oxidative stress, inflammation, impaired immune function, alteration in cellular and energy metabolism, inhibition in cell proliferation, tissue degeneration, abnormal organ development and dysfunction, alteration in biochemical parameters and even cause genotoxicity and carcinogenicity'.

Epilogue

'If we save our oceans, we save our world.'
SIR DAVID ATTENBOROUGH

I'M BACK AT THE CLIFFTOPS AT FLAMBOROUGH, gazing out across a seascape that is both familiar yet fleeting. A conspiracy of wind and waves harries the burnished blue surface, rendering each moment as unique as a human fingerprint. It was here among the boulders and pools of the shore and the sprung turf of the headlands that my passion for the sea was kindled. Since that time as a boy almost half a century ago, I've devoted my life to the study of the sea, diving both literally and metaphorically into the world beneath the waves.

Contemplating the immensity of the scene before me, I feel small, insignificant, yet this is just one panoramic pixel of the Earth's great marine canopy. The scale of the ocean makes the challenge of understanding it too great for one person, in one lifetime. It's a job for all of humanity, for scientists and laypeople, for artists and naturalists, for the expert and the beginner, as I was as a boy, all those years ago. Moreover, it's a job that is essential if we are to protect our oceans because – make no mistake – if we fail in this endeavour, the future is bleak indeed.

The winds that touch my face are chill and fresh, sweeping down from the north. The breeze is suffused with the memory of Arctic ice, but this polar powerhouse, like its southern counterpart, is weakening. These twin hearts of ocean circulation are showing symptoms of a kind of thermal angina, weakening perceptibly under the assault of rising global temperatures. For now, the currents of gelid water from the poles bring their bounty of minerals and oxygen to invigorate the lower latitudes, and the ocean clings on.

If I had the exquisite sense of smell enjoyed by a seabird such as a petrel, perhaps I'd scent among the gusts the signature bouquet of far-off phytoplankton, blooming in unimaginable numbers in the late Arctic summer. Amid the relentless dynamism of the diatoms and dinoflagellates in the Barents Sea swim members of an iconic species in one of its last remaining fastnesses. Atlantic cod were once so plentiful that they supported a huge fishing industry and were a staple for rich and poor alike throughout Europe and North America. Perhaps it was partly of cod that the head of the Woods Hole Oceanographic Institute was thinking when in the 1950s he declared the sea's riches to be inexhaustible. And my, haven't we run with that vision?

We've since learned, to our cost, that the seas are far from illimitable, that the ocean cannot prosper in the face of either our rapacious greed or our squalid indifference. Cod were the bellwether for the consequences of industrial fishing, their jaw-dropping collapse the signal that something was very, very wrong. The remaining survivors of this slaughter exist now in anything like their former numbers only at the margins of our reach. Theirs is just one cautionary tale of many. In this book, I have told sobering tales of demise, depauperation and destruction, of marine life sacrificed at the altar of progress and profit. I've described underwater mausoleums shorn of fauna, of tropical waves washing across crumbling bleached corals, the mass retreat of mangroves, of kelp and of sea ice. I've related cliffs that once hosted raucous carnivals of tumbling, white spindrift in avian form, now stark and silent.

My bookcase is home to some of the most beautiful and haunting writing on nature. Each work, in common with almost every contemporary portrayal of the natural world, is suffused with concerns and cautions, and an exhortation: the time to act is now. Yet the warning passes unheeded, and why? Ignorance, indifference, short-term thinking and economic anxieties all play their role; fear of change permits the blinkered acceptance of an unsustainable status quo. Society's lassitude is encouraged by huckstering populist politicians and their allies, the proud-to-be-stupid shouters who infest social media, each marinated in their silos of idiocy, their asininity spewing like sewage outfalls into the channels of public discourse.

Yet if in response conservationists simply offer gloom, if all we can do is glower and chide, then it's perhaps little wonder that the audience turns away. Make no mistake, I have felt despondency myself. Among much else, I've seen seal pups, agonisingly beyond reach, being slowly garrotted by discarded fishing gear, suffering their inexorable fate in silent pain. I've witnessed a magical, technicolour reef give way to a dead, bone-white simulacrum, interred in a winding sheet of algae. I've stood at urban waterfronts the world over feeling disconsolate at the sight of taupe-coloured waves beating a dull tattoo of defeat against plastic-choked shores.

I experienced my nadir in the otherwise wonderful city of Mumbai on a visit to the predawn fish market at Sassoon Docks. I mingled with people as they bustled around scant baskets of fish for sale, with smatterings still being lifted through blue-grey cumuli of diesel fumes from boats at the quayside. Within their woven caskets, sightless, clouding eyes stared up at electric floodlights that, in turn, lit the slicked surface of a sickly sea, laminated in oil. Among the lizardfish known as Bombay duck and disc-shaped pomfret were juvenile sharks, pitifully small, stacked in stiff pyramids of death. At the end of an aisle, a majestic devil ray lay prostrate on the grimy market floor.

I can't adequately explain why Mumbai's fish market affected me quite as it did; after all, there are innumerable such places around the world, many of them far worse than this. Perhaps it was because the Arabian Sea was so obviously offering mere remnants, shoddy returns for the many boats trying to eke a living. Or maybe it was the pollution, the waters thereabouts seeming as dead as the market's offerings. Whatever it was, the experience lingered; it was as if I'd received a visit from the ghost of oceans yet to come foretelling the end of that which I love so dearly.

In the days that followed, the solace that I usually gain from proximity to the sea was nowhere to be found. I felt paralysed by a sense of despair to the point that I felt like giving up. If it's going to hell anyway, what's the point? I was scheduled to go diving later that week and almost cancelled, such were my feelings. But I did go, and it proved the perfect antidote. As I descended into the sea's embrace, and reconnected with marine life, my despondency gave way to determination. This book is the outcome of that change in attitude, and while there's no hiding the fact that the oceans are suffering, their fate still lies within our hands. There is every reason to continue the fight, every reason for optimism. In what remains of this book, I'll set out why this is the case.

* * *

The scale of the challenge that faces us is daunting, admittedly, but we are not helpless: we know how to fix the problems, and we have the expertise and the capacity to restore the oceans. It's a struggle that we *must* win, and that we absolutely *can* win. Indeed, realisation of this is the critical first step. Doom-laden pronouncements have fostered a growing fatalism, most worryingly among the young. A recent survey asked children and young adults across ten countries about their feelings on the environment. Three quarters of them reported fear or dread for the future. They felt betrayed by older generations yet powerless to change what, for them, seems inescapably apocalyptic. It's a perspective that was encapsulated for me in a conversation I had with a backpacker in the Galapagos. She'd embarked on what she portrayed as a funerary grand tour, telling me that she was travelling now because 'the world's going to burn'. Although I empathise, I have no plans to resign my fate, or that of my two children, to becoming pieces of bespoke charcoal. If apathy replaces agency, if we simply give up, then nothing that I relate below matters and disaster is inevitable.

Hope is the antidote to this. Just as Robert Paine described the ochre sea star as the keystone species of his Pacific rocky shore communities, so hope is the keystone motif for the future; it's what sustains the entire conservation ecosystem. If we can instil hope, we might harness people's concerns in a positive way, rather than leaving them prey to a perception of powerlessness. Individually, people can take responsibility for their own actions, making informed choices about everything from the plastic that they use to the fish that they buy. Collectively, we can drive shifts in society's attitudes, compel business to operate responsibility, and elect politicians who are prepared to push this most important of issues up the agenda.

Beyond this, ever more people are getting involved in the fight for the future of the seas. Around the world, community-driven projects are springing up at an unprecedented rate and participation has skyrocketed. New seagrass meadows and mangrove forests are being planted, turning the tide on decades of destruction. In the tropics, bespoke frames, festooned with fragments of live coral, are helping to repopulate damaged sections of coral reef. Another kind of reef is being fostered in temperate seas by the annual sowing of countless millions of oysters.

Armies of volunteers are also at work loosening plastic's stranglehold on our shores, each year collecting mountains of trash from the coasts of more than a hundred countries. In a similar way, divers have mobilised worldwide under the aegis of organisations such as Ghost Divers and the Ocean Defenders Alliance, recovering hundreds of tonnes of lost or discarded fishing gear from the seabed.

While it's true that an ounce of prevention is worth a pound of cure, the spectacle of degraded marine environments has already stirred millions of people to action worldwide. Ideally, the kind of remediation that their efforts provide wouldn't be needed, but in many places, it so plainly is. The immediate reward on offer to these people is the satisfaction of a job well done, of a clean shore, of rows of newly planted seagrass seedlings repopulating a damaged seabed. However, the impacts go far beyond this, sending a message both to polluters and to society at large. Such local initiatives turn people into stakeholders and promote marine conservation at a relatable scale. Small though each individual programme may be, together they can change the course of our relationship to the ocean.

Our efforts to conserve are not always successful, but setbacks can inform us how to do things better next time. The overwhelming majority of marine conservation initiatives have a relatively short pedigree and for much of this time, we've been learning by doing. This has been bolstered through collaboration, through sharing knowledge across networks of participants, and through the integration of local knowledge with cutting-edge science. We're now reaping the rewards, becoming more adept, more efficient at restoration.

* * *

Innovation is the hallmark of our species. When faced with a challenge, we're characteristically good at coming up with solutions. Science lies at the heart of this. During my career as a biologist, I've had the privilege of working with marine researchers from across different disciplines, on projects from Antarctic krill to coral reef ecology and much else in between. The work of these researchers is often unheralded, taking place out of the glare of public attention, yet hope blossoms from each new discovery. The conservation of complex marine ecosystems is founded on two key elements: a detailed knowledge of what makes them tick, and the technological capacity to address the challenges that they face. Thanks to the efforts of researchers in both of these regards, we now have an unprecedented opportunity to bring positive change.

Not so long ago, the ocean was a black box, its inhospitable immensity concealing its mysteries. That situation has been transformed in recent times; a spate of scientific advances are yielding the ocean's secrets. We now have at our disposal remote sensing, telemetry, drones and autonomous submersibles to map the ocean and its inhabitants. Alongside these, environmental DNA, or eDNA, sampling allows us to compile inventories of species from the tiny fragments of their genetic code that they leave behind as they move around. As data floods in, AI is being used to analyse it, finding patterns and detecting change. Taken together, these approaches mean that we can assess not only what's out there but gather detailed and almost instantaneous information on the health of the ocean and its inhabitants. This, in turn, means that we can develop more dynamic and targeted approaches to preserving and restoring the oceans.

Alongside this, technological innovations offer hope in myriad contexts, and – happily – are too numerous to list. Nevertheless, a few selections from what is a long list include the ease with which we can now genotype organisms to learn the genetic basis of resilience, taking some of the guesswork and risk out of restoration efforts. Biodegradable nets are beginning to replace the deathless synthetic fishing gear that's currently used. It's now possible to imagine a future free from both the microplastic pollution and the deadly ghost fishing perpetrated by the 50,000 tonnes of gear that is lost or discarded each year. Meanwhile, more sustainable fishing practices may soon be achieved using bots, offering alternatives to practices such as dredging. Instead of blindly destroying acres of seabed in search of shellfish, a bot could pluck them one by one, leaving the environment untouched. Whereas fisherbots remain a work in progress, robots are currently hard at work in restoration projects, bringing their tireless precision to the job of planting thousands of seagrass seeds in the seabed, and to distributing coral fragments on damaged reefs.

Clamping down on illegal fishing would also represent a vital win in the fight for the ocean's recovery. Currently, operators can evade detection by switching off their identification systems at sea and then landing their spoils at ports where officials are happy to look the other way, allowing illegally caught fish to enter the market and blend anonymously with legitimate catches. Blockchains, tamper-proof electronic ledger systems, provide a solution to the latter. Using this, every single fish can be tracked by inspectors through the entire supply network from the moment of capture to its appearance on a supermarket shelf. The same information will also be immediately accessible by consumers, who can scan a barcode to allow them to make the best choices about the fish they're buying.

Perhaps my favourite example in this fight comes from a new and creative strategy to counter illegal fishers, simply because it is being led by one of the animals that has been so badly damaged by their activities. As one of the most iconic, and most threatened, of our marine species, albatrosses are closely monitored by the scientific community. Large numbers of these birds carry miniature tracking devices that relay their position to data loggers, and these are now being tweaked to detect radar signals.

Switching off identification systems is one thing for the skippers of illegal boats, but they're reluctant to do the same with their radar systems in the hostile isolation of the high seas. This is where the albatrosses come in. The birds are attracted to fishing boats in the hopes of scraps. As they approach, the albatrosses' tracking devices pick up and register the precise locations of boats from their radar. This information is then relayed to a website for access by fisheries inspectors. A proof-of-concept study reported that over the course of six months, the birds located several hundred vessels, more than a third of which had their identification systems switched off while in international waters. Given the enormous distances over which albatrosses range in the southern hemisphere, the presence of these avian sentinels means that there's now nowhere for illegal fishers to hide.

* * *

Much of what I've discussed so far involves action and intervention, invoking the notion that the sea must be managed intensively to survive. Though there are instances where this is unavoidable, for the most part the best thing for us to leave well alone. The sea has demonstrated, time and again, its breathtaking ability to regenerate.

I've described the recovery of Atlantic fish stocks during the two world wars, but this is one example among many.

It's hard to imagine a more severe test for any environment than the nuclear maelstrom that was visited upon Bikini Atoll in the mid twentieth century. Over a period of 12 years, a total of 23 atomic bombs were detonated there, the largest of which was 1,000 times more powerful than that dropped on Hiroshima. Each explosion in this delicate coral habitat almost instantaneously heated seawater to 55,000°C, obliterated massive sections of reef, and caused a firestorm of radioactive fallout. When marine biologists returned to Bikini a few years ago, they reported that despite the destruction, and the pernicious radiation, the corals of Bikini have come rampaging back, with huge shoals of fish swirling around the regrowing reefs.

Meanwhile, the Black Sea was once a byword for ecosystem collapse. By the 1990s, large parts of it were lifeless and the remainder was afflicted with plagues of invasive comb jellies, but little else. It was an ecological disaster, driven by a combination of overfishing alongside massive inputs of pollution from Eastern bloc countries during the Soviet era. Within 20 years of this toxic tap being turned off, sections of what was once known as the world's most polluted sea have come back from the dead.

The roll call of marine recoveries provides proof of the extraordinary ability of the sea to bounce back on its own. Sometimes, however, nature needs a helping hand. The Moratorium agreed by the International Whaling Commission in 1982 is one example of this. Characteristically, we waited until the eleventh hour to act, but still the ban changed the fortunes of whale species. Humpback and sperm whale populations, in particular, have responded gratifyingly well, to the point that humpback numbers are back to pre-hunting levels. Other persecuted marine mammals, including sea otters and Antarctic fur seal, have shown similarly impressive recoveries.

Northern elephant seals were almost wiped out by hunting in the late nineteenth century. By 1892, only around 20 individuals remained, confined to a single island; 132 years later, those 20 have mushroomed to 220,000.

The return of one of the sea's top predators to UK waters in recent times after an absence of 60 years is another cause for celebration. Atlantic bluefin tuna are an extraordinary sight, their streamlined, muscular form a paean to power and grace. The largest of them can be twice as long as a person is tall and half a tonne in muscly weight, yet they're rapier quick. That they grace our oceans now after decades of intense persecution is akin to a miracle. During the latter stages of last century and the early years of this, bluefins came close to outright extermination. Enormous fleets of boats hounded them, and when they'd slaughtered one shoal, they sent spotter planes up to locate more. Atlantic tunas were killed in their millions until by the late 1990s, the populations of this magnificent creature had sunk close to oblivion. Even when the International Union for Conservation of Nature (IUCN) listed them as an endangered species, the killing went on. In 2006, the responsible regulatory authority, the International Commission for the Conservation of Atlantic Tunas (ICCAT) finally began to get its act together. Quotas were cut, there was a clampdown on illegal fishing, a ban on spotter planes, and a minimum size limit was enforced to preserve juveniles. Less than two decades on, the results are little short of astonishing. The population of bluefins has quadrupled, and from being listed as endangered, they have now been reclassified as a species of least concern.

It remains essential that we guard against complacency. The fishing lobby fought tuna conservation plans at every step of the way and even demonstrated the extent of their pernicious reach by persuading the United Nations to vote against protection. Now that bluefin tuna stocks have rebounded, the lobby chafes at the restrictions that prevent them from resuming their war on nature. The same is true of recovering populations of just about every overexploited marine fish species. It's clear that for some in the industry, marine life has no intrinsic value beyond hard currency and more animals simply mean more profits. Heartening as the tuna recovery has been, their numbers remain far below historical levels and it's critical that public pressure is maintained as a bulwark against the fishing industry's lethal intent.

Tunas are among just a handful of exalted creatures worldwide that attract a specific focus for regulation. What many such species share in common is their profitability. For the millions of other less lucrative but equally valuable species to gain some of the protective limelight, a piecemeal approach isn't practical, and that's where marine protected areas (MPAs) come in. By setting aside conservation zones within the sea, entire communities of marine animals gain sanctuary, and the critical functions of their habitat are preserved. Achieving the goal of protecting 30% of the sea in this way by 2030, mandated by the UN's Global Biodiversity Framework, would represent a huge step in safeguarding the future of our oceans. At the time of writing, four years on from that 2022 agreement, it's fair to say that we're not on course to achieve this. Nevertheless, the coverage of MPAs around the world has increased exponentially during the course of this century, from around 3 million square kilometres in 2000 to around ten times that amount in 2025. If the current rate of growth is maintained, we'll hit the 30% target in the mid-2030s. Disappointing as this may be, if there's a silver lining it's that momentum is gathering.

That said, hitting 30% isn't a panacea on its own. While a clear target is a good thing, there is a danger of prioritising quantity over quality. To yield the greatest benefits, MPAs need to be situated according to the best scientific advice. Areas of high species diversity, and those that contain critical resources, such as breeding grounds and nursery habitats, must be a high priority. Sadly, a recurring theme in conservation is the siting of protected areas in places that were no good for anything else, and it's important that cynical ploys by the political classes don't derail conservation goals. As an indication of just how important siting is, it's been suggested that if we were to focus on protecting the most critical areas, we'd gain 90% of the predicted benefits of MPAs by setting aside just 21% of the ocean. If, on the other hand, we place those zones in a less targeted manner, we'll need far more than the 30% cover we're aiming for.

Simply declaring an MPA means little unless restrictions on fishing are enforced. An analysis of data from MPAs around the world shows an average increase in fish biomass within fully protected zones of more than 400%, in stark contrast to the 20% gains obtained in partly protected equivalents. Achieving this will need the firm commitment of governments and buy-in from local people. Money underpins all of this. The total price of establishing and then enforcing MPAs, and restoring areas that have been damaged, will be considerable. Whether the cost is high or not depends on your perspective. The annual worldwide bill for a comprehensive network of MPAs that meets the 30% criterion will be in the region of $20 to $40 billion. Expensive? It's less than half the gross annual profits of companies such as Shell and ExxonMobil, and represents around 1% of global military spending.

This is not, as is sometimes projected, a zero-sum game in which a greener world is a poorer world. Those of us with a passion for the oceans and the desire to fight for their future need to communicate not only the importance of protecting them but also the benefits of doing so. MPAs are investments that provide enormous returns.

I realise that hard-headed investors with an eye on the bottom line might fail to be impressed by something supposedly so intangible as biophilia, the profound, innate joy that we gain through our connection to nature. But even if we put aside the exhilarating prospect of rebuilding thriving marine ecosystems, even if we disregard the accompanying ecosystem services that enrich us, the business case is undeniable.

Back in 2009, the World Bank, historically hardly a champion of environmental causes, pointed out that the global fishing industry was costing itself $83 billion a year through its lunatic mismanagement. Restoring fish populations through a worldwide network of MPAs would mean that far more fish would be caught with far less effort. Meanwhile, ecotourism provides vastly greater – and more reliable – income for coastal communities than relentless fishing of depleted populations through activities such as whale watching, seabird tours and diving. It is important to acknowledge that these benefits would not accrue immediately – a timeline of five to ten years is more reasonable – but the costs of the short-term approaches that have hitherto dominated our thinking are everywhere to be seen. We stand at a crossroads in time; will our legacy to future generations be a resurgent, healthy ocean, or its depauperate shade?

* * *

Conflict is a mainstay of modern global politics, and in an atmosphere of bellicosity and bloodshed, and the decline in the rules-based order, the prospects of securing the international cooperation to secure the future of the ocean may seem bleak. Yet even as sabres rattle, agreements have been reached, suggesting that a moral instinct survives. Over the course of the last half-century, there has been a swathe of hugely important agree-ments. Against the backdrop of the Cold War, the groundbreaking United Nations Convention on the Law of the Sea (UNCLOS) in 1982 was a historic achievement, establishing a legal framework governing our relationship with the oceans and enshrining the principle that the high seas are the common heritage of mankind. Four decades on, the High Seas Treaty that grew from it provides for the first ever legally binding agreement of its kind.[91] International waters, representing two thirds of the ocean, are to gain new protection, with key provisions for setting aside marine protected areas and the agreement that any activity that occurs on the high seas must be consistent with conservation objectives.[92]

The High Seas Treaty dramatically improves the chances that we'll be able to achieve the 30x30 target and signals a global commitment to ocean health. Nevertheless, a long struggle lies ahead to ensure that the treaty, ratified in 2025, is obeyed in both letter and spirit, that its provisions are implemented and enforced. The onus is on all of us to make sure that this opportunity is not missed.

Chemical pollution remains a pervasive problem in marine environments, but the picture is improving in many places thanks to more stringent treatment of wastewater, the shift away from leaded fuels and the tightening of laws on the production of hazardous materials. The Stockholm Convention, which entered into force in 2004, has been instrumental in driving down the production of some of the worst chemicals. While Stockholm deals largely with land-based pollution, the International Convention for the Prevention of Pollution from Ships established the principle at sea. One of the effects of MARPOL, as it's known, has been to drive

---

91 More formally, the treaty is known as the Agreement on the Conservation and Sustainable Use of Marine Biological Diversity of Areas Beyond National Jurisdiction.

92 International waters, or the high seas, are those beyond the 200-nautical mile limits of countries' exclusive economic zones.

down the frequency of oil spills from tankers to less than a tenth of what they were before. As a measure of the effectiveness of these measures, levels of contamination in many parts of the globe show long-term decreasing trends.

Agreements have also been made to preserve pristine polar environments. The Antarctic Treaty, which came into force in 1961, paved the way for cooperation and conservation in the far south. Then, in 2021 an agreement was reached between nations to protect the Arctic from commercial fishing. Perhaps the most remarkable thing about the Central Arctic Ocean Fisheries Agreement was that it was the first agreement to protect an area from commercial fishing before it had even begun.

The news isn't always positive, however. Discussions on the UN's proposed Global Plastics Treaty failed to reach agreement in 2024 after two years of negotiation. Talks have since resumed, and we can only hope that the determination that carried the High Seas Treaty to its conclusion despite years of deadlock and dispute will prevail here too. As a guide to just how challenging these processes can be, it took more than 20 years of back and forth before the World Trade Organization Agreement on Fisheries Subsidies was ratified in 2025.

There is one agreement, however, that a decade ago shone like a beacon of hope but whose flames now flicker and threaten to die. Its failures cast a pall over the progress that has been made in other areas because it concerns the single greatest threat faced by the planet: climate change. The Paris Agreement of 2015 aimed to limit warming to less than 2°C, and ideally less than 1.5°C, above pre-industrial levels by compelling member states to cut emissions of greenhouse gases. The results, however, have been underwhelming; bluntly, the Paris Agreement is weak as an instrument of change. Any commitments that nations make are voluntary, and there are no enforceable penalties for non-compliance. As a result, global emissions – and global temperatures – are still climbing.

It is not irrational to retain hope and determination, even as we acknowledge the climate crisis. Optimism comes in many forms, not least of which is the massive reduction in the prices of renewable energy and battery storage. Whether we like to admit it or not, the single most important metric in determining people's attitudes to cutting emissions is cost. We want to save the planet, but seemingly only without affecting our wallets. Until recently, fossil fuels provided a far cheaper source of energy than renewables, but this is changing. The price per megawatt hour of solar energy has fallen by 90% in the last ten years alone, while that of onshore wind is now 70% lower.[93] This is partly the result of technological improvements to the designs of wind turbines and photovoltaic cells, and partly because of economies of scale: the more you make of something, the cheaper it becomes to make. As renewable costs come down, we enter a cycle where more uptake drives falling prices, which then drives more uptake.

Alongside cost, another factor that has held back the transition to renewables is the cost and capacity of electricity storage that's needed to maintain steady supplies when the sun doesn't shine, or the wind doesn't blow. Here, too, there's good news. The price of the batteries we need has fallen by 97% this century and capacity is increasing exponentially. The environmental and humanitarian costs of sourcing lithium remain a major concern, but if we can get this right, the end of fossil fuels for electricity generation is in sight.

None of this negates the difficulties that we face. We're still locked in a cycle of rabid overconsumption and we're still pumping

---

93 This is the LCOE (levelised cost of energy) price, an industry measure that accounts for the cost of generating energy as well as the cost of building and maintaining energy plants.

out massive volumes of greenhouse gases. That we have to cut emissions drastically, and soon, is obvious, and has been for some time, but the pace of change is desperately, agonisingly slow. Nonetheless, as William Ellery Channing said, 'Difficulties are meant to rouse, not discourage,' and I remain optimistic. History shows us that we can move quickly when provoked – the Montreal Protocol that addressed ozone depletion in the 1980s is one example of this – and we're better equipped than we've ever been to make the changes necessary.

Those changes will come, even if our political and economic inertia mean that we wait – so very characteristically – until the very last moment to make them. What can we do in the here and now? The priority must be to redouble our efforts to protect and conserve the oceans. Doing so sends the important message to those who are losing hope that tomorrow matters. At the same time, protecting marine diversity helps to build resilience against the effects of climate change, potentially buying us a little time in the epoch-defining struggle for the future of the planet. Safeguarding marine ecosystems and restoring others that have become depleted provides a powerful and natural means of sucking up carbon. We now know the immense potential that environments like seagrass meadows and mangrove forests have in this regard, but the entirety of the ocean is a vast carbon sink.

Out beyond the bay, I watch as one of Flamborough's fishing cobles bobs its way around the headland, a view broken momentarily by a train of gannets arrowing towards their feeding grounds far out to sea. It's a scene that has changed little over the course of my lifetime, and despite the exotic places to which I've travelled during my career, I will always return to Flamborough. But it's not simply my love of the sea – this sea – that brings me back. It was here, at the top of East Scar cliff, that a few years ago I scattered my mother's ashes. This was a special place to her, as it is to me. Perhaps, when the day comes that my children perform the same rite for me, we'll have changed the course of our relationship with the marine environment, and my existence will disperse over a resurgent sea. I cherish that hope.

Acknowledgements

I'VE BEEN ENORMOUSLY FORTUNATE TO HAVE THE support of many wonderful people throughout the writing process. My spectacular agent, Max Edwards, is as fabulous as he is tall. Thank you, Max, for your encouragement, insight and patience. I'm extremely grateful, too, to the dedicated team of people at my publishers, Bonnier. Particular mentions must go to Rik Ubhi for championing the project in the first place, Justine Taylor for her expertise and unyielding belief, and Sarah Braybrooke for her vision and leadership. In addition, I'd like to thank Susan Pegg for her editorial brilliance. Over the course of the three years that it took to bring this book to fruition, I have been lucky enough to have the opportunity to discuss initial thoughts and to share drafts with a trusted group of people, whose advice and support have sustained me. My heartfelt thanks go to Callum Steven, Robyn Scott, Stella Encel, Tony Wu and Karl Stanley.

Index, photographs, species glossary and further reading can be found on the author's website at https://ashleyward.net/saltwater/